Lecture Notes in Computer Science 3351

Commenced Publication in 1973
Founding and Former Series Editors:
Gerhard Goos, Juris Hartmanis, and Jan van Leeuwen

T0232828

Giuseppe Persiano   Roberto Solis-Oba (Eds.)

Approximation and Online Algorithms

Second International Workshop, WAOA 2004
Bergen, Norway, September 14-16, 2004
Revised Selected Papers

 Springer

Volume Editors

Giuseppe Persiano
Università degli Studi di Salerno
Dipartimento di Informatica ed Applicazioni
Via S. Allende, 43, 84081 Baronissi (SA), Italy
E-mail: giuper@dia.unisa.it

Roberto Solis-Oba
The University of Western Ontario
Department of Computer Science
London, Ontario, N6A 5B7, Canada
E-mail: solis@csd.uwo.ca

Library of Congress Control Number: 2005921102

CR Subject Classification (1998): F.2.2, G.2.1-2, G.1.2, G.1.6, I.3.5, E.1

ISSN 0302-9743
ISBN 3-540-24574-X Springer Berlin Heidelberg New York

Springer is a part of Springer Science+Business Media

springeronline.com

© Springer-Verlag Berlin Heidelberg 2005
Printed in Germany

Typesetting: Camera-ready by author, data conversion by Scientific Publishing Services, Chennai, India
Printed on acid-free paper   SPIN: 11389811   06/3142   5 4 3 2 1 0

Preface

The 2nd Workshop on Approximation and Online Algorithms (WAOA 2004) focused on the design and analysis of algorithms for online and computationally hard problems. Both kinds of problems have a large number of applications arising from a variety of fields. WAOA 2004 took place in Bergen, Norway, from September 14 to September 16, 2004. The workshop was part of the ALGO 2004 event which also hosted ESA, WABI, IWPEC, and ATMOS.

Topics of interests for WAOA 2004 were: applications to game theory, approximation classes, coloring and partitioning, competitive analysis, computational finance, cuts and connectivity, geometric problems, inapproximability results, mechanism design, network design, routing, packing and covering, paradigms, randomization techniques, and scheduling problems. In response to our call we received 47 submissions. Each submission was reviewed by at least 3 referees, who judged the paper on originality, quality, and consistency with the topics of the conference. Based on the reviews, the Program Committee selected 21 papers. This volume contains the 21 selected papers and the two invited talks given by Yossi Azar and Klaus Jansen.

We thank all the authors who submitted papers to the workshop and we also kindly thank the local organizers of ALGO 2004.

November 2004

G. Persiano
R. Solis-Oba

WAOA 2004
September 14–16 2004, Bergen, Norway

Program Co-chairs

Giuseppe Persiano	Università di Salerno, Italy
Roberto Solis-Oba	University of Western Ontario, Canada

Program Committee

Yossi Azar	Tel Aviv University, Israel
Evripidis Bampis	University of Evry, France
Thomas Erlebach	ETH, Zurich, Switzerland
Klaus Jansen	University of Kiel, Germany
Christos Kaklamanis	University of Patras, Greece
Marek Karpinski	Rheinische Friedrich-Wilhelms Universität, Germany
Claire Kenyon	Ecole Polytechnique, France
Samir Kuller	University of Maryland, USA
David Peleg	Weizmann Institute of Science, Israel
Giuseppe Persiano	Università di Salerno, Italy
Martin Skutella	Max-Planck-Institut für Informatik, Germany
Roberto Solis-Oba	University of Western Ontario, Canada

Sponsoring Institutions

Dipartimento di Informatica ed Applicazioni "R.M. Capocelli", Università di Salerno, Italy

Referees

Ernst Althaus	Han Hoogeveen	Micha Sharir
Vincenzo Auletta	Alex Kesselman	Maxim Sviridenko
Holger Bast	Yoo Ah Kim	Kavitha Telikepalli
Chandra Chekuri	Stavros Kolliopoulos	Ralf Thöle
Amir Epstein	Martin Loehnertz	Justin Wan
Leah Epstein	Maren Martens	Deshi Ye
Sandor Fekete	Paolo Penna	Hu Zhang
Amos Fiat	Kirk Pruhs	
Alexei Fishkin	Dr. Raghavachari	
Olga Gerber	Yossi Richter	

Table of Contents

Online Packet Switching

Yossi Azar[1],[*]

School of Computer Science, Tel Aviv University,
Tel Aviv, 69978, Israel
azar@tau.ac.il

Abstract. We discuss packet switching for single-queue, multi-queue buffers and CIOQ buffers. We evaluate the algorithms by competitive analysis. We also mention the zero-one principle that applies to general switching networks.

1 Introduction

Overview: Packet routing networks, most notably the Internet, have become the preferred platform for carrying data of all kinds. Due to the steady increase of network traffic, and the fact that Internet traffic volume tends to constantly fluctuate, Quality of Service (QoS) networks, which allow prioritization between different traffic streams have gained considerable attention within the networking community. As network overloads become frequent, intermediate switches have to cope with increasing amounts of traffic, while attempting to pass forward more "valuable" packets, where values correspond to the required quality of service for each packet. We can measure the quality of the decisions made within a network by considering the total value of packets that were delivered to their destination.

Traditionally, the performance of queuing systems has been studied within the stability analysis framework, either by a probabilistic model for packet injection (queuing theory, see e.g. [11, 18]) or an adversarial model (adversarial queuing theory, see e.g. [5, 12]). In stability analysis packets are assumed to be identical, and the goal is to determine queue sizes such that no packet is ever dropped. However, real-world networks do not usually conform with the above assumptions, and it seems inevitable to drop packets in order to maintain efficiency. As a result, the competitive analysis framework, which avoids any assumptions on the input sequence and compares the performance of online algorithms to the optimal solution, has been adopted recently for studying throughput maximization problems.

Single-Queue: In general we assume that all packets have a fixed size and each is associated with a value. We are given a FIFO queue with bounded capacity. At each time step new packets arrive to the end of the queue, and the packet at the head of the queue is transmitted. The goal is to maximize the total value of

---

[*] Research supported in part by the Israel Science Foundation.

G. Persiano and R. Solis-Oba (Eds.): WAOA 2004, LNCS 3351, pp. 1–5, 2005.

transmitted packets. We distinguish between two models: preemptive and non-preemptive. The former allows to discard packets stored in the queue, while the latter does not, i.e. whenever a packet is accepted to the queue it has to be eventually transmitted. In both cases a packet can be dropped at its arrival.

Aiello *et al.* [2] initiated the study of different queuing policies for the 2-value non-preemptive model in which each packet has a value of either 1 or $\alpha > 1$. Andelman *et al.* [4] later showed tight bounds for this case. The preemptive 2-value single-queue model was initially studied by Kesselman and Mansour [14], followed by Lotker and Patt-Shamir [17] who showed almost tight bounds. The general preemptive single-queue model, where packets can take arbitrary values, was investigated by Kesselman *et al.* [13], who proved that the natural greedy algorithm is 2-competitive (specifically $2\alpha/(1 + \alpha)$-competitive where $\alpha \geq 1$ is the ratio between the largest value to the smallest one).

The natural greedy preemptive admission control strategy for a single queue studied in [13] is defined as follows: Enqueue a new packet if the queue is not full, or a packet with the smallest value in the queue has a lower value than the new packet. In the latter case a smallest value packet is discarded.

Kesselman *et al.* [15] were the first to show a preemptive algorithm whose competitiveness is strictly below 2, followed by Bansal *et al.* [10] who presented a 1.75-competitive algorithm.

Multi-Queue: The multi-queue QoS switching model that was originally introduced in [7]. In this model we have a switch with m incoming FIFO queues with bounded capacities and one output port. At each time step new packets arrive to each of the queues. Additionally, at each time step the switch selects one non-empty queue and transmits the packet at the head of the queue through the output port. As before, the goal is to maximize the total value of transmitted packets.

The results for a single queue were generalized for multi-queue switches with arbitrary number of input queues in [7], by a general reduction from the multi-queue model to the single-queue model. Specifically, a 4-competitive algorithm is presented in [7] for the weighted multi-queue switch problem. An improved 3-competitive algorithm was shown in [9]. The 3-competitive algorithm is the natural greedy which works as follows. Use the greedy single-queue policy in all m incoming queues to handle admission control. At each time step, transmit the packet with the largest value among all packets at the head of the queues.

The multi-queue switch model has been also investigated for the special case of unit-value packets, which corresponds to IP networks. First, the result from [7] shows that any algorithm that transmits any packet if exists is 2-competitive. Albers and Schmidt [3] showed that any greedy algorithm for the unit-value problem is not better than 2-competitive. In addition, they introduced a deterministic 1.89-competitive algorithm for this problem; A randomized 1.58-competitive algorithm was previously shown in [7]. Recently, a deterministic 1.58-competitive algorithm was shown in [6], for the case where the size of the queues is quite large compared with their number.

CIOQ Switch: To date, the most general switching model that has been studied using competitive analysis is CIOQ (Combined Input and Output Queued) architecture. A CIOQ switch with speedup $S \geq 1$ is an $N \times N$ switch, with N input ports and N output ports. The internal fabric that connects the input and output FIFO queues is S times faster than the queues. A switching policy for a CIOQ switch consists of two components. First, an admission control policy to determine the packets stored in the bounded-capacity queues. Second, a scheduling strategy to decide which packets are transferred from input queues to output queues through the intermediate fabric at each time step. The goal is to maximize the total value of packets transmitted from the switch.

The online problem of maximizing the total throughput of a CIOQ switch was initiated by Kesselman and Rosén in [16]. For the special case of unit-value packets, they proved that the greedy algorithm is 2-competitive for a speedup of 1 and 3-competitive for any speedup. For the general case they obtained non-constant bounds of $4S$ and $\log \alpha$, where α is the ratio between the largest and smallest values. Recently, a constant (about 9) competitive algorithm was presented in [8] for the general CIOQ model.

Multiple-Node Networks: The simplest network one may consider has a topology of a line of length k, where node i is connected to node $i + 1$ by a unidirectional link, and contains a fixed-size FIFO queue to store the packets waiting to be transmitted. At each time step new packets may arrive online to the network nodes, each is associated with a value and a destination node. Additionally, each node can transmit the packet at the head of its queue to the next node. The goal is to maximize the total value of packets that were delivered to their destination. The case $k = 1$ corresponds to single-queue. The unweighted version of the line model, in which all packets have unit value, was investigated by Aiello *et al.* [1] who proved that the greedy algorithm is $O(k)$-competitive. This was generalized and improved by [9] that showed that the natural greedy algorithm is $(k + 1)$-competitive for the weighted problem.

The Zero-One Principle: While different techniques were used to analyze algorithms in various switching models, there was one common property: analysis of 2-value sequences, in which packets can take only 2 distinct values, was always substantially easier compared with arbitrary packet sequences. Moreover, many results are known only for restricted value sequences, since handling the state of a system containing packets with arbitrary values is significantly more involved. Motivated by this, [9] introduced the zero-one principle for switching networks. This principle applies to all *comparison-based* switching algorithms, that base their decisions on the relative order between packet values. The principle says that in order to prove that an algorithm achieves c-approximation it is sufficient to prove that it achieves c-approximation with respect to sequences composed solely of 0's and 1's, where ties between packets with equal values may be broken arbitrarily. One might have assumed that without loss of generality there are no 0-value packets in the input sequence since those could have been dropped. Indeed, the optimal solution may ignore all 0-value packets, however, the comparison-based algorithm may not, since it only regards the relative order between values. The zero-one principle was applied, among others, to get the 3-competitive algorithm for the multi-queue switch as well as the $k+1$ -competitive algorithm for the line.

References

1. W. Aiello, R. Ostrovsky, E. Kushilevitz, and A. Rosén. Dynamic routing on networks with fixed-size buffers. In *Proc. 14th ACM-SIAM Symp. on Discrete Algorithms*, pages 771–780, 2003.
2. W. A. Aiello, Y. Mansour, S. Rajagopolan, and A. Rosén. Competitive queue policies for differentiated services. In *Proceedings of the IEEE INFOCOM 2000*, pages 431–440.
3. S. Albers and M. Schmidt. On the performance of greedy algorithms in packet buffering. In *Proc. 36th ACM Symp. on Theory of Computing*, pages 35–44, 2004.
4. N. Andelman, Y. Mansour, and A. Zhu. Competitive queueing policies for QoS switches. In *Proc. 14th ACM-SIAM Symp. on Discrete Algorithms*, pages 761–770, 2003.
5. M. Andrews, B. Awerbuch, A. Fernández, J. Kleinberg, T. Leighton, and Z. Liu. Universal stability results for greedy contention-resolution protocols. In *Proc. 37th IEEE Symp. on Found. of Comp. Science*, pages 380–389, 1996.
6. Y. Azar and M. Litichevskey. Maximizing throughput in multi-queue switches. In *Proc. 12th Annual European Symposium on Algorithms*, pages 53–64, 2004.
7. Y. Azar and Y. Richter. Management of multi-queue switches in QoS networks. In *Proc. 35th ACM Symp. on Theory of Computing*, pages 82–89, 2003.
8. Y. Azar and Y. Richter. An improved algorithm for CIOQ switches. In *Proc. 12th Annual European Symposium on Algorithms*, pages 65–76, 2004.
9. Y. Azar and Y. Richter. The zero-one principle for switching networks. In *Proc. 36th ACM Symp. on Theory of Computing*, 2004. 64–71.
10. N. Bansal, L. Fleischer, T. Kimbrel, M. Mahdian, B. Schieber, and M. Sviridenko. Further improvements in competitive guarantees for QoS buffering. In *Proc. 31st International Colloquium on Automata, Languages, and Programming*, pages 196–207, 2004.
11. A. Birman, H. R. Gail, S. L. Hantler, Z. Rosberg, and M. Sidi. An optimal service policy for buffer systems. *Journal of the Association Computing Machinery (JACM)*, 42(3):641–657, 1995.
12. A. Borodin, J.Kleinberg, P. Raghavan, M. Sudan, and D. Williamson. Adversarial queuing theory. In *Proc. 28th ACM Symp. on Theory of Computing*, pages 376–385, 1996.
13. A. Kesselman, Z. Lotker, Y. Mansour, B. Patt-Shamir, B. Schieber, and M. Sviridenko. Buffer overflow management in QoS switches. In *Proc. 33rd ACM Symp. on Theory of Computing*, pages 520–529, 2001.
14. A. Kesselman and Y. Mansour. Loss-bounded analysis for differentiated services. In *Proc. 12th ACM-SIAM Symp. on Discrete Algorithms*, pages 591–600, 2001.
15. A. Kesselman, Y. Mansour, and R. van Stee. Improved competitive guarantees for QoS buffering. In *Proc. 11th Annual European Symposium on Algorithms*, pages 361–372, 2003.

16. A. Kesselman and A. Rosén. Scheduling policies for CIOQ switches. In *Proceedings of the 15th Annual ACM Symposium on Parallel Algorithms and Architectures*, pages 353–362, 2003.

17. Z. Lotker and B. Patt-Shamir. Nearly optimal fifo buffer management for diffserv. In *Proc. 21st ACM Symp. on Principles of Distrib. Computing*, pages 134–143, 2002.

18. M. May, J. C. Bolot, A. Jean-Marie, and C. Diot. Simple performance models of differentiated services for the internet. In *Proceedings of the IEEE INFOCOM 1999*, pages 1385–1394.

Approximation Algorithms for Mixed Fractional Packing and Covering Problems

Klaus Jansen

Institut für Informatik und Praktische Mathematik, Universität Kiel,
Olshausenstr. 40, 24098 Kiel, Germany
kj@informatik.uni-kiel.de*

We study general mixed fractional packing and covering problems (MPC_ϵ) of the following form: Given a vector $f : B \to \mathrm{I\!R}_+^M$ of M nonnegative continuous convex functions and a vector $g : B \to \mathrm{I\!R}_+^M$ of M nonnegative continuous concave functions, two M - dimensional nonnegative vectors a, b, a nonempty convex compact set B and a relative tolerance $\epsilon \in (0, 1)$, find an approximately feasible vector $x \in B$ such that $f(x) \leq (1 + \epsilon)a$ and $g(x) \geq (1 - \epsilon)b$ or find a proof that no vector is feasible (that satisfies $x \in B$, $f(x) \leq a$ and $g(x) \geq b$).

The fractional packing problem with convex constraints, i.e. to find $x \in B$ such that $f(x) \leq (1 + \epsilon)a$, is solved in [4, 5, 8] by the Lagrangian decomposition method in $O(M(\epsilon^{-2} + \ln M))$ iterations where each iteration requires a call to an approximate block solver $ABS(p, t)$ of the form: find $\hat{x} \in B$ such that $p^T f(\hat{x}) \leq (1 + t)\Lambda(p)$ where $\Lambda(p) = \min_{x \in B} p^T f(x)$. Furthermore, Grigoriadis et al. [6] proposed also an approximation algorithm for the fractional covering problem with concave constraints, i.e. to find $x \in B$ such that $g(x) \geq (1 - \epsilon)b$, within $O(M(\epsilon^{-2} + \ln M))$ iterations where each iteration requires here a call to an approximate block solver $ABS(q, t)$ of the form: find $\hat{x} \in B$ such that $q^T g(\hat{x}) \geq (1 - t)\Lambda(q)$ where $\Lambda(q) = \max_{x \in B} q^T g(x)$. Both algorithms solve also the corresponding min-max and max-min optimization variants within the same number of iterations. Furthermore, the algorithms can be generalized to the case where the block solver has arbitrary approximation ratio [7, 8, 9].

Further interesting algorithms for the fractional packing and fractional covering problem with linear constraints were developed by Plotkin et al. [13] and Young [15]. These algorithms have a running time that depends linearly on the width - an unbounded function of the input instance. Several relatively complicated techniques were proposed to reduce this dependence. Garg and Könemann [3] described a nice algorithm for the fractional packing problem with linear constraints that needs only $O(M\epsilon^{-2} \ln M)$ iterations.

* Research of the author was supported in part by EU Thematic Network APPOL, Approximation and Online Algorithms, IST-2001-30012, by EU Project CRESCCO, Critical Resource Sharing for Cooperation in Complex Systems, IST-2001-33135 and by DFG Project, Entwicklung und Analyse von Approximativen Algorithmen für Gemischte und Verallgemeinerte Packungs- und Überdeckungsprobleme, JA 612/10-1. Part of this work was done while visiting the Department of Computer Science at ETH Zürich.

G. Persiano and R. Solis-Oba (Eds.): WAOA 2004, LNCS 3351, pp. 6–8, 2005.

For the mixed packing and covering problem (with linear constraints), Plotkin et al. [13] proposed also approximation algorithms where the running time depends on the width. Young [16] described an approximation algorithm for a special mixed packing and covering problem with linear constraints and special convex set $B = \mathbb{R}^N_+$. The algorithm has a running time of $O(M^2 \epsilon^{-2} \ln M)$. Recently, Fleischer [1] gave an approximation scheme for the optimization variant (minimizing $c^T x$ such that $Cx \geq b$, $Px \leq a$ and $x \geq 0$ where a, b, and c are nonnegative integer vectors and P and C are nonnegative integer matrices).

Young [16] posed the following interesting open problem: find an efficient width-independent Lagrangian-relaxation algorithm for the mixed packing and covering problem (with linear constraints): find $x \in B$ such that $Px \leq (1 + \epsilon)a$, $Cx \geq (1 - \epsilon)b$, where P, C are nonnegative matrices, a, b are nonnegative vectors and B is a polytope that can be queried by an optimization oracle (given a vector c, return $x \in B$ minimizing $c^T x$) or some other suitable oracle.

New results: We found an approximation algorithm for the general mixed problem with M convex and M concave functions f_m, g_m that uses an suitable oracle of the form: find $\hat{x} \in B$ such that $p^T f(\hat{x}) \leq \sum_{m=1}^M p_m$ and $q^T g(\hat{x}) \geq \sum_{m=1}^M q_m$ [10]. The algorithm uses $O(M \epsilon^{-2} \ln(M \epsilon^{-1}))$ iterations or coordination steps, where in each iteration an oracle of the form above is called. Recently we found an improved width-independent Lagrangian-relaxation algorithm for the general mixed problem [11]. The algorithm uses a variant of the Lagrangian or price directive decomposition method. This is an iterative strategy that solves (MPC_ϵ) by computing a sequence of triples (p, q, x) as follows. A coordinator uses the current vector $x \in B$ to compute two price vectors $p = p(x) \in \mathbb{R}^M_+$ and $q = q(x) \in \mathbb{R}^M_+$ with $\sum_{m=1}^M p_m + q_m = 1$. Then the coordinator calls here a feasibility oracle to compute a solution $\hat{x} \in B$ of the block problem $BP(p, q, t)$

$$\text{find } \hat{x} \in B \text{ s.t. } p^T f(\hat{x})/(1 + t) \leq q^T g(\hat{x})(1 + t) + 2\bar{p} - 1,$$

(where $t = \Theta(\epsilon)$ and $\bar{p} = \sum_{m=1}^M p_m$) and makes a move from x to $(1 - \tau)x + \tau \hat{x}$ with an appropriate step length $\tau \in (0, 1)$. Such a iteration is called a coordination step. In case $\bar{p}$ is close to $1/2$, we use a slightly different block problem $BP'(p, q, t)$ of the form:

$$\text{find } \hat{x} \in B \text{ s.t. } p^T f(\hat{x})/(1 + 8t) \leq q^T g(\hat{x})(1 + 8t) + (2\bar{p} - 1 - t).$$

Our main result is the following: There is an approximation algorithm that for any given accuracy $\epsilon \in (0, 1)$ solves the general mixed fractional packing and covering problem (MPC_ϵ) within

$$N = O(M(\epsilon^{-2} \ln \epsilon^{-1} + \ln M))$$

iterations or coordination steps, where each of which requires a call to the block problem $BP(p, q, t)$ or $BP'(p, q, t)$.

Independently, Khandekar and Garg [2] proposed an approximation algorithm for the general mixed problem that uses $O(M \epsilon^{-2} \ln M)$ iterations or coordination steps.

References

1. L. Fleischer, A fast approximation scheme for fractional covering problems with variable upper bounds, *Proceedings of the 15th ACM-SIAM Symposium on Discrete Algorithms*, SODA 2004.
2. N. Garg and R. Khandekar, personal communication, 2004.
3. N. Garg and J. Könemann, Fast and simpler algorithms for multicommodity flow and other fractional packing problems, *Proceedings of the 39th IEEE Annual Symposium on Foundations of Computer Science*, FOCS 1998, 300-309.
4. M.D. Grigoriadis and L.G. Khachiyan, Fast approximation schemes for convex programs with many blocks and coupling constraints, *SIAM Journal on Optimization*, 4 (1994), 86-107.
5. M.D. Grigoriadis and L.G. Khachiyan, Coordination complexity of parallel price-directive decomposition, *Mathematics of Operations Research*, 2 (1996), 321-340.
6. M.D. Grigoriadis, L.G. Khachiyan, L. Porkolab and J. Villavicencio, Approximate max-min resource sharing for structured concave optimization, *SIAM Journal on Optimization*, 41 (2001), 1081-1091.
7. K. Jansen and L. Porkolab, On preemptive resource constrained scheduling: polynomial-time approximation schemes, *Proceedings of the 9th International Conference on Integer Programming and Combinatorial Optimization*, IPCO 2002, LNCS 2337, 329-349.
8. K. Jansen and H. Zhang, Approximation algorithms for general packing problems with modified logarithmic potential function, *Proceedings of the 2nd IFIP International Conference on Theoretical Computer Science*, TCS 2002, Montreal, Kluwer Academic Publisher, 2002, 255-266.
9. K. Jansen, Approximation algorithms for the general max-min resource sharing problem: faster and simpler, *Proceedings of the 9th Scandinavian Workshop on Algorithm Theory*, SWAT 2004, LNCS 3111, 311-322.
10. K. Jansen, Approximation algorithms for mixed fractional packing and covering problems, *Proceedings of the 3rd IFIP Conference on Theoretical Computer Science*, TCS 2004, Toulouse, Kluwer Academic Publisher, 2004, 223-236.
11. K. Jansen, Improved approximation algorithms for mixed fractional packing and covering problems, unpublished manuscript.
12. J. Könemann, Fast combinatorial algorithms for packing and covering problems, Diploma Thesis, Max-Planck-Institute for Computer Science Saarbrücken, 2000.
13. S.A. Plotkin, D.B. Shmoys, and E. Tardos, Fast approximation algorithms for fractional packing and covering problems, *Mathematics of Operations Research*, 20 (1995), 257-301.
14. J. Villavicencio and M. D. Grigoriadis, Approximate Lagrangian decomposition with a modified Karmarkar logarithmic potential, *Network Optimization*, Lecture Notes in Economics and Mathematical Systems, 450 (1997), 471-485.
15. N.E. Young, Randomized rounding without solving the linear program, *Proceedings of the 6th ACM-SIAM Symposium on Discrete Algorithms* SODA 1995, 170-178.
16. N.E. Young, Sequential and parallel algorithms for mixed packing and covering, *Proceedings of the 42nd Annual IEEE Symposium on Foundations of Computer Science*, FOCS 2001, 538-546.

Minimum Sum Multicoloring on the Edges of Planar Graphs and Partial k-Trees*
(Extended Abstract)

Dániel Marx

Department of Computer Science and Information Theory,
Budapest University of Technology and Economics,
Budapest H-1521, Hungary
dmarx@cs.bme.hu

Abstract. The edge multicoloring problem is that given a graph G and integer demands $x(e)$ for every edge e, assign a set of $x(e)$ colors to edge e, such that adjacent edges have disjoint sets of colors. In the *minimum sum edge multicoloring problem* the *finish time* of an edge is defined to be the highest color assigned to it. The goal is to minimize the sum of the finish times. The main result of the paper is a polynomial time approximation scheme for minimum sum multicoloring the edges of planar graphs and partial k-trees.

1 Introduction

In this paper we study an edge multicoloring problem that is motivated by applications in scheduling. We are given a graph with an integer demand $x(e)$ for each edge e. A *multicoloring* is an assignment of a set of $x(e)$ colors to each edge e such that the colors assigned to adjacent edges are disjoint. In multicoloring problems the usual aim is to minimize the *makespan* of the coloring, that is, the total number of different colors used. However, in this paper a different optimization goal is studied, which is related to minimizing the average completion time in scheduling problems. Given a multicoloring, the *finish time* of an edge is defined to be the highest color assigned to it. In the *minimum sum edge multicoloring problem* the goal is to minimize the sum of the finish times.

An application of edge coloring is to model dedicated scheduling of biprocessor tasks. The vertices correspond to the processors and each edge $e = uv$ corresponds to a job that requires $x(e)$ time units of simultaneous work on the two preassigned processors u and v. The colors correspond to the available time slots: by assigning $x(e)$ colors to edge e, we select the $x(e)$ time units when the job corresponding to e is executed. A processor cannot work on two jobs at the same time, this corresponds to the requirement that a color can appear at most

---

* Research is supported in part by grants OTKA 44733, 42559 and 42706 of the Hungarian National Science Fund.

G. Persiano and R. Solis-Oba (Eds.): WAOA 2004, LNCS 3351, pp. 9–22, 2005.
© Springer-Verlag Berlin Heidelberg 2005

once on the edges incident to a vertex. The finish time of edge e corresponds to the time slot when job e is finished, therefore minimizing the sum of the finish times is the same as minimizing the sum of completion times of the jobs. Using the terminology of scheduling theory, we minimize the mean flow time, which is a well-studied optimization goal in the scheduling literature. Such biprocessor tasks arise when we want to schedule file transfers between processors [4] or the mutual diagnostic testing of processors [7]. Note that it is allowed that a job is interrupted and continued later: the set of colors assigned to an edge does not have to be consecutive, hence our problem models preemptive scheduling.

In [11] it is shown that for trees the problem is NP-hard, but admits a polynomial time approximation scheme (PTAS). In this paper we extend the PTAS to partial k-trees and planar graphs. The problem is NP-hard for partial 2-trees and planar bipartite graphs even in the unit demand case [9], hence the approximation schemes given in this paper cannot be improved to exact polynomial time algorithms (assuming P $\neq$ NP).

Recently, the vertex coloring version of minimum sum multicoloring was investigated by several papers [2, 5, 6, 10]. In [5, 6] PTAS is given for the vertex coloring version of the problem in the case when the graph is a tree, partial k-tree, or a planar graph. The line graph of a bounded degree partial k-tree has bounded treewidth, hence the PTAS of [5] for vertex coloring partial k-trees can be used for the edge coloring of bounded degree partial k-trees. However, in Section 3 we present a more efficient, linear time PTAS for edge coloring such graphs. In Section 4 a PTAS is given for general partial k-trees by reducing the problem to the bounded degree case. In Section 5 the PTAS is extended to planar graphs: using standard techniques (the layering method of Baker [1]) we show how the PTAS for partial k-trees can be used for planar graphs.

2   Preliminaries

The problem considered in this paper is the edge coloring version of minimum sum multicoloring, which can be stated formally as follows:

Minimum Sum Edge Multicoloring (SEMC)
Input: A graph $G(V, E)$ and a *demand* function $x \colon E \to \mathbb{N}$.
Output: A *multicoloring* $\Psi \colon E \to 2^{\mathbb{N}}$ such that $|\Psi(e)| = x(e)$ for every edge e, and $\Psi(e_1) \cap \Psi(e_2) = \emptyset$ if e_1 and e_2 are adjacent in G.
Goal: The *finish time* of edge e in coloring Ψ is the highest color assigned to it, $f_\Psi(e) = \max\{c : c \in \Psi(e)\}$. The goal is to minimize $f_\Psi(G) = \sum_{e \in E} f_\Psi(e)$, the *sum* of the coloring Ψ.

We extend the notion of finish time to a set E' of edges by defining $f_\Psi(E') = \sum_{e \in E'} f_\Psi(e)$. Given a graph G and a demand function $x(e)$ on the edges of G, the minimum sum that can be achieved is denoted by $\mathrm{OPT}(G, x)$.

As in [5, 6, 11], we divide the infinite color spectrum into geometrically increasing layers. For some $\epsilon > 0$ and integer $\ell \geq 0$, the (ϵ, ℓ)-*decomposition* divides the set of colors into *layers* $L_0, L_1, \ldots$ and *zones* $Z_0, Z_1, \ldots, Z_\ell$. The layers are

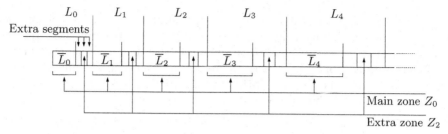

Fig. 1. The decomposition of the colors into layers ($\ell = 3$)

of geometrically increasing size: layer L_i contains the range of colors from q_i to $q_{i+1} - 1$, where $q_i = \lfloor (1 + \epsilon)^i \rfloor$. (If $q_i = q_{i+1}$, then layer L_i is empty). Denote by $Q_i = |L_i| = q_{i+1} - q_i$ the size of the ith layer. The total size of layers L_0, L_1, ..., L_i is $q_{i+1} - 1$. Later we will use that $(1 + 2\epsilon)q_i \geq q_{i+1} - 1$ for every $i \geq 1$. That is, if we replace a color from layer L_i with another color from L_i, then the new color is at most $(1 + 2\epsilon)$ times larger than the original.

Layer L_i is divided into two parts: the first $\frac{1}{1+\epsilon\ell}Q_i$ colors form the *main block* of layer L_i and the remaining $\frac{\epsilon\ell}{1+\epsilon\ell}Q_i$ colors the *extra block* (see Fig. 1). The main block of layer L_i is denoted by $\overline{L}_i$. The union of the main block of every layer L_i is the *main zone* Z_0. Divide the extra block of every layer L_i into ℓ equal parts: these are the ℓ *extra segments* of L_i. The union of the jth extra segment of every layer L_i forms the jth *extra zone* Z_j. Each extra zone contains $\frac{\epsilon}{1+\epsilon\ell}Q_i$ colors from layer L_i. We ignore rounding problems here. It can be shown that by defining the zones carefully, one can achieve the following:

Lemma 1 ([11]). *For given ℓ and $\epsilon \leq \frac{1}{2\ell}$, the (ϵ, ℓ)-decomposition of the colors has the following properties:*

(a) *For every $c \geq 1$, zone Z_0 contains at least c colors not greater than $\lfloor (1+\epsilon\ell)c \rfloor$.*
(b) *For every $c \geq 1$ and $1 \leq j \leq \ell$, zone Z_j contains at least c colors not greater than $(2/\epsilon) \cdot c$.*

Given a multicoloring Ψ, the operation (ϵ, ℓ)-*augmentation* creates a multicoloring Φ the following way. Consider the (ϵ, ℓ)-decomposition of the colors, and if $\Psi(e)$ contains color c, then let $\Phi(e)$ contain instead the cth color from the main zone Z_0. By Lemma 1a, $f_\Phi(e) \leq \lfloor (1 + \epsilon\ell)f_\Psi(e) \rfloor$, thus this operation increases the sum by at most a factor of $(1 + \epsilon\ell)$. After the augmentation, the colors of the extra zones are not used, only the colors of the main zone.

A *tree decomposition* of $G(V, E)$ is a tree $T(U, F)$ together with a bag $U_x \subseteq V$ for each $x \in U$ such that

- for every $uv \in E$, there is an $x \in U$ with $u, v \in U_x$, and
- for every $v \in V$, the bags containing v induce a connected subtree of T.

The *width* of the decomposition is $\max_{x \in U} |U_x| - 1$, and the *treewidth* of a graph is the smallest width that its tree decompositions can have. A *partial k-tree* is a graph with treewidth at most k. For background on partial k-trees and treewidth, the reader is referred to [3].

In [11], the following scaling property is proved for trees: if the demand of every edge is increased by at most a factor of $1 + \epsilon$, then the sum increases by at most a factor of $1 + \epsilon$. As an application of the layer and zones defined above, we prove a weaker property for partial k-trees: a $1 + \epsilon$ increase of demand causes an at most $1 + 2(k + 1)\epsilon$ increase of the sum (Lemma 3). First we need a special orientation, which will be used later as well.

Lemma 2. *Let G be a partial k-tree. There is a proper $k + 1$-coloring of the vertices of G, and an orientation of the edges of G such that the outdegree of every vertex is at most k, and the outneighbors of a vertex have distinct colors.*

Proof. Consider a tree decomposition $T(U, F)$ of G. Assume that the tree T is rooted. For each vertex v of G, consider those nodes of T whose bags contain v, and let f_v be the highest such node (i.e., the node that is closest to the root).

The vertices can be colored with $k + 1$ colors such that the vertices in the same bag have distinct colors (that is, we take a coloring of the chordal graph induced by the decomposition).

Let $e = uv$ be an edge of G. The trees T_u and T_v intersect, hence either node $f_u \in U$ of u is descendant of $f_v \in U$ of v, or vice versa. If f_u is the descendant of f_v, then we direct e from u to v, otherwise we direct it from v to u. If $f_u = f_v$, then direct the edge uv arbitrarily. Notice that all the outneighbors of u are contained in the bag of f_u: if a subtree intersects T_u, and it contains an ancestor of f_u, then it has to contain f_u as well. Now it is clear that u has at most k outneighbors, and they have distinct colors.

Lemma 3. *Let (G, x) be an instance of SEMC and let x' be a demand function with $x'(e) \leq (1 + \epsilon) \cdot x(e)$ for every edge e. If G is a partial k-tree, then $\mathrm{OPT}(G, x') \leq (1 + 2(k + 1)\epsilon) \cdot \mathrm{OPT}(G, x)$.*

Proof. Let Ψ be an optimum coloring of (G, x), and let Ψ' be the $(2\epsilon, k + 1)$-augmentation of Ψ. By Lemma 1, we have $f_{\Psi'}(e) \leq (1 + 2(k+1)\epsilon)f_\Psi(e)$ for every edge e.

Consider the coloring and orientation given by Lemma 2. Let $e_1, \ldots, e_\ell$ be those edges that enter v, assume that $f_{\Psi'}(e_1) \leq f_{\Psi'}(e_2) \leq \cdots \leq f_{\Psi'}(e_\ell)$. Edge e_i requires $x'(e_i) - x(e_i) \leq \epsilon x(e_i)$ extra colors to satisfy demand function x'. If vertex v has color c, then we use extra zone Z_c to give additional colors to these edges. The first $\epsilon x(e_1)$ colors of Z_c are given to edge e_1, the next $\epsilon x(e_2)$ colors are given to e_2. It is clear that no conflict arises with the assignment of these new colors. If two edges conflict in zone Z_c, then their end vertices have the same color c. By Lemma 2, this is only possible if they enter the same vertex, but in this case the construction ensures that the edges receive different colors.

We show that these additional colors do not increase the finish time of the edges. The finish time of e_i is clearly at least $\sum_{j=1}^{i} x(e_i)$ in Ψ'. The last color given to edge e_i is color $\sum_{j=1}^{i} \epsilon x(e_i)$ from extra zone Z_c. However, by Lemma 1b, zone Z_c contains at least $\sum_{j=1}^{i} \epsilon x(e_i)$ colors below $\frac{2}{2\epsilon} \cdot \sum_{j=1}^{i} \epsilon x(e_i) = \sum_{j=1}^{i} x(e_i) \leq f_{\Psi'}(e_i)$. Therefore it remains true that $f_{\Psi'}(e) \leq (1 + 2(k+1)\epsilon)f_\Psi(e)$ for every edge e.

3   Bounded Degree Partial k-Trees

In this section we show that SEMC admits a linear time PTAS for bounded degree partial k-trees. If G is a partial k-tree with maximum degree D, then its line graph has treewidth at most $(k+1)D - 1$ (see [3–Lemma 32]). Therefore the PTAS given by Halldórsson and Kortsarz [5] for the minimum sum vertex coloring of partial k-trees can be used to edge color bounded degree partial k-trees. However, the time complexity of their algorithm is $n^{O(k^2/\epsilon^5)}$. We show here that for line graphs of bounded degree partial k-trees the running time can be reduced to linear.

The main idea of the PTAS in [5] is that one can assign a polynomial number of color sets to each vertex such that the graph has a good approximate coloring where every vertex uses one of the sets assigned to it. This is proved by using probabilistic arguments and by transforming the solution to a standard form. The best coloring using the selected color sets can be found in polynomial time with the standard dynamic programming algorithm of partial k-trees. In [11] a similar path is taken to give a linear time PTAS for SEMC in bounded degree trees: it is shown that a constant number of color sets can be assigned to the edges such that there is good approximation with these sets. In this case the probabilistic argument of [5] can be replaced by a simple greedy algorithm. Here we construct a constant number of color sets for the edges of bounded degree partial k-trees, yielding a linear time PTAS. The construction depends on some strong results concerning the asymptotics of the chromatic index.

3.1   Approximating the Makespan

As explained in the introduction, in this paper our goal is to minimize the sum of finish times, and we are not interested in minimizing the makespan. However, (as in [5] and [11]) the algorithm for minimizing the sum requires the use of some results on approximating the makespan.

Minimizing the makespan is closely related to the chromatic index problem. Given a graph G and a demand function x on the edges, let (G, x) denote the multigraph obtained by replacing every edge e by $x(e)$ parallel edges, and let $\chi'(G, x)$ be the chromatic index of (G, x). It is clear that the minimum makespan with demand x is $\chi'(G, x)$.

Shannon [12] has shown that $\chi'(G) \leq \lfloor 3\Delta(G)/2 \rfloor$ for every multigraph G, where $\Delta(G)$ is the maximum degree of G. Since $\Delta(G)$ is a lower bound on the fractional chromatic index $\chi'^*(G)$, this implies that $\chi'(G) \leq 3\chi'^*(G)/2$. The following theorem shows that better bounds can be given for $\chi'(G)$ if $\chi'^*(G)$ is large:

Theorem 1 (Kahn, [8]).
For every $\gamma > 0$, there exists $D(\gamma)$ such that for any multigraph G with $\chi'^*(G) > D(\gamma)$, we have $\chi' < (1 + \gamma)\chi'^*(G)$.

We use this result to construct a constant number of color sets that can approximate the makespan:

Lemma 4. *For every $\epsilon > 0$ and integers $\Delta, t > 0$, there is a family of color sets $\mathcal{C}_{t,\Delta,\epsilon}$ such that if the maximum degree of G is Δ, and graph G with demand $x(e)$ has a coloring with makespan t, then it has a coloring with makespan at most $(1+\epsilon)t$ such that every edge receives a color set from $\mathcal{C}_{t,\Delta,\epsilon}$. Moreover, the size of $\mathcal{C}_{t,\Delta,\epsilon}$ depends only on ϵ and Δ.*

Proof. Set $\gamma = \epsilon/6$, and let $x_1(e) = \lfloor 4D(\gamma)\Delta/(\gamma t) \cdot x(e)\rfloor$, where $D(\gamma)$ is from Theorem 1. It follows that $\chi'^*(G, x_1) \leq 4D(\gamma)\Delta/(\gamma t) \cdot \chi'^*(G, x) \leq 4D(\gamma)\Delta/\gamma \cdot \chi'(G, x)/t \leq 4D(\gamma)\Delta/\gamma$. Let $x_2(e) = \lceil 4D(\gamma)\Delta/(\gamma t) \cdot x(e)\rceil$. In x_2 every edge requires at most one color more than in x_1, therefore by Vizing's theorem, x_2 can be satisfied with $\Delta + 1$ new colors: $\chi'^*(G, x_2) \leq \chi'^*(G, x_1) + \Delta + 1$. If $\chi'^*(G, x_2) > 2D(\gamma)\Delta/\gamma$, then we can apply Theorem 1 to show that $\chi'(G, x_2) \leq (1 + \gamma)\chi'^*(G, x_2) \leq (1 + \gamma)(\chi'^*(G, x_1) + \Delta + 1) \leq (1 + \gamma)^2\chi'^*(G, x_1) \leq (1 + \gamma)^2 4D(\gamma)\Delta/\gamma$. If $\chi'^*(G, x_2) \leq 2D(\gamma)\Delta/\gamma$, then $\chi'(G, x_2) \leq 3D(\gamma)\Delta/(\gamma)$ follows simply from $\chi' \leq 3\chi'^*/2$. Now if we repeat every color $\lceil \gamma t/(4D(\gamma)\Delta)\rceil$ times in the coloring of (G, x_2), then we obtain a coloring of (G, x) with at most $\chi'(G, x_2) \cdot \lceil \gamma t/(4D(\gamma)\Delta)\rceil \leq \chi'(G, x_2) \cdot (1+\gamma) \cdot \gamma t/(4D(\gamma)\Delta) \leq (1+\gamma)^3 t \leq (1+\epsilon)t$ colors (in the first inequality we assume that t is sufficiently large). Let $\mathcal{C}_{t,\Delta,\epsilon}$ contain those color sets that can be obtained from a subset of $\{1, \ldots, \lfloor(1 + \epsilon)4D(\gamma)\Delta/\gamma\rfloor\}$ by repeating every color $\lceil \gamma t/(4D(\gamma)\Delta)\rceil$ times. What we have shown is that there is an $(1 + \epsilon)$-approximate coloring where every edge uses a set from $\mathcal{C}_{t,\Delta,\epsilon}$. The size of $\mathcal{C}_{t,\Delta,\epsilon}$ is $2^{\lfloor(1+\epsilon)4D(\gamma)\Delta/\gamma\rfloor} \leq 2^{O((1+\epsilon)D(\epsilon/4)\Delta/\epsilon)}$, which depends only on ϵ and Δ. $\qquad\square$

3.2   Approximating the Sum

To give a linear time PTAS for SEMC on almost bounded degree partial k-trees, first we construct a constant number of color sets for each edge. Moreover, we give a slightly more general algorithm for SEMC that works on "almost bounded degree" partial k-trees as well. A *pendant edge* is an edge that has a degree one end vertex, an *almost bounded degree* graph is a graph that has bounded degree after deleting the pendant edges. A non-pendant edge will be called a *core edge*, and the number of core edges incident to a vertex will be called the *core degree* of the vertex. Deleting the pendant edges of G gives the *core G_0* of G. The treewidth of the line graph of an almost bounded degree partial k-tree is not necessarily bounded, hence the vertex coloring algorithm of [5] cannot be used for such graphs.

Lemma 5. *For every $\epsilon_0 > 0$ and integers $k, y, D > 0$ there is a family of color sets $\mathcal{D}_{k,y,D,\epsilon_0}$ such that for every partial k-tree G with maximum core degree D and every demand function there is a $(1 + \epsilon_0)$-approximate coloring where every core edge e with demand y receives a color set from $\mathcal{D}_{k,y,D,\epsilon_0}$. Moreover, the size of $\mathcal{D}_{k,y,D,\epsilon_0}$ depends only on k, D, and ϵ_0.*

Proof. Let Ψ be an optimum coloring of the instance (G, x). With a series of transformations, we modify Ψ to a special form where only a constant number of possible color sets can appear at a given core edge. Moreover, the transformations will increase the sum only by a factor of $1 + \epsilon_0$. Most of the ideas in this proof are taken from [5].

Set $\epsilon := \epsilon_0/(10D)$ and let Ψ' be obtained by performing an $(\epsilon, D + 2)$-augmentation on Ψ. By Lemma 1a, we have $f_{\Psi'}(G) \le (1 + (D + 2)\epsilon)f_\Psi(G)$.

The first step is to ensure that every core edge e uses the main zone only between $\epsilon x(e)/2$ and $2x(e)/\epsilon$. This means that e can use the main zone only from layer $\lfloor \log_{1+\epsilon} \epsilon x(e)/2 \rfloor$ to layer $\lceil \log_{1+\epsilon} 2x(e)/\epsilon \rceil$, that is, only from at most $\log_{1+\epsilon}((2x(e)/\epsilon)/(\epsilon x(e)/2)) + 2 = \log_{1+\epsilon} 4/\epsilon^2 + 2 = O(1/\epsilon \cdot \log 1/\epsilon)$ layers.

By Vizing's Theorem, one can assign a type $1, \ldots, D + 1$ to every core edge e such that adjacent edges have different types. If $f_{\Psi'(e)} > 2x(e)/\epsilon$ for a core edge e of type j, then modify $\Psi'(e)$ to be the first $x(e)$ colors of zone Z_j. By Lemma 1b, Z_j contains at least $x(e)$ colors not greater than $2x(e)/\epsilon$, therefore the modification does not increase the finish time of edge e.

If $\Psi'(e)$ contains colors from the main zone below $\epsilon x(e)/2$, then delete these colors and let $\Psi'(e)$ contain instead the first $\epsilon x(e)/2$ colors from zone Z_j. There are at least $\epsilon x(e)/2$ colors in Z_j below $2/\epsilon \cdot \epsilon x(e)/2 = x(e)$. Since the finish time of e is at least $x(e)$, hence this modification does not increase the finish time of e.

For every core edge e, define $x_i(e) := |\Psi'(e) \cap \overline{L}_i|$ the number of colors used by e in the main block of layer i. In the core graph G_0, the multicoloring problem with demand $x_i(e)$ has a solution with makespan $|\overline{L}_i|$: for example, $\Psi'(e) \cap \overline{L}_i$ is such a coloring. Therefore by Lemma 4, there is a coloring Ψ_i with makespan $\lfloor (1 + \frac{\epsilon}{2})|\overline{L}_i| \rfloor$ such that $\Psi_i(e) \in C_{|\overline{L}_i|, D, \epsilon/2}$. We let Ψ_i determine how the colors are used in the main block of layer $\overline{L}_i$. That is, if Ψ_i assigns color c to edge e, then we assign the cth color of $\overline{L}_i$ to e. However, Ψ_i uses $\lfloor (1 + \frac{\epsilon}{2})|\overline{L}_i| \rfloor$ colors, so we run out of the colors of $\overline{L}_i$. Extra zone Z_{D+2} is used to provide $\lfloor \frac{\epsilon}{2}|\overline{L}_i| \rfloor$ additional colors. We use the first $\lfloor \frac{\epsilon}{2}|\overline{L}_1| \rfloor$ colors of Z_{D+2} when recoloring $\overline{L}_1$, the next $\lfloor \frac{\epsilon}{2}|\overline{L}_2| \rfloor$ colors for the recoloring of $\overline{L}_2$, and so on. This means that when we recolor $\overline{L}_i$, only the first $\sum_{j=1}^{i} \lfloor \frac{\epsilon}{2}|\overline{L}_1| \rfloor$ colors of Z_{D+2} are used. Zone Z_{D+2} contains that many colors below $\frac{2}{\epsilon} \sum_{j=1}^{i} \lfloor \frac{\epsilon}{2}|\overline{L}_j| \rfloor \le \sum_{j=1}^{i} |\overline{L}_j|$ (Lemma 1b). Therefore we do not use colors above layer $\overline{L}_i$, and the last color of every edge remains in the same layer. As noted in Section 2, this implies that the finish time is increased by at most a factor of $1 + 2\epsilon$ for every edge.

In the previous paragraph we have recolored all the core edges at the same time, thus no conflict arises between the core edges, but there can be conflicts between a core edge and a pendant edge. However, at each vertex v, the number of colors used by the core edges from the main block of layer i did not increase. In fact, it is possible that it was decreased since in the recoloring extra zone Z_{D+2} was also used. Therefore there remains enough colors in $\overline{L}_i$ to satisfy the requirements of the pendant edges attached to v. This means that the last color of each pendant edge will remain in the same layer, hence the finish time of a pendant edge increases by at most a factor of $(1 + 2\epsilon)$. Thus the finish time of every edge is at most $(1 + 2\epsilon)(1 + (D + 2)\epsilon) \le (1 + \epsilon_0)$ times higher than in Ψ, and the coloring obtained is $(1 + \epsilon_0)$-approximate.

It is easy to show that only a constant number of different color sets can arise for a given edge e. This follows from the facts that e uses the main zone only in a constant number of different layers, and in each layer it has a constant number of different possibilities. □

The PTAS uses standard dynamic programming methods to find the best coloring with the selected color sets. However, there is a slight twist: the pendant edges have to be handled differently. The idea is that if the core edges incident to v are given a coloring, then this determines what is the best way of coloring the pendant edges incident to v. Details omitted.

Theorem 2. *For every $\epsilon_0 > 0$, and integers $k, D > 0$, there is a linear time algorithm that gives a $(1 + \epsilon_0)$-approximate solution to the* SEMC *problem for partial k-trees with maximum core degree D.* □

4   General Partial k-Trees

We prove that SEMC admits a linear time PTAS for partial k-trees. The main idea is to modify the graph to be an almost bounded degree graph. After these modifications the algorithm of Theorem 2 can be used, and we show that the coloring for the almost bounded degree graph can be transformed into a coloring of the original graph with only a small increase of the sum.

Theorem 3. *For every $\epsilon > 0$ and integer $k > 0$, there is a linear time algorithm that gives a $(1 + \epsilon_0)$-approximate solution to the* SEMC *problem for every partial k-tree G and demand function x_0.*

Proof. We show how to find a $(1 + K\epsilon)$-approximate solution in linear time, where K is a constant depending only on k. This implies that there is a $(1 + \epsilon)$-approximation algorithm for every k and $\epsilon > 0$. The algorithm consists of a series of phases, in the following we describe these phases.

Phase 1: Rounding the Demands. Let $x(e)$ be the smallest q_i that is not smaller than $x_0(e)$. Since $q_{i+1} \le (1 + \epsilon)^{i+1} \le (1 + \epsilon)(q_i + 1)$, thus $x(e) \le (1 + \epsilon)x_0(e)$. Therefore by Lemma 3, this modification increases the minimum sum by at most a factor of $1 + 2(k + 1)\epsilon$, hence a $(1 + O(\epsilon))$-approximation for (G, x) is also a $(1 + O(\epsilon))$-approximation for (G, x_0). An edge e with demand q_i will be called a *class i* edge (if $x(e) = q_i$ for more than one i, then take the smallest i).

Assume we have a subset of the edges that are incident to the same vertex, and there are exactly $c(i)$ edges of class i among them. The best way to color these edges is to order them by increasing demand size, and color them in this order. Therefore the sum of these edges in every coloring is at least

$$\sum_{i=1}^{\infty} \sum_{j=1}^{c(i)} \left(\sum_{k=1}^{i-1} c(k)q_k + jq_i \right) = \sum_{i=1}^{\infty} \left(c(i) \sum_{k=1}^{i-1} c(k)q_k + q_i \sum_{j=1}^{c(i)} j \right) =$$

$$\sum_{i=1}^{\infty} \left(c(i) \sum_{k=1}^{i-1} c(k)q_k + q_i(c(i) + 1)c(i)/2 \right) \qquad (1)$$

The parentheses in the first expression contains the finish time of the jth edge of class i. The first term is the contribution of the edges with class less than i, the second term is the contribution of the first j edges of class i.

Phase 2: Classifying the edges. Consider the $(k+1)$-coloring of the vertices and the orientation of the edges given by Lemma 2. The edges of the graph are divided into *large edges*, *small edges*, and *frequent edges*. It will be done in such a way that at most $D := 6/\epsilon^7$ large edges enter every vertex. Denote by $n(v, i)$ the number of class i edges entering v. Let $N(v)$ be the largest i such that $n(v, i) > 0$ and set $F := 6/\epsilon^5$. Let e be a class i edge entering v. If $n(v, i) > F$, then e is a *frequent edge*. If $n(v, i) \le F$ and $i \le N(v) - 1/\epsilon^2$, then e is a *small edge*. Otherwise, if $n(v, i) \le F$ and $i > N(v) - 1/\epsilon^2$, then e is a *large edge*. Clearly, at most $F \cdot 1/\epsilon^2 = 6/\epsilon^7 = D$ large edges can enter v: for each class $N(v)$, $N(v) - 1$, $\ldots$, $N(v) - 1/\epsilon^2 + 1$, there are at most F such edges.

Phase 3: Splitting the edges. The graph is split at the tail of every small and frequent edge. That is, if $e = \overrightarrow{uv}$ is a small or frequent edge, then add a new vertex u', and replace $\overrightarrow{uv}$ by $\overrightarrow{u'v}$. The resulting graph G' is an almost bounded degree graph. Deleting the pendant edges deletes every small and frequent edge, therefore only the large edges remain. We have seen that at most D large edges enter every node, and the outdegree of every vertex is at most k, thus the degree of the remaining graph is at most $D + k$. Therefore graph G' can be colored with the algorithm of Theorem 2 in linear time. This gives a $(1+\epsilon)$-approximate coloring Ψ_1 of G'. Coloring Ψ_1 can be used as a coloring for the original graph G', but in this case there might be conflicts: a small or frequent edge might be in conflict with the edges incident to its tail vertex. In the rest of the proof, we transform Ψ_1 into a proper coloring of G in such a way that the sum of the coloring does not increase too much. We will distinguish between the following 4 types of conflicts, they will be handled separately in the phases to follow:

1. Conflicts involving a small edge $\overrightarrow{uv}$.
2. Conflict between two frequent edges $\overrightarrow{uv_1}$ and $\overrightarrow{uv_2}$.
3. Conflict between a frequent edge $\overrightarrow{uv_1}$ and a large edge $\overrightarrow{uv_2}$.
4. Conflict between a frequent edge $\overrightarrow{uv}$ and a frequent or large edge $\overrightarrow{wu}$.

Phase 4: Small Edges. In this phase we resolve the conflicts of the first type. Consider the $(\epsilon, k+1)$-augmentation of the coloring Ψ_1, this results in a $(1+O(\epsilon))$-approximate coloring Ψ_2. We modify Ψ_2 in such a way that the small edges use only the extra zones. More precisely, if the head of a small edge e has color $r \in \{1, 2, \ldots, k+1\}$, then e is recolored using the colors in Z_r. Since the extra zones contain only a very small fraction of the color spectrum, the recoloring can significantly increase the finish time of the small edges, but not more than by a factor of $2/\epsilon$ (Lemma 1b). However, we show that the total demand of the small edges entering v is so small compared to the demand of the largest edge entering v, that their total finish time will be negligible, even after this large increase.

By definition, the largest edge entering v has demand $q_{N(v)}$. Let S_v be the set of small edges entering v. Let r be the color of vertex v. Color the edges in S_v one after the other, in the order of increasing demand size, using only the colors in Z_r. Call the resulting coloring Ψ_3. We claim that $f_{\Psi_3}(S_v) \le \epsilon q_{N(v)}$ for every node v, thus transforming Ψ_2 into Ψ_3 increases the total sum by at most $\sum_{v \in G} f_{\Psi_3}(S_v) \le \epsilon \sum_{v \in G} q_{N(v)} \le \epsilon f_{\Psi_2}(G)$ and $f_{\Psi_3}(G) \le (1+\epsilon) f_{\Psi_2}(G)$ follows. To give an upper bound on $f_{\Psi_3}(S_v)$, we assume the worst case, that is, $n(v, i) = F$ for every $i \le N(v) - 1/\epsilon^2$. Imagine first that the small edges are colored using the full color spectrum, not only with the colors of zone Z_r. Assume that the small edges are colored in the order of increasing demand size, and consider a class m edge e. In the coloring, only edges of class not greater than m are colored before e. Hence the finish time of e is at most

$$\sum_{i=0}^{m} n(v, i) q_i \le F \sum_{i=0}^{m} (1+\epsilon)^i \le 6(1+\epsilon)/\epsilon^6 \cdot (1+\epsilon)^m$$

$$\le 14/\epsilon^6 \cdot \frac{1}{2}(1+\epsilon)^m \le 14/\epsilon^6 \cdot \lfloor (1+\epsilon)^m \rfloor = 14/\epsilon^6 \cdot q_m.$$

That is, the finish time of an edge is at most $14/\epsilon^6$ times its demand (in the second inequality, we used $\sum_{i=0}^{m}(1+\epsilon)^i = ((1+\epsilon)^{m+1} - 1)/\epsilon < (1+\epsilon)^{m+1}/\epsilon$). Therefore the total finish time of the small edges is at most $14/\epsilon^6$ times the total demand, which is

$$\frac{14}{\epsilon^6} \sum_{i=0}^{N(v)-1/\epsilon^2} n(v, i) q_i \le \frac{84}{\epsilon^{11}} \sum_{i=0}^{N(v)-1/\epsilon^2} (1+\epsilon)^i \le \frac{85}{\epsilon^{12}} (1+\epsilon)^{N(v)-1/\epsilon^2}$$

$$\le \frac{85}{\epsilon^{12}} \cdot 2^{-1/\epsilon} \cdot (1+\epsilon)^{N(v)} \le \frac{\epsilon^2}{2} \cdot \frac{1}{2}(1+\epsilon)^{N(v)} \le \frac{\epsilon^2}{2} \cdot \lfloor (1+\epsilon)^{N(v)} \rfloor = \frac{\epsilon^2}{2} \cdot q_{N(v)}.$$

(In the third inequality we use $(1+\epsilon)^{1/\epsilon} \ge 2$, in the fourth inequality it is assumed that ϵ is sufficiently small that $2^{1/\epsilon} \ge 4 \cdot 85/\epsilon^{14}$ holds.) However, the small edges do not use the full color spectrum, only the colors in zone Z_r. By Lemma 1b, zone Z_r contains at least c colors up to $2/\epsilon \cdot c$, thus every finish time in the calculation above should be multiplied by at most $2/\epsilon$. Therefore the sum of the small edges is at most $2/\epsilon \cdot \epsilon^2/2 q_{N(v)} \le \epsilon q_{N(v)}$ as claimed, and Ψ_3 is a $(1 + O(\epsilon))$-approximate coloring.

Phase 5: Reordering the frequent edges. Now we have a coloring Ψ_3 of G such that the only type of conflict that is possible is between a frequent edge $\overrightarrow{uv}$ and an edge incident to u. In this section, we resolve the conflicts of the second type. Consider the $F(v, i)$ frequent edges of class i that enter v. We randomly reorder the color sets assigned to these edges. That is, take a uniformly distributed random permutation σ of the class i frequent edges entering v, and set $\Psi_4(e) = \Psi_3(\sigma(e))$ for these edges. Since all the class i edges have the same demand, a class i edge can receive the color set assigned to another edge of class i. Moreover, since the reordered edges have the same head v, the only conflicts that can arise are of types 2–4. We repeat the reordering independently for every v and i, and set $\Psi_4(e) = \Psi_3(e)$ for every small and large edge.

We argue that after the reordering, with high probability there are only relatively few type 2 conflicts, and these conflicts can be resolved with only a small increase of the sum. A frequent edge $\overrightarrow{uv_1}$ that conflicts with another frequent edge $\overrightarrow{uv_2}$ will be called a *marked* edge. Denote by $M(v, i)$ the number of class i marked edges entering v. The following claim bounds the expected number of marked edges:

Claim. For every v and i, $\mathrm{E}[M(v, i)] \le \epsilon^4 F(v, i)$.

Proof. Consider frequent edges $\overrightarrow{uv_1}$ of class i_1 and $\overrightarrow{uv_2}$ of class i_2. We bound the probability that they conflict. Let $a_1, \ldots, a_{F(v_1, i_1)}$ be the class i_1 frequent edges entering v_1 ordered by increasing finish time. For every $1 \le j \le F(v_1, i_1)$, denote by A_j the interval between the first and last color of $\Psi_3(a_1)$, i.e., $A_j = [\min\{c : c \in \Psi_3(a_j)\}, \max\{c : c \in \Psi_3(a_j)\}]$. Since the algorithm of Theorem 2 colors the frequent (pendant) edges entering v_1 one after the other, thus the intervals A_j are disjoint. The intervals B_j are similarly defined for $1 \le j \le F(v_2, i_2)$, but now we consider the class i_2 edges entering v_2. Let A (resp. B) be the interval between the first and last color of $\Psi_4(\overrightarrow{uv_1})$ (resp. $\Psi_4(\overrightarrow{uv_2})$). If $\overrightarrow{uv_1}$ and $\overrightarrow{uv_2}$ conflict in Ψ_4, then A and B intersect. By bounding the probability that $A \cap B \ne \emptyset$, we give a bound on the probability that the two edges are in conflict.

Because of the way the color sets were reordered, interval A can be any of the intervals A_j with probability $1/F(v_1, i_1)$, and similarly for the interval B. Since the reordering at vertices v_1 and v_2 are independent, $(A, B) = (A_{j_1}, B_{j_2})$ with probability $[F(v_1, i_1)F(v_2, i_2)]^{-1}$ for any pair (A_{j_1}, B_{j_2}). We show that the number of pairs (A_{j_1}, B_{j_2}) such that $A_{j_1} \cap B_{j_2} \ne \emptyset$ is at most $F(v_1, i_1) + F(v_2, i_2)$. The proof is by induction on $F(v_1, i_1) + F(v_2, i_2)$, the total number of intervals. If either $F(v_1, i_1)$ or $F(v_2, i_2)$ is zero, then the claim trivially holds. Assume without loss of generality that the right end point of A_1 is not greater than the right end point of B_1. If we remove interval A_1, then the number of non-disjoint pairs is decreased by at most one, since A_1 can intersect only B_1. Therefore the number of intervals is decreased by one, the number of non-disjoint pairs is decreased by at most one, completing the induction. Therefore there are at most $F(v_1, i_1) + F(v_2, i_2)$ non-disjoint pairs, and the probability that $A \cap B \ne \emptyset$ is at most

$$\frac{F(v_1, i_1) + F(v_2, i_2)}{F(v_1, i_1)F(v_2, i_2)} = \frac{1}{F(v_1, i_1)} + \frac{1}{F(v_2, i_2)} \le \epsilon^5/3.$$

Since the outdegree of u is at most k, thus edge $\overrightarrow{uv_1}$ can be in type 2 conflict with at most $k - 1$ possible edges. Therefore $\overrightarrow{uv_1}$ is marked with probability at most $(k - 1)\epsilon^5/3 \le \epsilon^4$. Thus the expected number of class i marked edges entering v is at most $\epsilon^4 F(v, i)$. □

Take an $(\epsilon, k+1)$-augmentation of Ψ_4. If vertex v has color r, then the marked edges entering v are recolored using extra zone Z_r. The edges are recolored in increasing order of demand size: first the class 1 edges, etc. It can be shown that the expected total sum of the recolored marked edges is at most $2/\epsilon \cdot \epsilon^4 = 2\epsilon^3$ times the total sum of the frequent edges: this follows from Lemma 1b, from the claim, and from the convexity of the expression (1). Thus by Markov's Inequality, with probability at least $1 - 2\epsilon^2$, the transformation increases the sum by at most a factor of $(1 + \epsilon)$ and the resulting coloring Ψ_5 is $(1 + O(\epsilon))$-approximate.

Phase 6: Resolving type 3 conflicts. We show that with high probability there are only few type 3 conflicts in Ψ_5, therefore they can be resolved with only a small increase of the sum. If a frequent edge $\overrightarrow{uv_1}$ conflicts with a large edge $\overrightarrow{uv_2}$, then we remove the conflicting colors from the large edge. We say that a large edge $\overrightarrow{uv}$ is *marked* if more than $\epsilon x(\overrightarrow{uv})/2$ colors were removed from it. Take an $(\epsilon, k + 1)$-augmentation of Ψ_5, the removed colors of the marked edges are replaced using the extra zones. This results in a coloring Ψ_6, where every edge e lost at most $\epsilon x(\overrightarrow{uv})/2$ colors. By an argument similar to the one in Phase 5, it can be shown that with high probability Ψ_6 is an $(1 + O(\epsilon))$-approximate coloring. Furthermore, by Lemma 3 the size of the color sets can be increased by a factor of $(1 + \epsilon)$ at the cost of increasing the sum by a factor of $1 + 2(k + 1)\epsilon$. Therefore we get a $(1 + O(\epsilon))$-approximate coloring Ψ_7 that satisfies all the demands.

Phase 7: Low frequent edges. In this phase we ensure that if a frequent edge e is conflicting, then edge e uses only colors above $2x(e)/\epsilon$. This will help us in resolving the type 4 conflicts in Phase 8.

If a frequent edge e has finish time at most $8x(e)/\epsilon^2$ in Ψ_7, then e will be called a *low* edge. Clearly, there can be at most $8/\epsilon^2$ low edges of a given class entering vertex v. First we ensure that the low frequent edges are not conflicting. Let $e_1, \ldots, e_m$ be the low edges entering node v, ordered by increasing demand size. Let us perform an $(\epsilon, k + 1)$-augmentation on Ψ_7. If v has color r, then we use the colors in zone Z_r to recolor the low edges: the first $x(e_1)$ colors in Z_r are given to e_1, the next $x(e_2)$ colors are given to e_2 and so on. In the resulting coloring Ψ_8 the finish time of the low edges can increase. However, we show that this increase is negligible compared to the total sum of the frequent edges.

Denote by $F(v, m)$ the number of class m frequent edges entering v, and by $\ell(v, m) \le 8/\epsilon^2$ the number of class m low frequent edges entering v. Since $F(v, m) \ge 6/\epsilon^5$ whenever $\ell(v, m)$ is non-zero, we have $\ell(v, m) \le 2\epsilon^3 F(v, m)$. By Lemma 1b and the convexity of the expression (1), this means that the total sum of the low edges, even if they use only an extra zone, is at most $2/\epsilon \cdot 2\epsilon^3 < \epsilon$ times the sum of the frequent edges. Thus Ψ_8 is a $(1 + O(\epsilon))$-approximate coloring.

It is still possible that a conflicting frequent edge e uses colors up to $2x(e)/\epsilon$, but in this case its finish time is at least $8x(e)/\epsilon^2$. We perform once again an $(\epsilon, k + 1)$-augmentation on Ψ_8. Consider the conflicting frequent edges entering v in increasing order of demand size, and if we encounter an edge e that uses colors up to $2x(e)/\epsilon$, then these colors are replaced using the next available colors of zone Z_r (r is the color of vertex v). More precisely, we want to ensure that every color of e is at least $2x(e)/\epsilon$, hence we assign to it the next available color above $2x(e)/\epsilon$. At the point when edge e is recolored, only colors up to $2x(e)/\epsilon$ have been migrated to zone Z_r, since the edges before e have demand not greater than

$x(e)$. Therefore the new colors are not larger than the smallest $4x(e)/\epsilon$ colors of zone Z_r, and these colors are at most $(2/\epsilon) \cdot 4x(e)/\epsilon = 8x(e)/\epsilon^2$. We have ensured that every frequent edge has finish time greater than $8x(e)/\epsilon^2$, thus the recoloring does not increase the sum. Therefore we obtain a coloring Ψ_9 with $f_{\Psi_9}(G) = f_{\Psi_8}(G)$ where every frequent edge e uses only colors above $2x(e)/\epsilon$.

Phase 8: Resolving type 4 conflicts. First perform an $(\epsilon, k+1)$-augmentation on Ψ_9. Let $e = \overrightarrow{uv}$ be a frequent edge that conflicts with some of the edges entering u. Let the color of vertex u be r. There are at most $x(e)$ colors that are used by both e and an edge entering u. We resolve these conflicts by recoloring the edges entering u in such a way that they use the first at most $x(e)$ colors in zone Z_r instead of the colors in $\Psi_9(e)$. It is clear that if this operation is applied for every frequent edge e, then the resulting coloring Ψ_{10} does not contain type 4 conflicts.

In Phase 7 we have ensured that a conflicting frequent edge uses only colors above $2x(e)/\epsilon$. Therefore if an edge $\overrightarrow{wu}$ is recolored, then it has finish time at least $2x(e)/\epsilon$, otherwise it does not conflict with frequent edge $e = \overrightarrow{uv}$. On the other hand, by Lemma 1b, zone Z_r contains at least $x(e)$ colors up to $2x(e)/\epsilon$, thus the recoloring does not add colors above that. Therefore the finish time of $\overrightarrow{wu}$ is not increased. Thus in this phase the sum is only increased by the $(\epsilon, k+1)$-augmentation. Therefore finally we obtained a proper $(1 + O(\epsilon))$-approximate coloring Ψ_{10}. $\qquad\square$

5   Planar Graphs

In [5] the PTAS for minimum sum multicoloring the vertices of partial k-trees is extended to planar graphs using the layering method of Baker [1]. In this section we follow a similar path to obtain a linear time PTAS for SEMC on planar graphs. To present the algorithm, we will need the definition of t-outerplanar graphs:

Definition 1 (t-outerplanar). *An embedding of graph $G(V, E)$ is 1-outerplanar (or simply outerplanar), if it is planar, and all vertices lie on the exterior face. For $t \geq 2$, an embedding of a graph G is t-outerplanar, if it is planar, and when all vertices on the outer face are deleted, then a $(t-1)$-outerplanar embedding of the resulting graph is obtained. A graph is t-outerplanar, if it has a t-outerplanar embedding. A t-outerplanar embedding divides the vertices into t layers: layer ℓ_1 contains the vertices on the outer face, while for $i \geq 2$, layer ℓ_i contains those vertices that are on the outer face after deleting layers $\ell_1, \ldots, \ell_{i-1}$.*

Theorem 4. *For every $\epsilon > 0$, there is a linear time $(1 + \epsilon)$-approximation algorithm for SEMC on planar graphs.*

Proof. A planar embedding of G can be found in linear time. Assume that this embedding is t-outerplanar for some integer t. Set $K = 4/\epsilon^2$. Denote by E_i the edges connecting layer i and $i + 1$. For $1 \leq i \leq K$, let G_i be the subgraph of G spanned by the edges in E_i, E_{i+K}, E_{i+2K}, ... Each E_i spans a 2-outerplanar graph, hence each G_i is 2-outerplanar. A theorem of Bodlaender [3–Theorem 83] assures that a t-outerplanar graph has treewidth at most $3t - 1$, therefore we can use Theorem 3 to find a $(1 + \epsilon)$-approximate coloring Ψ_i for each graph G_i in linear time. Notice that the edges in the graphs G_i are pairwise disjoint. Therefore there is a $1 \leq i' \leq K$ such that the minimum sum on $G_{i'}$ is at most $\mathrm{OPT}(G, x)/K$, and consequently, the cost of $\Psi_{i'}$ is at most $(1 + \epsilon)\mathrm{OPT}(G, x)/K$.

Delete the edges of $G_{i'}$ from G, the resulting graph is clearly K-outerplanar. Therefore it has treewidth at most $3K - 1$, and Theorem 3 can be used to find a $(1 + \epsilon)$-approximate coloring Ψ_1. Perform an $(\epsilon, 1)$-augmentation of Ψ_1, this increases the sum by at most a factor of $(1 + \epsilon)$. We use extra zone Z_1 to color the edges in $G_{i'}$. If $\Psi_{i'}$ assigns color c to edge e, then assign to e the cth color of Z_1. By Lemma 1b, the sum of the edges in $G_{i'}$ will be at most $2/\epsilon$ times the cost of $\Psi_{i'}$, that is, at most $2/\epsilon \cdot (1 + \epsilon)\mathrm{OPT}(G, x)/K \leq \epsilon\mathrm{OPT}(G, x)$. Therefore the resulting coloring is a $(1 + O(\epsilon))$-approximate coloring for G.   □

References

1. B. S. Baker. Approximation algorithms for NP-complete problems on planar graphs. *J. Assoc. Comput. Mach.*, 41(1):153–180, 1994.
2. A. Bar-Noy, M. M. Halldórsson, G. Kortsarz, R. Salman, and H. Shachnai. Sum multicoloring of graphs. *J. Algorithms*, 37(2):422–450, 2000.
3. H. L. Bodlaender. A partial k-arboretum of graphs with bounded treewidth. *Theoret. Comput. Sci.*, 209(1-2):1–45, 1998.
4. E. G. Coffman, Jr., M. R. Garey, D. S. Johnson, and A. S. LaPaugh. Scheduling file transfers. *SIAM J. Comput.*, 14(3):744–780, 1985.
5. M. M. Halldórsson and G. Kortsarz. Tools for multicoloring with applications to planar graphs and partial k-trees. *J. Algorithms*, 42(2):334–366, 2002.
6. M. M. Halldórsson, G. Kortsarz, A. Proskurowski, R. Salman, H. Shachnai, and J. A. Telle. Multicoloring trees. *Inform. and Comput.*, 180(2):113–129, 2003.
7. J. A. Hoogeveen, S. L. van de Velde, and B. Veltman. Complexity of scheduling multiprocessor tasks with prespecified processor allocations. *Discrete Appl. Math.*, 55(3):259–272, 1994.
8. J. Kahn. Asymptotics of the chromatic index for multigraphs. *J. Combin. Theory Ser. B*, 68(2):233–254, 1996.
9. D. Marx. Complexity results for minimum sum edge multicoloring. Manuscript.
10. D. Marx. The complexity of tree multicolorings. In *Mathematical Foundations of Computer Science 2002 (Warsaw-Otwock)*, pages 532–542. Springer, Berlin, 2002.
11. D. Marx. Minimum sum multicoloring on the edges of trees. In *1st International Workshop on Approximation and Online Algorithms (WAOA) 2003*, volume 2909 of *Lecture Notes in Computer Science*, pages 214–226. Springer, Berlin, 2004.
12. C. E. Shannon. A theorem on coloring the lines of a network. *J. Math. Physics*, 28:148–151, 1949.

Online Bin Packing with Resource Augmentation

Leah Epstein[1],* and Rob van Stee[2],**

[1] School of Computer Science, The Interdisciplinary Center, Herzliya, Israel
lea@idc.ac.il.
[2] Centre for Mathematics and Computer Science (CWI), Amsterdam,
The Netherlands
Rob.van.Stee@cwi.nl.

Abstract. In competitive analysis, we usually do not put any restrictions on the computational complexity of online algorithms, although efficient algorithms are preferred. Thus if such an algorithm were given the entire input in advance, it could give an optimal solution (in exponential time). Instead of giving the algorithm more knowledge about the input, in this paper we consider the effects of giving an online bin packing algorithm larger bins than the offline algorithm it is compared to. We give new algorithms for this problem that combine items in bins in an unusual way and give bounds on their performance which improve upon the best possible bounded space algorithm. We also give general lower bounds for this problem which are nearly matching for bin sizes $b \geq 2$.

1   Introduction

In this paper we investigate the bin packing problem, one of the oldest and most thoroughly studied problems in computer science [1, 2]. In particular, we investigate this problem using the resource augmentation model, where the online algorithm has bins of size $b \geq 1$ and is compared to an offline algorithm that has bins of size 1. We show improved upper bounds and general lower bounds for this problem.

Problem Definition. In the *classical bin packing* problem, we receive a sequence σ of *items* $p_1, p_2, \ldots, p_N$. Each item has a fixed *size* in $(0, 1]$. In a slight abuse of notation, we use p_i to indicate both the ith item and its size. We have an infinite supply of *bins* each with *capacity* 1. Each item must be assigned to a bin. Further, the sum of the sizes of the items assigned to any bin may not exceed its capacity. A bin is *empty* if no item is assigned to it, otherwise it is *used*. The goal is to minimize the number of bins used.

---

* Research supported by Israel Science Foundation (grant no. 250/01).
** Research supported by the Netherlands Organization for Scientific Research (NWO), project number SION 612-30-002.

G. Persiano and R. Solis-Oba (Eds.): WAOA 2004, LNCS 3351, pp. 23–35, 2005.
© Springer-Verlag Berlin Heidelberg 2005

In the *resource augmentation* model [7, 9], one compares the performance of a particular algorithm $\mathcal{A}$ to that of the optimal offline algorithm (denoted by OPT) in an unfair way. The optimal offline algorithm uses bins of capacity one, where $\mathcal{A}$ is allowed to use bins of capacity $b > 1$. The goal is still to minimize the *number* of bins used.

In the *online* versions of these problems, each item must be assigned in turn, without knowledge of the next items. Since it is impossible in general to produce the best possible solution when computation occurs online, we consider approximation algorithms. Basically, we want to find an algorithm that incurs cost which is within a constant factor of the minimum possible cost, no matter what the input is. This constant factor is known as the asymptotic performance ratio.

The resource augmentation model was introduced due to the following drawback of standard competitive analysis. Competitive analysis compares the performance of an online algorithm, which must pack each item upon arrival, to that of an omniscient and all-powerful offline algorithm that gets the input as a set. Resource augmentation gives more power to the online algorithm, making the analysis more general.

A bin-packing algorithm uses *bounded space* if it has only a constant number of bins available to accept items at any point during processing. These bins are called *open* bins. Bins which have already accepted some items, but which the algorithm no longer considers for packing are *closed* bins. While bounded space algorithms are sometimes desirable, it is often the case that unbounded space algorithms can achieve lower performance ratios.

We define the asymptotic performance ratio more precisely. For a given input sequence σ, and a fixed bin size b, let $\text{cost}_{\mathcal{A},b}(\sigma)$ be the number of bins (of size b) used by algorithm $\mathcal{A}$ on σ. Let $\text{cost}(\sigma)$ be the minimum possible cost to pack items in σ using bins of size 1. The *asymptotic performance ratio* for an algorithm $\mathcal{A}$ is defined to be

$$R_{\mathcal{A},b}^{\infty} = \limsup_{n \to \infty} \max_{\sigma} \left\{ \frac{\text{cost}_{\mathcal{A},b}(\sigma)}{\text{cost}(\sigma)} \,\middle|\, \text{cost}(\sigma) = n \right\}.$$

The *optimal asymptotic performance ratio* is defined to be $R_{\text{OPT},b}^{\infty} = \inf_{\mathcal{A}} R_{\mathcal{A},b}^{\infty}$. Our goal is to find for all values of b ($b \geq 1$) an algorithm with asymptotic performance ratio close to $R_{\text{OPT},b}^{\infty}$.

Previous Results. The classic online bin packing problem was first investigated by Ullman [12]. He showed that the FIRST FIT algorithm has performance ratio $\frac{17}{10}$. This result was then published in [5]. Johnson [6] showed that the NEXT FIT algorithm has performance ratio 2. Yao showed that REVISED FIRST FIT has performance ratio $\frac{5}{3}$. Currently the best known lower bound is 1.54014, due to van Vliet [13].

Define $u_1 = 2, u_{i+1} = u_i(u_i - 1) + 1$, and $h_\infty = \sum_{i=1}^{\infty} \frac{1}{u_i - 1} \approx 1.69103$. Lee and Lee showed that the HARMONIC algorithm, which uses bounded space, achieves a performance ratio arbitrarily close to h_∞ [8]. They further showed that no bounded space online algorithm achieves a performance ratio less than h_∞ [8]. In addition, they developed the REFINED HARMONIC algorithm, which they showed to have a performance ratio of $\frac{273}{228} < 1.63597$. The next improvements were MODIFIED HARMONIC and MODIFIED HARMONIC 2. Ramanan, Brown, Lee and Lee showed that these algorithms have performance ratios of $\frac{538}{333} < 1.61562$ and $\frac{239091}{148304} < 1.61217$, respectively [10]. Currently, the best known upper bound is 1.58889 due to Seiden [11].

Bin packing with resource augmentation was first studied by Csirik and Woeginger [3]. They give an optimal bounded space algorithm. Naturally, its asymptotic performance ratio is strictly decreasing as a function of the bin size of the online algorithm. Some preliminary general lower bounds for bin packing with resource augmentation were given in [4]. In Section 5, we will compare them to our new lower bounds.

Our Results. In this paper, we present new algorithms for the online bin packing problem in the resource augmentation model. We introduce a general method which extends many previously studied algorithms for bin packing. This method takes the online bin size b as a parameter. We study four instances of the general method, each of our algorithms performs well for a different interval of values of b. By partitioning the interval $[1, 2)$ in four sub-intervals and using the most appropriate algorithm on each sub-interval, we give upper bounds that improve upon the bounds from [3] on the entire interval. That is, these algorithms are better than the best possible bounded space algorithm.

Our analysis technique extends the general packing algorithm analysis technique developed by Seiden [11]. Specifically, unlike previous algorithms which pack the relatively small items by a very simple heuristic (Next-Fit, or any fit), we combine small items with large items in the same bins in order to achieve good performance (see the algorithms SMH and TMH).

We also show new lower bounds for this model, by using improved sequences. We omit detailed descriptions of these sequences in this extended abstract. For $b \geq 2$, our lower bounds show that the best bounded space algorithm is very close to optimal (among unbounded space algorithms).

2   The HARMONIC Algorithm and Variations

In this section we discuss the important HARMONIC algorithm [8] and possible variations on it. In the next section we will discuss the specific variations on HARMONIC that we have used in this paper.

The fundamental idea of these algorithms is to first classify items by size, and then pack an item according to its class (as opposed to letting the exact size influence packing decisions). For the classification of items, we need to partition the interval $(0, 1]$ into subintervals. The standard HARMONIC algorithm uses $n - 1$ subintervals of the form $(1/(i + 1), 1/i)$ for $i = 1, \ldots, n - 1$ and one final subinterval $(0, 1/n]$. Each bin will contain only items from one subinterval (type).

Items in subinterval i are packed i to a bin for $i = 1, \ldots, n-1$ and the items in interval n are packed in bins using NEXT FIT.

A disadvantage of HARMONIC is that items of type 1, that is, the items larger than $1/2$, are packed one per bin, possibly wasting a lot of space in each single bin. To avoid this large waste of space, later algorithms used two extra interval endpoints, of the form $\Delta > 1/2$ and $1 - \Delta$. Then, some small items can be combined in one bin together with an item of size $\in (1/2, \Delta]$. Items larger than Δ are still packed one per bin as in HARMONIC. These algorithms furthermore use parameters α^i ($i = 3, \ldots, n$) which represent the fraction of bins allocated to type i where the algorithm will reserve space for items $\in (1/2, \Delta]$. The remaining bins with items of type i still contain i items per bin.

Example. MODIFIED HARMONIC (MH) is defined by $n = 38$ (the number of intervals) and $\Delta = 419/684$.

$$\alpha^2 = \tfrac{1}{9}; \quad \alpha^3 = \tfrac{1}{12}; \quad \alpha^4 = \alpha^5 = 0;$$

$$\alpha^i = \frac{37 - i}{37(i+1)}, \qquad \text{for } 6 \le i \le 36;$$

$$\alpha^{37} = \alpha^{38} = 0.$$

The results of [10] imply that the asymptotic performance ratio of MH is at most $\frac{538}{333} < 1.61562$. (In the original definition, Δ was used to denote $1 - \Delta$.)

In the current paper, we will use as interval endpoints the points of the form b/i (as long as they are below 1) instead of $1/i$, since items in $(b/(i+1), b/i]$ can be placed exactly i to a bin in an (online) bin of size b. Moreover, sometimes we will also use points of the form $\Delta, b - \Delta, 1 - b/2$ as interval endpoints, in order to combine items from different types where they would otherwise waste much space.

Note that for $b \in [1, 2)$ we always have $b/2 \le 1$. We now consider an algorithm $\mathcal{A}$ that uses n basic intervals (some might be subdivided further):

$$I_{\mathcal{A}}^1 = (b/2, 1]$$
$$I_{\mathcal{A}}^j = (b/(j+1), b/j] \qquad j = 2, \ldots, n-1$$
$$I_{\mathcal{A}}^n = (0, b/n]$$

In case Δ is used as an endpoint, the interval $I_{\mathcal{A}}^1 = (b/2, 1]$ is partitioned into two subintervals, which will be denoted by $I_{\mathcal{A}}^{\Delta(2)} = (b/2, \Delta]$ and $I_{\mathcal{A}}^{\Delta(1)} = (\Delta, 1]$. ($\Delta$ will always be chosen larger than $b/2$.) We will use two versions of algorithms, that are determined by whether they use $b - \Delta$ or $1 - b/2$ as an additional endpoint. We denote the largest possible size of an item of the smallest type by ε. This is b/n unless $I_{\mathcal{A}}^n$ is divided further into two subintervals.

Version 1. We use the endpoint $b - \Delta$ (but not the endpoint $1 - b/2$). Let j_Δ be the integer such that $b/(j_\Delta + 1) < b - \Delta \le b/j_\Delta$. Then $I_{\mathcal{A}}^{j_\Delta}$ is partitioned into two subintervals, which will be denoted by $I_{\mathcal{A}}^{\Delta(4)} = (b/(j_\Delta + 1), b - \Delta]$ and $I_{\mathcal{A}}^{\Delta(3)} = (b - \Delta, b/j_\Delta]$.

Version 2. We use the endpoint $1 - b/2$ (but not the endpoint $b - \Delta$). Let j_Δ be an integer such that $b/(j_\Delta + 1) < 1 - b/2 \le b/j_\Delta$. In this version we always take $n \ge j_\Delta$.

If $n \ge j_\Delta + 1$, then $I_\mathcal{A}^{j_\Delta}$ is partitioned into two subintervals, which will be denoted by $I_\mathcal{A}^{\Delta(4)} = (b/(j_\Delta + 1), 1 - b/2]$ and $I_\mathcal{A}^{\Delta(3)} = (1 - b/2, b/j_\Delta]$.

Otherwise $n_\mathcal{A} = j_\Delta$ and $I_\mathcal{A}^n$ is partitioned into the two subintervals $I_\mathcal{A}^{\Delta(4)} = (0, 1 - b/2]$ and $I_\mathcal{A}^{\Delta(3)} = (1 - b/2, b/n]$.

In both versions, the intervals are disjoint and cover $(0, 1]$. $\mathcal{A}$ assigns each item a *type* depending on its size. An item of size s has type $\tau_\mathcal{A}(s)$ where $\tau_\mathcal{A}(s) = j \Leftrightarrow s \in I_\mathcal{A}^j$. Note that either $2 \le j \le n$ ($j \ne j_\Delta$) or $j = \Delta(i)$ for some $1 \le i \le 4$.

Note that if we place an item from the interval $I_\mathcal{A}^{\Delta(2)}$ in a bin, the amount of space left over is at least $b - \Delta$. If possible, we would like to use this space to pack more items. To accomplish this, we assign each item a color, *red* or *blue*. $\mathcal{A}$ attempts to pack red items with type $I_\mathcal{A}^{\Delta(2)}$ items. For both versions, all items of types $2, \ldots, j_\Delta - 1$ and $\Delta(k), k = 1, 2, 3$ (where applicable) are blue. Other items can be either red or blue.

To assign colors to items, the algorithm uses two sets of counters, $e_{j_\Delta}, \ldots, e_n$ and $s_{j_\Delta}, \ldots, s_n$, all of which are initially zero. The counter s_{j_Δ} counts the number of bins for items of type $\Delta(4)$, and the counter s_i keeps track of the total number of bins in which we packed items of type i for $i = j_\Delta + 1, \ldots, n$. The counters e_i are defined analogously, but only count the number of bins containing *red* items of type $\Delta(4)$ or i. These bins are also called red themselves.

For $j_\Delta \le i \le n$, $\mathcal{A}$ maintains the invariant $e_i = \lfloor \alpha^i s_i \rfloor$, i.e. the fraction of bins with type i items that contain red items is approximately α^i. Recall that α^i is defined only for $j_\Delta \le i \le n$. For each such interval, at least one item can fit in a bin together with an item of size at most Δ in a bin of size b. Moreover, for version 2 we combine only relatively small items with items of interval $\Delta(2)$, so in most cases several items fit together with the $\Delta(2)$ item.

We now describe how blue and red items are packed. The packing of blue items is simple. For $i < n$, the number of items with sizes in $(b/(i+1), b/i]$ which fit in a bin of size b is i. Blue items with such sizes are placed i in a bin, as in the HARMONIC algorithm. Note that the type of such an item is either i or $\Delta(k)$ for some $1 \le k \le 4$. Small items (type n) which are colored blue are packed into separate bins using NEXT FIT, again as in the HARMONIC algorithm.

For the red items, we consider the two versions of algorithms defined before separately.

Version 1. One red item of type $\Delta(4)$ can be combined with an item of type $\Delta(2)$. We define $\gamma_{j_\Delta} = 1$. For $j_\Delta < j < n$, the number of red items we will assign together with a type $\Delta(2)$ item is $\gamma_j = \lfloor j(b - \Delta)/b \rfloor$. For type n, we treat the remaining space of $b - \Delta$ in bins containing an item of type $\Delta(2)$ as a bin, and use NEXT FIT to place red items in such bins. Clearly we can fill at least $b - \Delta - b/n$ of this space by small red items.

Version 2. If $n = j_\Delta$, it means that we combine only the smallest interval with items of type $\Delta(2)$. Then we can assign at least $b - \varepsilon = 3b/2 - 1$ to blue items bins, and $b - \Delta - \varepsilon = 3b/2 - \Delta - 1$ to red items bins. If $n > j_\Delta$, all the amounts are defined as for version 1, except for $\gamma_{j_\Delta} = \lfloor (b - \Delta)/(1 - b/2) \rfloor$.

We explain more precisely the method by which red items are packed with type $\Delta(2)$ items. When a bin is opened, it is assigned to a *group*. If $\varepsilon = b/n$, the bin groups are named:

$$\Delta(1), \Delta(3), \Delta(4), 2, 3, , \ldots, j_\Delta - 1, j_\Delta + 1, \ldots n; \tag{1}$$
$$(\Delta(2), i), \qquad \text{for } \alpha^i \neq 0, \ j_\Delta \leq i \leq n; \tag{2}$$
$$(\Delta(2), *); \tag{3}$$
$$(*, i), \qquad \text{for } \alpha^i \neq 0, \ j_\Delta \leq i \leq n; \tag{4}$$

If $\varepsilon = 1 - b/2$, i.e. the smallest interval was partitioned, the bin groups are named:

$$\Delta(1), \Delta(3), \Delta(4), 2, 3, \ldots, n - 1; \tag{5}$$
$$(\Delta(2), \Delta(4)); \tag{6}$$
$$(\Delta(2), *); \tag{7}$$
$$(*, \Delta(4)); \tag{8}$$

Bins from groups in (1) and (5) contain only blue items of the type they is named after. The closed bins all contain the maximum number of items they can have (explained earlier).

If the smallest interval was not partitioned, then for $j_\Delta \leq i < n$, a closed bin in group $(\Delta(2), i)$ contains one type $\Delta(2)$ item and γ_i type i items, and a closed bin in group $(\Delta(2), n)$ contains one type $\Delta(2)$ item and red items of total size at least $b - \Delta - b/n$. If the smallest interval was partitioned, a closed bin in group $(\Delta(2), \Delta(4))$ contains red items of total size at least $3b/2 - \Delta - 1$. There is at most one open bin in any of these groups.

The group $(\Delta(2), *)$ contains bins which hold a single blue item of type $\Delta(2)$. These bins are all open, as we hope to add red items to them later.

The group $(*, j)$ contains bins which hold only red items of type j. Again, these bins are all open, but only one has fewer than γ_j items if $j < n$. For $j = n$ only one bin can contain total size of less than $b - \Delta - \varepsilon$ of red items of the last interval. We will try to add a type $\Delta(2)$ item to these bins if possible.

We call bins in the last two group classes ((3) and (4), or (7) and (8)) *indeterminate*. Essentially, the algorithm tries to minimize the number of indeterminate bins, while maintaining all the aforementioned invariants. I.e. we try to place red and type $\Delta(2)$ items together whenever possible; when this is not possible we place them in indeterminate bins in hope that they can later be combined.

On arrival of an item, it gets the same color as the previous item of the same type, if it can also fit into the same bin. Otherwise, we update the bins counter, and according to the counters, decide which color it gets.

3   Algorithms in This Paper

After describing the general framework, we now describe the specific algorithms that we have designed. They are all instances of the general algorithm above.

Generalized Modified Harmonic (GMH). This algorithm has the same structure as the regular MODIFIED HARMONIC, i.e. $n = 38$, and the same values of α^i. The only difference is that the variable Δ is adjusted to ensure that $\Delta \in (b/2, 1)$ for $b \in [1, 2)$.

Specifically, we let Δ grow linearly with the bin size until it reaches the value 1 for a bin size of 2, i.e. $\Delta = 419/684 + 265(b-1)/684$. We applied this algorithm on the interval $[1, 6/5)$. This algorithm is of version 1 as we only modify Δ.

Convenient Modified Harmonic (CMH). On the interval $[6/5, 4/3)$, we focus on the items that could be packed together in one offline bin together with items of type 1, that is, items that are just larger than $b/2$. This was done specifically to handle the *greedy* input sequence, which starts with an item just larger than $b/2$ and repeatedly adds an item of the form $b/i_j + \varepsilon$ such that all items together fit into a bin of size b.

Our algorithm is of version 1 and does the following. Let $k = \lfloor 1/(1 - b/2) \rfloor$. This means that the largest items that can be packed together with an item of size $b/2$ in a single bin of size 1 are in the interval $(1/(k+1), 1/k]$ (possibly not every size in this interval can be so packed). Let $\Delta = (k-1)b/k$. Note that in the interval of b we consider, we always have $k = 3$ and hence $\Delta = 2b/3$. Note that $b - \Delta = b/k$ and therefore $I^{\Delta(3)} = \emptyset$.

Our choice of Δ ensures that items of type $\Delta(2)$, with sizes in $(b/2, (k-1)b/k]$, can be packed very well together with items of type k, with sizes in $(b/(k+1), b/k]$, in our case this is $(b/4, b/3]$. In the discussed interval we have $b/2 + b/4 < 1$, so in the optimal packing such items could also be together in one bin. The choice of $n = 38$ is as in GMH and the values α^i are chosen by experimenting. The values we used are

$$\alpha^3 = \tfrac{1}{8}; \quad \alpha^4 = \tfrac{1}{10}; \quad \alpha^5 = 0;$$
$$\alpha^i = \frac{37 - i}{37(i + 4)}, \qquad \text{for } 6 \le i \le 36;$$
$$\alpha^{37} = \alpha^{38} = 0.$$

Small Modified Harmonic (SMH). On the interval $[4/3, 12/7 \approx 1.7143)$, it becomes more important how to pack smaller items (relative to b). We define $\Delta = 1$, and $n = 12$. Thus $I^{\Delta(1)} = \emptyset$. Note that we use the second version of the algorithm, which means that in marked contrast to all other previously defined variations on Harmonic that the authors are aware of, we do not take $\alpha_{12} = 0$, that is, we pack some of the smallest items together with the large items.

We illustrate the reason. Consider a bin of size $3/2$. Taking $\Delta = 1$ leaves a space of $1/2$ in a bin. This space could be used to accommodate an item of size $b/3 = 1/2$. However, items of size in $(b/4, b/3]$, when packed three to a bin, occupy at least $3b/4 = 9/8 > 1$. Considering an offline packing we can see that such items do not fit together with an items of type $\Delta(2)$. Therefore there is no reason to improve their packing which is already relatively good.

However, items that do fit together with type $\Delta(2)$ items do need to be packed more carefully (partly red and partly blue), including the ones from the last interval, since they can be combined in an offline packing. We determine the largest item type that OPT could pack together with an item in $(b/2, 1]$ (i.e. the smallest i such that $b/i \leq 1 - b/2$). Larger items are packed according to Harmonic, while a fraction of these smaller items are reserved to be packed together with an item of type $\Delta(2)$, i.e. in $(b/2, 1]$.

We explain how to fix the values α^i for this algorithm in Section 4.

Tiny Modified Harmonic (TMH). On the interval $[12/7, 2)$, it turns out that it is crucial to pack the smallest items better than with Harmonic. All other items are packed in their own bins according to Harmonic. We use the second version of the algorithm. We use $\Delta = 1$ (so $I^{\Delta(1)} = \emptyset$) and let $n = j_\Delta$.

In other words, we determine the number of intervals that we use in such a way that $1 - b/2 \in (b/(n+1), b/n]$. The smallest interval boundary of the form b/i is just larger than $1 - b/2$ (or equal to it). This ensures that in the optimal packing, only items of the smallest type could be packed together with large items with size in $(b/2, 1]$. We use $\alpha_{j_\Delta} = (2b - 2)/(4 - b)$.

It would be possible to improve very slightly using the algorithm SMH with more intervals, but the number of intervals required grows without bound as b approaches 2, and it becomes infeasible to calculate all the patterns.

4   Analysis

An algorithm for a given bin size b can be used without change for any bin size $c \geq b$, and will have the same performance ratio since for any given sequence, the offline optimal packing and the cost of the algorithm remain unchanged. This means that the function $R^\infty_{\text{OPT},b}$ is monotonically decreasing in b. This property allows us to give bounds on an interval by *sampling* a large but finite number of points. An upper bound for the bin size b holds for $b + \gamma$ for any $\gamma > 0$. A lower bound for the bin size b holds for a bin size $b - \gamma$ for any $\gamma > 0$.

Weighting functions. The type of algorithm described in Section 2 can be analyzed using the method of weighting systems developed in [11]. The full generality of weighting systems is not required here, so we adopt a slightly different notation than that used in [11], and restrict ourselves to a subclass of weighting systems.

A weighting system for an algorithm $\mathcal{A}$ is a pair $(W_\mathcal{A}, V_\mathcal{A})$. $W_\mathcal{A}$ and $V_\mathcal{A}$ are *weighting functions* which assign each item p a real number based on its size. The weighting functions for an algorithm $\mathcal{A}$ are defined as follows.

If $\varepsilon = 1 - b/2$, the only value of α^i which is not zero is α^{j_Δ}. The weighting functions are defined as follows.

Type of item p	$W_\mathcal{A}(p)$	$V_\mathcal{A}(p)$
$\Delta(1)$	1	1
$\Delta(2)$	1	0
$k \in \{2, 3, \ldots, j_\Delta - 1\}$	$1/k$	$1/k$
$\Delta(3)$	$1/j_\Delta$	$1/j_\Delta$
$\Delta(4)$	$\dfrac{p(1 - \alpha^{j_\Delta})}{3b/2 - 1 - \Delta\alpha^{j_\Delta}}$	$\dfrac{p}{3b/2 - 1 - \Delta\alpha^{j_\Delta}}$

For the cases that $\varepsilon = b/n$ we define the functions differently.

Type of item p	$W_\mathcal{A}(p)$	$V_\mathcal{A}(p)$
$\Delta(1)$	1	1
$\Delta(2)$	1	0
$k \in \{2, 3, \ldots, j_\Delta - 1\}$	$1/k$	$1/k$
$\Delta(3)$	$1/j_\Delta$	$1/j_\Delta$
$\Delta(4)$	$\dfrac{1 - \alpha^j}{\gamma_{j_\Delta}\alpha_\Delta^j + j_\Delta(1 - \alpha_\Delta^j)}$	$\dfrac{1}{\gamma_{j_\Delta}\alpha_\Delta^j + j_\Delta(1 - \alpha_\Delta^j)}$
$k \in \{j_\Delta + 1, \ldots, n - 1\}$	$\dfrac{1 - \alpha^k}{\gamma_k\alpha^k + k(1 - \alpha^k)}$	$\dfrac{1}{\gamma_k\alpha^k + k(1 - \alpha^k)}$
n	$\dfrac{p(1 - \alpha^n)}{b - b/n - \Delta\alpha^n}$	$\dfrac{p}{b - b/n - \Delta\alpha^n}$

The following lemma follows directly from Lemma 4 of [11]:

Lemma 1. *For all σ, we have*

$$\mathrm{cost}_\mathcal{A}(\sigma) \leq \max\left\{\sum_{p \in \sigma} W_\mathcal{A}(p), \sum_{p \in \sigma} V_\mathcal{A}(p)\right\} + O(1).$$

So the cost to $\mathcal{A}$ can be upper bounded by the weight of items in σ, and the weight is independent of the order of items in σ.

We give a short intuitive explanation of the weight functions and Lemma 1: Consider the final packing created by an algorithm $\mathcal{A}$ on some input σ. In this final packing, let r be the number of bins containing red items, let b_1 be the number of type $\Delta(2)$ items, and let b_2 be the number of bins containing blue items of type other than $\Delta(2)$. The total number of bins is just $\max\{r, b_1\} + b_2 = \max\{r + b_2, b_1 + b_2\}$. We have chosen our weighting functions so that $\sum_{p \in \sigma} W_\mathcal{A}(p) = b_1 + b_2 + O(1)$ and $\sum_{p \in \sigma} V_\mathcal{A}(p) = r + b_2 + O(1)$. In both $W_\mathcal{A}$ and $V_\mathcal{A}$, the weight of a blue item of type other than $\Delta(2)$ is just the fraction of a bin that it occupies. $W_\mathcal{A}$ counts type $\Delta(2)$ items, but ignores red items. $V_\mathcal{A}$ ignores type $\Delta(2)$ items, but counts bins containing red items. For a formal proof, we refer the reader to [11].

Let f be some function $f : (0, 1] \mapsto \mathbb{R}^+$.

Definition 1. $P(f)$ *is the mathematical program: Maximize* $\sum_{x \in X}^{n} f(x)$ *subject to* $\sum_{x \in X} x \leq 1$, *over all finite sets of real numbers X. In an abuse of notation, we also use $P(f)$ to denote the value of this mathematical program.*

Intuitively, given a weighting function f, $P(f)$ upper bounds the amount of weight that can be packed in a single bin. It is shown in [11] that the performance ratio of $\mathcal{A}$ is upper bounded by $\max\{P(W_{\mathcal{A}}), P(V_{\mathcal{A}})\}$. The value of P is easily determined using a branch and bound procedure very similar to those in [11, 4].

Choice of values α^i for SMH. To choose the values of α^i in the algorithm SMH we use the following idea. We would like to balance the total weight of two particular offline packings. The first offline packing contains one item of interval $\Delta(2)$ and smaller items of type i (here the weight is maximized by considering the weight function $W_{\mathcal{A}}$). The second offline packing contains only items of type i, and we use $V_{\mathcal{A}}$ to determine the maximum weight.

In order to balance these weights, we define the *expansion* of type i to be the maximum ratio of weight to size of an item of type I. Let $EV(i)$ be the expansion according to $V_{\mathcal{A}}$ and $EW(i)$ be the expansion according to $W_{\mathcal{A}}$. We would like to have

$$EV(i) = 1 + (1 - b/2)EW(i).$$

This implies $\alpha^i = (S - b/2)/(S - s + 1 - b/2)$, where S is the minimal occupied area in a closed bin containing blue items of type i and s is the minimal occupied area by red items of interval I in a closed bin.

Note that this computation is not entirely accurate for all types, as it is not always possible to fill a bin of size 1 or of size $1 - b/2$ completely with items of the largest expansion. However, the interval which affects the asymptotic performance ratio the most is $(0, \varepsilon]$.

Analysis of TMH. The simple structure of TMH allows an analytical solution. For this algorithm, we do not need to solve mathematical programs, but can instead calculate the asymptotic worst case performance directly, as follows.

For all types but the smallest and the largest, the weight of an item of size x is at most x. The reason for this is that they are packed according to Harmonic, and TMH can fit at least the same number of items per bin as OPT can. To get a bin of weight more than 1, there must be some items of the first or the last type.

The upper bound of the last interval is $1 - b/2$, denoted by ε. Only items in this interval can be packed together with a type $\Delta(2)$ item in one bin.

Recall that the algorithm uses a parameter $\alpha = \alpha^{j\Delta}$ that determines how the small items are packed. The algorithm maintains the invariant that a fraction α of the bins containing small items are red and have room for a type $\Delta(2)$ item. The total size of all the small items in such a bin is at least $b - 1 - \varepsilon$. The rest of these bins are blue and contain a volume of at least $b - \varepsilon$. There are two cases.

Case 1. There is no item of type $\Delta(2)$. If TMH uses k bins to pack all items of type $\Delta(4)$ (the last type), then αk bins are red and contain a minimum volume of $b - 1 - \varepsilon$ each; $(1 - \alpha)k$ bins contain a minimum total volume of $b - \varepsilon$ of small blue items each. Thus k bins contain a total volume of at least $\alpha k(b - 1 - \varepsilon) + (1 - \alpha)k(b - \varepsilon) = k(b - \alpha - \varepsilon)$, in other words each bin contains on average a volume of at least $b - \alpha - \varepsilon$. The worst case is that all the items are small. Since an offline bin can contain one unit of such items, this gives an asymptotic performance ratio of $1/(b - \alpha - \varepsilon)$. Note that this is consistent with the definition of the function $V_\mathcal{A}$ for this case.

Case 2. There is an item of type $\Delta(2)$. We are interested in the case that its weight is 1, i.e. in the weights according to the function $W_\mathcal{A}$. The large item is of size at least $b/2$. The weight in a bin that contains such an item is maximized by filling up the bin with items of type $\Delta(4)$. The remaining space in the offline bin is exactly $1 - b/2$. In this case, TMH only needs "to pay" for the blue bins. It packs a volume of $k(b - \alpha - \varepsilon)$ using only $(1 - \alpha)k$ blue bins. The total weight according to $W_\mathcal{A}$ is $1 + \frac{1-\alpha}{b-\alpha-\varepsilon}(1 - b/2)$.

Balancing the weights gives that the best choice is $\alpha = \frac{2b-2}{4-b}$ and a ratio of $1/(b - \alpha - \varepsilon) = (2b - 8)/(3b^2 - 10b + 4)$.

5   Results

As mentioned in Section 4, we can determine valid upper and lower bounds on the asymptotic performance ratio for this problem on any interval by sampling a finite number of points. In fact, since we have given an analytical solution for the algorithm TMH, it is not necessary to do any sampling for the upper bound on the interval $(12/7, 2]$.

On the remaining intervals, we used a computer program to solve the associated mathematical program $\mathcal{P}$ for many specific values of b (we sampled integer multiples of $\frac{1}{1000}$) and whichever algorithm is used for that value of b.

We also used a computer program to generate lower bounds for 1,000 values of b between 1 and 2. There were some values of b where all lower bound sequences that we used gave a worse lower bound than we had already found for some higher value of b. However, a lower bound of c for a value b_0 also implies a lower bound of c for all values $1 \le b \le b_0$ as stated before. Therefore, whenever we found a lower bound that was worse than one that was found for some higher value of b, we instead use this higher bound. This explains the small intervals in the graph where the lower bound is constant.

Our results are summarized in the two Figures 1 and 2. The horizontal axis is the size of the online bin, and the vertical axis is the asymptotic performance ratio. For comparison, we have included the graph of the bounded space upper bound (which matches the bounded space lower bound).

It can be seen that for all bin sizes between 1 and 2, we have given substantial improvements on the bounded space algorithm, which was the best known algorithm for this problem so far. The lower bounds from [4] were also significantly improved: for instance for $b = 6/5$, the lower bound was improved from less than 1.18 to above 1.34, and for $b \ge 3/2$, the previous lower bound was less than 0.8.

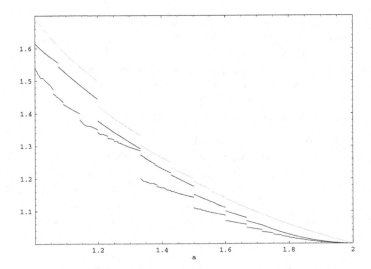

Fig. 1. The lower bound (lowest graph), upper bound (middle), and bounded space bound (highest). Horizontal axis is size of online bin, vertical axis is asymptotic performance ratio

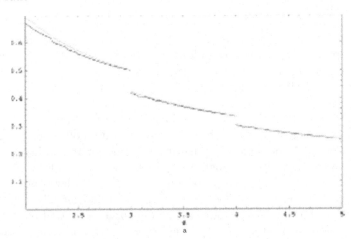

Fig. 2. The lower bound (lowest graph) and bounded space bound on $[2, 5]$. Axes as in previous figure

References

1. Edward G. Coffman, Michael R. Garey, and David S. Johnson. Approximation algorithms for bin packing: A survey. In D. Hochbaum, editor, *Approximation algorithms*. PWS Publishing Company, 1997.
2. János Csirik and Gerhard J. Woeginger. On-line packing and covering problems. In *A. Fiat and G. J. Woeginger, editors,* Online Algorithms: The State of the Art, volume 1442 of *Lecture Notes in Computer Science*, pages 147–177. Springer-Verlag, 1998.

3. János Csirik and Gerhard J. Woeginger. Resource augmentation for online bounded space bin packing. *Journal of Algorithms*, 44(2):308–320, 2002.
4. Leah Epstein, Steve S. Seiden, and Rob van Stee. New bounds for variable-sized and resource augmented online bin packing. In P. Widmayer, F. Triguero, R. Morales, M. Hennessy, S. Eidenbenz, and R. Conejo, editors, *Proc. 29th International Colloquium on Automata, Languages, and Programming (ICALP)*, volume 2380 of *Lecture Notes in Computer Science*, pages 306–317. Springer, 2002.
5. Michael R. Garey, Ronald L. Graham, and Jeffrey D. Ullman. Worst-case analysis of memory allocation algorithms. In *Proceedings of the Fourth Annual ACM Symposium on Theory of Computing*, pages 143–150. ACM, 1972.
6. David S. Johnson. Fast algorithms for bin packing. *Journal Computer Systems Science*, 8:272–314, 1974.
7. Bala Kalyanasundaram and Kirk Pruhs. Speed is as powerful as clairvoyance. *Journal of the ACM*, 47:214–221, 2000.
8. C. C. Lee and D. T. Lee. A simple online bin packing algorithm. *Journal of the ACM*, 32:562–572, 1985.
9. Cynthia Phillips, Cliff Stein, Eric Torng, and Joel Wein. Optimal time-critical scheduling via resource augmentation. *Algorithmica*, pages 163–200, 2002.
10. P. Ramanan, D. J. Brown, C. C. Lee, and D. T. Lee. Online bin packing in linear time. *Journal of Algorithms*, 10:305–326, 1989.
11. Steve S. Seiden. On the online bin packing problem. *Journal of the ACM*, 49(5):640–671, 2002.
12. Jeffrey D. Ullman. The performance of a memory allocation algorithm. Technical Report 100, Princeton University, Princeton, NJ, 1971.
13. Andre van Vliet. An improved lower bound for online bin packing algorithms. *Information Processing Letters*, 43:277–284, 1992.

A PTAS for Delay Minimization
in Establishing Wireless Conference Calls

Leah Epstein[1,*] and Asaf Levin[2]

[1] School of Computer Science, The Interdisciplinary Center, Herzliya, Israel
lea@idc.ac.il.
[2] Department of Statistics, The Hebrew University, Jerusalem, Israel
levinas@mscc.huji.ac.il

Abstract. A prevailing feature of mobile telephony systems is that the location of a mobile user may be unknown. Therefore, when the system is to establish a call between users, it may need to search, or page, all the cells that it suspects the users may be located in, in order to find the cells where the users currently reside. The search consumes expensive wireless links which motivates search techniques that page as few cells as possible.

We consider cellular systems with n cells and m mobile users roaming among the cells. The location of the users is uncertain and is given by m probability distribution vectors. Whenever the system needs to find specific users, it conducts a search operation lasting at most d rounds. In each round the system may check an arbitrary subset of cells to see which users are located there. The problem of finding a single user is known to be polynomially solvable. Whereas the problem of finding any constant number of users (at least 2) in any fixed (constant) number of rounds (at least two rounds) is known to be NP-hard. In this paper we present a simple polynomial-time approximation scheme for this problem with a constant number of rounds and a constant number of users. This result improves an earlier $\frac{e}{e-1} \sim 1.581977$-approximation of Bar-Noy and Malewicz.

1 Introduction

ESTABLISHING WIRELESS CONFERENCE CALLS UNDER DELAY CONSTRAINTS PROBLEM (EWCC) is concerned with establishing a conference call involving $m + 1$ users (from which one has a static position and the other m users have dynamic locations) in a cellular network. The main property of a cellular network is that the users are roaming. This places another step in the process of establishment of the conference call. I.e., the system needs to find out to which cell each user is connected at the moment. Using historical data the system has a certain probability vector for each user that describes the probability that the

---

* Research supported by Israel Science Foundation (grant no. 250/01).

G. Persiano and R. Solis-Oba (Eds.): WAOA 2004, LNCS 3351, pp. 36–47, 2005.

system will find the user in each cell. We assume that each user is connected to exactly one cell in the system and that the locations of the different users are independent random variables.

In order to find a set of users the system may page a subset of cells. Each cell in this subset returns a complete and accurate list of all the users that are connected to it. We assume that the search lasts a short period and during this period users do not move from one cell to another. The search strategy is to page a certain subset of cells looking for the users that participate in the conference call. After the system gets the answers from all the paged cells, it decides whether it needs to continue to the next round (i.e., the search did not find all the users) or it can stop the search (i.e., all the users have been already found). In order to ensure a reasonable quality of service there is an upper bound on the maximum number of rounds, denoted by d. We assume that the system must find all the users within d rounds. Therefore, if the system does not find all the participating users within the first $d-1$ rounds then in the last round it must page all the cells it did not page before. In this paper we follow Bar-Noy and Malewicz [1], and restrict our search strategy to *oblivious* algorithms, in which the subset of cells that is paged in round i does not depend on the actual users that the system found in round $1, 2 \ldots, i-1$. However, if the search process is completed at round i, the algorithm may stop. There are other search strategies that are known as *adaptive search strategies* in which the subset of cells that is paged in round i depends on the users that have been found so far. As noted in [1] the two versions coincide for the special case of two rounds.

The goal of EWCC is to minimize the expected number of cells that the system pages throughout the search.

If $d = 1$ then EWCC is trivial since the system must page all cells in the first round, and the solution costs n. If $m = 1$ EWCC can be solved in polynomial time using a simple dynamic programming [2, 3]. Bar-Noy and Malewicz [1] showed that EWCC is NP-hard for any pair of fixed values of m, d such that $m, d \geq 2$. They also presented an approximation algorithm with performance guarantee of $\frac{e}{e-1} \sim 1.581977$ for arbitrary values of d, m, and for the special (NP-hard) case where $d = m = 2$ they showed that their approximation algorithm is a $\frac{4}{3}$-approximation. Bar-Noy and Malewicz raised the open problem of the existence of a polynomial-time approximation scheme for EWCC. In this paper we give the first positive answer for this question by presenting a polynomial-time approximation scheme (PTAS) for the case of a fixed number of rounds and a fixed number of users (m and d are arbitrary constant integer values).

We now present a formal definition of EWCC. Denote the cell set by $C = \{1, 2, \ldots, n\}$, and the user set by $U = \{1, 2, \ldots, m\}$. For $i \in U$ and $j \in C$, denote by p_i^j the probability that user i is located at cell j. We assume that $p_i^j > 0$, $\forall i, j$. Given a positive matrix $P = \{p_i^j\}_{i,j}$ and a bound d on the number of rounds, a feasible solution is a partition $C_1, C_2, \ldots, C_d$ of C with the interpretation that in round k the system pages the cells in the subset C_k, unless it has already found all the users. A partition $C_1, C_2, \ldots, C_d$ induces probabilities $(P_k)_{k=1}^d$ where P_k

denotes the probability that the search will last for at least k rounds. I.e., P_k is the probability that $C_1 \cup C_2 \cup \cdots \cup C_{k-1}$ does not contain U. Then, the cost of $C_1, C_2, \ldots, C_d$ is $\sum_{k=1}^{d} P_i |C_i|$. The goal of EWCC is to find a partition of C that minimizes its cost.

We now give a more detailed expression for the cost of a partition for the special case of two rounds: $C_2 = C \setminus C_1$, denote $p_i(C_1) = \sum_{j \in C_1} p_i^j$, which is the probability to find user i in the first round. Therefore $P_2 = 1 - \prod_{i \in U} p_i(C_1)$, and the cost associated with the partition $C_1, C \setminus C_1$ is exactly $|C_1| + (n - |C_1|) \cdot \left(1 - \prod_{i \in U} p_i(C_1)\right)$.

We start the paper with a PTAS for two rounds, and later extend it to an arbitrary (constant) number of rounds d.

2   A PTAS for Two Rounds

In this section we present the main result of this paper; a polynomial time approximation scheme for EWCC when $d = 2$. We fix an optimal solution OPT. Our scheme is composed of two guessing steps. In these guessing steps we guess certain information about the structure of OPT. Each guessing step can be emulated via an exhaustive enumeration of all the possibilities for this piece of information. So our algorithm runs all the possibilities, and among them chooses the best solution achieved. In the analysis it is sufficient to consider the solution obtained when we check the right guess.

Given OPT, denote by OPT_1 the number of cells that OPT pages in the first round, and by α_i the probability that OPT does not find user i in the first round. Therefore, the cost of OPT denoted by $COST(OPT)$ is $COST(OPT) = OPT_1 + (n - OPT_1) \cdot \left(1 - \prod_{\ell \in U}(1 - \alpha_\ell)\right)$.

Recall that m is a constant, and let ε be a value such that $0 < \varepsilon < \frac{1}{(m+1)}$. If $n \leq m$, then EWCC can be solved in a constant time via exhaustive enumeration (since m is a constant), and therefore we assume that $n > m$. Denote the probability intervals $I_0 = (0, \frac{\varepsilon}{n^2}]$, and for $1 \leq i \leq \left\lceil \log_{1+\varepsilon}\left(\frac{n^2}{\varepsilon}\right)\right\rceil$,

$$I_i = \left(\frac{\varepsilon}{n^2}(1+\varepsilon)^{i-1}, \frac{\varepsilon}{n^2}(1+\varepsilon)^i\right].$$

Our first guessing step guesses for each $\ell \in U$, the index $i(\ell)$ such that $\alpha_\ell \in I_{i(\ell)}$. The following lemma is trivial:

Lemma 1. *The number of possibilities for the first guessing step is*

$$O\left(\left\lceil \log_{1+\varepsilon}\left(\frac{n^2}{\varepsilon}\right) + 2\right\rceil^m\right).$$

Therefore, performing an exhaustive enumeration for this guessing step can be done in polynomial time. We continue to analyze the iteration of this step in which we guess the right values that correspond to OPT. For all $\ell \in U$, we denote *the guess of α_ℓ by β_ℓ* to be the upper bound of $I_{i(\ell)}$; $\beta_\ell = \frac{\varepsilon}{n^2}(1+\varepsilon)^{i(\ell)}$.

The next step is to scale up the probabilities as follows: for all i, j define $q_i^j = p_i^j/\beta_i$ to be the *scaled probability of i and j*. We consider the vector $Q^j = (q_i^j)_{i \in U}$ of the scaled probabilities that the users are in cell j. We remove all cells with scaled probability larger than 1. Such cells cannot be paged in the second round, and therefore must be paged in the first round. We further assign a type and weight for each Q^j according to the following way. Let q_i^j be a maximum entry in Q^j, then we assign a weight $w^j = q_i^j$ to Q^j, and we define $\tilde{Q}^j = \left(q_\ell^j/w^j\right)_{\ell \in U}$. Note that $Q^j = w^j \cdot \tilde{Q}^j$. We define a set of intervals $\mathcal{J}$ as follows: $J_0 = (0, \varepsilon]$, and for all $k \geq 1$, $J_k = (\varepsilon \cdot (1+\varepsilon)^{k-1}, \varepsilon \cdot (1+\varepsilon)^k]$, and $\mathcal{J} = \{J_0, J_1, \ldots\}$. For each entry ℓ in $\tilde{Q}^j$, we find the interval from $\mathcal{J}$ that contains q_ℓ^j/w^j. We assign the type of Q^j to be the following vector. For each ℓ, compute a value t_ℓ such that $\frac{q_\ell^j}{w^j} \in J_{t_\ell}$, then the type of Q^j is the vector $(t_1, t_2, \ldots, t_m)$.

Lemma 2. *The number of possible types is $O\left(\left[\log_{1+\varepsilon}\left(\frac{1}{\varepsilon}\right) + 2\right]^m\right)$.*

Proof. To see this note that for all ℓ, j, we have $\frac{q_\ell^j}{w^j} \leq 1$. Therefore, it is enough to use the first $\lceil\log_{1+\varepsilon}\left(\frac{1}{\varepsilon}\right)\rceil + 1$ intervals in $\mathcal{J}$. The bound on the number of possibilities of types that our instance contains follows.

Note that the bound on the number of types is a constant (for fixed values of ε, m). Our second guessing step is to guess OPT_1 (since the first round is never skipped, this is an integer between 1 and n) and guess the *number* of cells from each type that OPT pages in the second round (this also gives the number of cells from each type that OPT pages in the first round).

Lemma 3. *The number of possible guesses is bounded by $O\left(n^{\left[\log_{1+\varepsilon}\left(\frac{1}{\varepsilon}\right)+2\right]^m}\right)$.*

Proof. The number of cells from each type is an integer between 0 and $OPT_2 = n - OPT_1 \leq n - 1$. This guess also implies a guess for OPT_2 and OPT_1.

Note that the number of possibilities for this guessing step is polynomial (for fixed values of ε, m).

Assume that for a type $T = (t_1, t_2, \ldots, t_m)$, OPT has $OPT_2(T)$ cells of type T that are paged in the second round. We sort the cells of type T according to their weight, and we assign the second round the $OPT_2(T)$ cells (among the cells of type T) that have the least weight. We apply this procedure for all the types T. We would like to ignore all invalid solutions. In order to be valid, the probability bounds β_ℓ must be satisfied, i.e. the sum of (non-scaled) probabilities must be in the interval $I_{i(\ell)}$. We slightly relax this requirement since a result of the scaling may shift the sum out of this interval. Instead, we disregard probabilities such that their scaled probability is in the interval $(0, \varepsilon]$, and we allow the sum of the other (non-scaled) probabilities to reside in the interval $[0, \beta_\ell(1 + \varepsilon)]$. Equivalently, the sum of scaled and rounded vectors of probability of chosen cells (ignoring the small components as explained above) should be such that no component exceeds $1 + \varepsilon$. For some guesses we obtain a candidate solution. Among all candidate solutions we output the one whose cost is minimized.

Lemma 4. *For a fixed number of users m, and for a constant $\varepsilon > 0$, the above scheme takes polynomial time.*

Proof. By Lemmas 1,2 and 3, the number of possibilities in the first step and in the second step is polynomial. The time to compute the resulting candidate solution for a single guess is clearly polynomial (i.e., finding a maximum value for each cell and finding its weight is polynomial, and the rest is simply sorting of the cells according to their weights), and the time to compute its cost is also polynomial. Therefore, the scheme has polynomial running time.

3   Analysis

We analyze the iteration of the first guessing step in which the guessed values of β_i $\forall i$ are the right guesses. We also assume that in the second guessing step we guess the right value of OPT_1 and the right number of cells of each type that OPT pages in the second round. We analyze the cost of the corresponding candidate solution.

Lemma 5. *The right set of guesses leads to a candidate solution.*

Proof. We have to show that for each user i, the sum of the probabilities (when we ignore cells whose scaled probability is at most ε) of finding user i in the second round is at most $\beta_i(1 + \varepsilon)$. For a type $T = (t_1, t_2, \ldots, t_m)$ with $t_i \geq 0$, OPT selects $OPT_2(T)$ cells of type T with sum of weights that is at least the sum of weights of the cells of type T that the candidate solution selects (note that the weights are not changed in the process of partitioning the cells into types). By definition of $\mathcal{J}$, the probabilities of having user i in a pair of cells of type T with the same weight, are within a multiplicative factor of $1 + \varepsilon$. Therefore, the contribution of type T cells to the probability that the candidate solution finds i during the second round, is at most $1 + \varepsilon$ times the contribution of type T cells to the probability that OPT finds i only during the second round. Since the probability that OPT finds i only during the second round is $\alpha_i \leq \beta_i$, the claim follows.

Lemma 6. *Consider a user i, then the probability that the candidate solution finds i during the second round (and not during the first round) is at most $\beta_i(1 + (m+1)\varepsilon)$.*

Proof. Consider a type $T = (t_1, t_2, \ldots, t_m)$. First, assume that $t_i \geq 1$. By Lemma 5, the contribution of type T cells to the probability that the candidate solution finds i during the second round, is at most $1 + \varepsilon$ times the contribution of type T cells to the probability that OPT finds i only during the second round.

Next, consider a type T such that $t_i = 0$. For such a type we define the *leader* of T to be the first entry of the type vector that relates to the largest interval (the interval which contains the point 1). There exists at least one such entry as in $\tilde{Q}^j$ there is at least one unit entry. Note that the sum of scaled probabilities of finding user ℓ of all the cells paged by OPT with a type such that ℓ is the leader, is at most $1 + \varepsilon$. Therefore, the total contribution of scaled probabilities of all the cells of any type T such that $t_i = 0$ and ℓ is acting as the leader of T is at most $(1 + \varepsilon)\varepsilon$. Summing over all ℓ (note that $\ell \neq i$), we get an increase of $(m - 1)\varepsilon(1 + \varepsilon)$ caused by the types where $t_i = 0$. In terms of the original probabilities (i.e., for each cell we multiply its probability by β_i) the types T such that $t_i = 0$ increase the probability of not finding user i in the first round by (an additive factor of) $(m - 1)(1 + \varepsilon)\varepsilon\beta_i$.

To conclude (the two above arguments) the probability that the candidate solution finds i during the second round (and not during the first round) is at most $\beta_i(1 + \varepsilon) + \beta_i(1 + \varepsilon)(m - 1)\varepsilon = \beta_i(1 + \varepsilon)(1 + (m - 1)\varepsilon) \leq \beta_i(1 + (m + 1)\varepsilon)$ (since $\varepsilon < 1/(m + 1)$).

We denote by $S = \{i \in U | \beta_i = \frac{\varepsilon}{n^2}\}$ the set of users with *small* probability of being left for the second round, and by $L = U \setminus S$ the set of users with *large* probability of being left for the second round.

Theorem 1. *The best candidate solution is a $(1 + \varepsilon)(1 + 2\varepsilon)(1 + (m + 1)\varepsilon)$-approximated solution.*

Proof. We will analyze the candidate solution that corresponds to the right guesses (with respect to the information used by the solution OPT). By Lemma 5, this is a candidate solution. The best candidate solution clearly outperforms this particular candidate solution.

For a user $i \in L$, the probability that OPT does not find i in the first round is α_i, whereas by Lemma 6 the probability that the candidate solution does not find i in the first round is at most $\beta_i(1 + (m + 1)\varepsilon)$. Since $i \in L$, we conclude that $\alpha_i \geq \frac{\beta_i}{1 + \varepsilon}$. Therefore, the probability that the candidate solution does not find i in the first round is at most $(1 + \varepsilon)(1 + (m + 1)\varepsilon)\alpha_i$.

For a user $i \in S$, the probability that the candidate solution does not find i in the first round is at most $\beta_i(1 + (m + 1)\varepsilon) = \frac{\varepsilon(1 + (m+1)\varepsilon)}{n^2} \leq \frac{2\varepsilon}{n^2}$, where the inequality follows as $\varepsilon < \frac{1}{(m+1)}$. Using the union bound we conclude that the probability that the candidate solution does not find at least one of the users in S is at most $\frac{2\varepsilon|S|}{n^2} \leq \frac{2\varepsilon m}{n^2} \leq \frac{2\varepsilon}{n}$, where the last inequality follows from the assumption $n > m$. In case this event happens we assign an extra cost of n (for the second round). This extra cost incurs an expected extra cost (an additive factor) of at most $\frac{2\varepsilon}{n} \cdot n = 2\varepsilon$. Since OPT costs at least 1, we will conclude that the users in S caused an increase of the approximation factor by a multiplicative factor of at most $1 + 2\varepsilon$.

We first assume that there is $\ell \in U$ such that $\beta_\ell(1 + (m + 1)\varepsilon)(1 + \varepsilon) \geq 1$. In this case $\alpha_\ell \geq \frac{1}{(1+(m+1)\varepsilon)(1+\varepsilon)^2}$, and therefore $COST(OPT) \geq OPT_1 + (n - OPT_1)\alpha_\ell \geq n\alpha_\ell \geq \frac{n}{(1+(m+1)\varepsilon)(1+\varepsilon)^2} \geq \frac{n}{(1+(m+1)\varepsilon)(1+\varepsilon)(1+2\varepsilon)}$. Note that the returned solution costs at most n, and therefore in this case the returned solution pays at most $(1+(m+1)\varepsilon)(1+\varepsilon)(1+2\varepsilon)COST(OPT)$. Therefore, we can assume that for all ℓ, $\beta_\ell(1 + (m + 1)\varepsilon)(1 + \varepsilon) < 1$.

We denote by $\tau = (1 + \varepsilon)(1 + (m + 1)\varepsilon)$. The cost of the candidate solution is at most:

$$OPT_1 + (n - OPT_1) \cdot \left(1 - \prod_{\ell \in U}(1 - \beta_\ell(1 + (m + 1)\varepsilon))\right) \tag{1}$$

$$\leq OPT_1 + (n - OPT_1) \cdot \left(1 - \prod_{\ell \in L}(1 - \beta_\ell(1 + (m + 1)\varepsilon))\right) + 2\varepsilon \tag{2}$$

$$\leq OPT_1 + (n - OPT_1) \cdot \left(1 - \prod_{\ell \in L}(1 - \tau\alpha_\ell)\right) + 2\varepsilon \tag{3}$$

$$\leq OPT_1 + (n - OPT_1)\tau \cdot \left(1 - \prod_{\ell \in L}(1 - \alpha_\ell)\right) + 2\varepsilon \tag{4}$$

$$\leq (\tau + 2\varepsilon)\left[OPT_1 + (n - OPT_1) \cdot \left(1 - \prod_{\ell \in L}(1 - \alpha_\ell)\right)\right] \tag{5}$$

$$\leq \tau(1 + 2\varepsilon)\left[OPT_1 + (n - OPT_1) \cdot \left(1 - \prod_{\ell \in L}(1 - \alpha_\ell)\right)\right] \tag{6}$$

$$\leq \tau(1 + 2\varepsilon)\left[OPT_1 + (n - OPT_1) \cdot \left(1 - \prod_{\ell \in U}(1 - \alpha_\ell)\right)\right] \tag{7}$$

$$= \tau(1 + 2\varepsilon)COST(OPT), \tag{8}$$

where (1) follows from Lemma 6 and the monotonicity of the goal function (increasing the probability of not finding a user in the first round only increases the solution cost). (2) follows as explained above since the users in S incur an additive increase of the expected cost by at most 2ε. (3) follows since for all $i \in L$, $\alpha_\ell(1 + \varepsilon) \geq \beta_\ell$. (4) follows because given a set of $|L|$ independent random events the probability of their union is multiplied by at most τ if we multiply the probability of each event in this set by τ. (5) and (6) follow by simple algebra (and by $OPT_1 \geq 1$). (7) follows since we deal with probabilities, and for each $\ell \in S$, $1 - \alpha_\ell \leq 1$, and therefore $\prod_{\ell \in L}(1 - \alpha_\ell) \geq \prod_{\ell \in U}(1 - \alpha_\ell)$. (8) follows from the fact that we consider the right guesses on OPT.

By Theorem 1 we conclude that,

Corollary 1. *The above scheme is a $[1 + 6m\varepsilon]$-approximation for all $\varepsilon > 0$.*

Proof. Since $\varepsilon < \frac{1}{(m+1)}$ and $m \geq 2$ we get $(1 + (m + 1)\varepsilon)(1 + \varepsilon)(1 + 2\varepsilon) \leq (1 + (m + 1)\varepsilon)(1 + 4\varepsilon) \leq 1 + (m + 9)\varepsilon \leq 1 + 6m\varepsilon$.

By setting $\varepsilon' = \frac{\varepsilon}{6m}$, and applying the above algorithm with ε' instead of ε, we get a $1 + \varepsilon$-approximation algorithm whose time complexity is polynomial for any fixed value of ε. Therefore, we proved the main result:

Theorem 2. *Problem EWCC with two rounds and a constant number of users has a polynomial time approximation scheme.*

4 Extension of the PTAS to Any Fixed Number of Rounds

In this section we show how the PTAS of the previous sections can be extended to provide a PTAS for EWCC when the number of rounds d is an arbitrary constant (the number of users is also a constant).

We fix an optimal solution OPT. Our scheme is again composed of two guessing steps.

Given OPT, denote by OPT_r the number of cells that OPT pages in the r-th round, and by α_i^r the probability for OPT to find user i exactly in the r-th round (i.e., OPT does not find i in the first $r-1$ rounds but finds i in the r-th round). Denote by $\pi_i^r = \sum_{s=r}^d \alpha_i^s$ the probability that OPT does not find i in the first $r-1$ rounds. Therefore, the cost of OPT denoted by $COST(OPT)$ is
$$COST(OPT) = \sum_{r=1}^d OPT_r \cdot (1 - \prod_{i \in U}(1 - \pi_i^r)).$$

Recall that m, d are constants, and let ε be a value such that $0 < \varepsilon < \frac{1}{(md+1)}$. If $n \leq md^2$, then EWCC can be solved in a constant time via exhaustive enumeration (as m and d are constants), therefore we assume that $n > md^2$. Similarly to the $d = 2$ case we denote the probability intervals $I_0 = (0, \frac{\varepsilon}{n^2}]$, and for $1 \leq i \leq \left\lceil \log_{1+\varepsilon}\left(\frac{n^2}{\varepsilon}\right)\right\rceil$, $I_i = \left(\frac{\varepsilon}{n^2}(1+\varepsilon)^{i-1}, \frac{\varepsilon}{n^2}(1+\varepsilon)^i\right]$.

Our first guessing step guesses for each $\ell \in U$ and $1 \leq r \leq d$, the index $i_r(\ell)$ such that $\alpha_\ell^r \in I_{i_r(\ell)}$. The following lemma is trivial:

Lemma 7. *The number of possibilities for the first guessing step is*

$$O\left(\left[\log_{1+\varepsilon}\left(\frac{n^2}{\varepsilon}\right) + 2\right]^{md}\right).$$

Therefore, performing an exhaustive enumeration for this guessing step can be done in polynomial time. We continue to analyze the iteration of this step in which we guess the right values that correspond to OPT. For all $\ell \in U$, we denote *the guess of α_ℓ^r* by β_ℓ^r to be the upper bound of $I_{i_r(\ell)}$; $\beta_\ell^r = \frac{\varepsilon}{n^2}(1+\varepsilon)^{i_r(\ell)}$.

The next step is to scale up the probabilities. Similarly in the $d = 2$ case we define $q_i^j(r) = p_i^j/(\beta_i^r)$ to be the scaled probability for user i to be found in cell j in round r. The matrix of cell j is $Q^j = (q_i^j(r))_{1 \leq i \leq m, 1 \leq r \leq d}$. For every matrix, each component larger than 1 is replaced by ∞ as this probability means that such cells cannot be paged in the relevant round. We further assign a type to each cell in the following way.

Let $q_i^j(r)$ be a maximum real entry in Q^j (if all entries are ∞, we can skip the current guess as it cannot lead to a valid solution), then we assign a weight $w^j = q_i^j(r)$ to Q^j, and we define $\tilde{Q}^j = \left(q_\ell^j(r)/w^j\right)_{\ell \in U, 1 \leq r \leq d}$. Note that $Q^j = w^j \cdot \tilde{Q}^j$. We define a set of intervals $\mathcal{J}$ as follows: $J_0 = (0, \varepsilon]$, and for all $k \geq 1$, $J_k = (\varepsilon \cdot (1+\varepsilon)^{k-1}, \varepsilon \cdot (1+\varepsilon)^k]$, and $\mathcal{J} = \{J_0, J_1, \ldots\}$. For each entry (ℓ, r) in $\tilde{Q}^j$, we find the interval from $\mathcal{J}$ that contains $q_\ell^j(r)/w^j$. We assign the type of Q^j to be the following matrix. For each (ℓ, r) of real probability, compute a value $t_{(\ell,r)}$

such that $\frac{q_\ell^j(r)}{w^j} \in J_{t_{(\ell,r)}}$. Entries of infinite probability are assigned $(t_{(\ell,r)} = \infty$. The type of Q^j is the matrix $(t_{(\ell,r)})_{1\leq\ell\leq m,1\leq r\leq d}$.

Our second guessing step is to guess OPT_r for $r = 1, 2, \ldots, d$ (since the first round is never skipped, OPT_1 is an integer between 1 and n, and the other values are integers between 0 and $n - 1$) and guess the *number* of cells from each type that OPT pages in each round.

Lemma 8. *The number of possible types is* $O\left(\left[\log_{1+\varepsilon}\left(\frac{1}{\varepsilon}\right) + 3\right]^{md}\right)$.

Proof. Each entry can have any of the values as in the two round case or infinity.

Lemma 9. *The number of possible guesses is bounded by the value*
$$O\left((n + 1)^{d\left[\log_{1+\varepsilon}\left(\frac{1}{\varepsilon}\right)+3\right]^{md}}\right).$$

Proof. For round $1 \leq r \leq d$, the number of cells from each type is an integer between 0 and $OPT_r \leq n$. Guessing the number of cells of each type in every round implies a guess of the OPT_r values.

Note that the number of possibilities for this guessing step is polynomial (for fixed values of ε, m, d).

Assume that for a type T, OPT has $OPT_r(T)$ cells of type T that are paged in the r-th round. We sort the cells of type T according to their weight, and we iterate the following: we initialize $r = d$ and assign the r-th round $OPT_r(T)$ cells (among the cells of type T) that have the least weight. We remove this set of cells, we decrease r by 1 and repeat until no more cells of type T exist. We apply this procedure for all the types T. We would like to ignore all invalid solutions. In order to be valid, the probability bounds β_ℓ^r must be satisfied, i.e. the sum of probabilities must be in the interval $I_{i_r(\ell)}$. We slightly relax this requirement since a result of the scaling may shift the sum out of this interval. Instead, we disregard probabilities such that their scaled probability is in the interval $(0, \varepsilon]$, and we require that the sum over all rounds from r to d, of the sum of the other (non-scaled) probabilities should reside in the interval $[0, \sum_{s=r}^{d} \beta_\ell^s(1 + \varepsilon)]$. For some guesses we obtain a candidate solution. Among all candidate solutions we output the one whose cost is minimized.

Lemma 10. *For a fixed number of users* m, *a fixed number of rounds* d, *and for a constant* $\varepsilon > 0$, *the above scheme takes a polynomial time.*

Proof. By Lemmas 7,8 and 9, the number of possibilities in the first step and in the second step is polynomial. The time to compute the resulting candidate solution for a single guess is clearly polynomial (i.e., finding a maximum value for each cell and finding its weight is polynomial, and the rest is simply sorting of the cells according to their weights), and the time to compute its cost is also polynomial. Therefore, the scheme takes a polynomial time.

We analyze the iteration of the first guessing step in which the guessed values of $\beta_i^r \, \forall i, r$ are the right guesses. We also assume that in the second guessing step we guess the right values of OPT_r for $r = 1, 2, \ldots, d$ and the right number of cells of each type that OPT pages in each round. We analyze the cost of the corresponding candidate solution.

Lemma 11. *The right set of guesses leads to a candidate solution.*

Proof. We have to show that for each user i and each round r, the sum of the probabilities (when we ignore cells whose scaled probability is at most ε) of not finding user i within the first $r - 1$ rounds is at most $\sum_{s=r}^{d} \beta_i^s (1 + \varepsilon)$. For a type T, OPT selects $\sum_{s=r}^{d} OPT_s(T)$ cells of type T with sum of weights that is at least the sum of weights of the cells of type T that the candidate solution selects (note that the weights are not changed in the process of partitioning the cells into types). By definition of $\mathcal{J}$, the probabilities of having user i in a pair of cells of type T with the same weight, are within a multiplicative factor of $1 + \varepsilon$. Therefore, the contribution of type T cells to the probability that the candidate solution does not find i during the first $r - 1$ rounds, is at most $1 + \varepsilon$ times the contribution of type T cells to the probability that OPT does not find i during the first $r - 1$ rounds. As the probability that OPT finds i only during the s-th round is $\alpha_i^s \leq \beta_i^s$, the claim follows.

Lemma 12. *Consider a user i and a round r, then the probability that the candidate solution does not find i during the first $r - 1$ rounds is at most $\sum_{s=r}^{d} \beta_i^s (1 + (md + 1)\varepsilon)$.*

Proof. Consider a type matrix T. A type with an ∞ entry for round s will have zero cells for that round. Otherwise assume first that $t_{(i,r)} \geq 1$. By Lemma 11, the contribution of type T cells to the probability that the candidate solution does not find i during the first $r-1$ rounds, is at most $1+\varepsilon$ times the contribution of type T cells to the probability that OPT does not find i during the first $r - 1$ rounds.

Next, consider a type T such that $t_{(i,r)} = 0$. For such a type we define the *leader* of T to be the first entry of the type matrix that relates to the largest real interval (that contains the point 1). There exists at least one entry like this, as there is at least one unit entry in $\tilde{Q}^j$. Note that the sum of scaled probabilities of finding user ℓ in round r' of all the cells paged by OPT in that round with a type such that ℓ is the leader, is at most $1 + \varepsilon$. Therefore, the total contribution of scaled probabilities of all the cells of any type T such that $t_{(i,r')} = 0$ and ℓ, r' is acting as the leader of T is at most $(1 + \varepsilon)\varepsilon$. Summing over all ℓ and r' (note that we may exclude the case $\ell = i, r' = r$), we get an increase of $(md-1)\varepsilon(1+\varepsilon)$ caused by the types where $t_{(i,r)} = 0$. In terms of the original probabilities (i.e., for each cell and round s we multiply its probability by β_i^s) the types T such that $t_{(i,r)} = 0$ increase the probability of not finding user i in the first $r - 1$ rounds by at most (an additive factor of) $(dm - 1)\varepsilon(1 + \varepsilon) \sum_{s=r}^{d} \beta_i^s$.

To conclude (the two above arguments) the probability that the candidate solution does not find i during the first $r-1$ rounds is at most $\sum_{s=r}^{d} \beta_i^s (1+\varepsilon) + \sum_{s=r}^{d} \beta_i^s (md - 1)\varepsilon(1 + \varepsilon) = \sum_{s=r}^{d} \beta_i^s (1 + (md + 1)\varepsilon)$ (since $\varepsilon < 1/(dm + 1)$).

.

Theorem 3. *The best candidate solution is a* $(1+\varepsilon)^2(1+(md+1)\varepsilon)$-*approximated solution.*

Proof. We will analyze the candidate solution that corresponds to the right guesses (with respect to the information used by the solution OPT). By Lemma 11, this is a candidate solution. The best candidate solution clearly outperforms this particular candidate solution.

For a user i, the probability that OPT does not find i in the first $r-1$ rounds is $\sum_{s=r}^{d} \alpha_i^s$, whereas by Lemma 12 the probability that the candidate solution does not find i in the first $r-1$ rounds is at most $\sum_{s=r}^{d} \beta_i^s(1+(md+1)\varepsilon)$. For all $\ell \in U$ and for all $s = 1,2,\ldots,d$, $\alpha_\ell^s(1+\varepsilon)+\frac{\varepsilon}{n^2} \geq \beta_\ell^s$ holds. This gives $\sum_{s=r}^{d} \alpha_i^s \geq \frac{\sum_{s=r}^{d}\beta_i^s}{1+\varepsilon} - \frac{\varepsilon(d-r+1)}{n^2}$. Therefore, the probability that the candidate solution does not find i in the first $r-1$ rounds is at most $(1+\varepsilon)(1+(md+1)\varepsilon)\sum_{s=r}^{d}\alpha_i^s+(1+\varepsilon)(1+(md+1)\varepsilon)\frac{\varepsilon d}{n^2}$. Since the above term bounds a probability we conclude that the probability that the candidate solution does not find i in the first $r-1$ rounds is at most $\min\{1, (1+\varepsilon)(1+(md+1)\varepsilon)\sum_{s=r}^{d}\alpha_i^s + (1+\varepsilon)(1+(md+1)\varepsilon)\frac{\varepsilon d}{n^2}\}$.

We denote by $\tau = (1+\varepsilon)(1+md\varepsilon)$. The cost of the candidate solution is at most:

$$OPT_1 + \sum_{r=2}^{d} OPT_r \cdot \left(1 - \prod_{\ell \in U}\left(1 - \sum_{s=r}^{d}\beta_\ell^s(1+md\varepsilon)\right)^+\right) \tag{9}$$

$$\leq OPT_1 + \sum_{r=2}^{d} OPT_r \cdot \left(1 - \prod_{\ell \in U}\left(1 - \tau\left(\sum_{s=r}^{d}\alpha_\ell^s + \frac{\varepsilon d}{n^2}\right)\right)^+\right) \tag{10}$$

$$\leq OPT_1 + \sum_{r=2}^{d} OPT_r \cdot \left(1 - \prod_{\ell \in U}\left(1 - \tau\sum_{s=r}^{d}\alpha_\ell^s\right)^+ + \frac{\varepsilon\tau md}{n^2}\right) \tag{11}$$

$$\leq OPT_1 + \sum_{r=2}^{d} OPT_r \cdot \left(1 - \prod_{\ell \in U}\left(1 - \tau\sum_{s=r}^{d}\alpha_\ell^s\right)^+\right) + \frac{\varepsilon\tau md^2}{n} \tag{12}$$

$$\leq OPT_1 + \sum_{r=2}^{d} OPT_r \cdot \left(1 - \prod_{\ell \in U}\left(1 - \tau\sum_{s=r}^{d}\alpha_\ell^s\right)^+\right) + \varepsilon\tau \tag{13}$$

$$\leq OPT_1 + \sum_{r=2}^{d} OPT_r \cdot \tau\left(1 - \prod_{\ell \in U}\left(1 - \sum_{s=r}^{d}\alpha_\ell^s\right)\right) + \varepsilon\tau \tag{14}$$

$$\leq \tau(1+\varepsilon)\left[OPT_1 + \sum_{r=2}^{d} OPT_r \cdot \left(1 - \prod_{\ell \in U}\left(1 - \sum_{s=r}^{d}\alpha_\ell^s\right)\right)\right] \tag{15}$$

$$= \tau(1+\varepsilon)COST(OPT), \tag{16}$$

where (9) follows from Lemma 12 and the monotonicity of the goal function (increasing the probability of not finding a user in the first rounds only increase the solution cost). (10) follows as explained above. (11) follows by simple algebra. (12) follows since $OPT_r \leq n$, $\forall n$. (13) follows from the assumption $n \geq md^2$. (14) follows because given a set of $|L|$ independent random events the probability of their union is multiplied by at most τ if we multiply the probability of each event in this set by τ. (15) follow by simple algebra (and by $OPT_1 \geq 1$). (16) follows from considering the right guesses on OPT.

Similar to the $d = 2$ case, we establish the following theorem:

Theorem 4. *Problem EWCC with a constant number of rounds and a constant number of users has a polynomial time approximation scheme.*

References

1. A. Bar-Noy and G. Malewicz, "Establishing wireless conference calls under delay constraints," *Journal of Algorithms*, **51**, 145-169, 2004.
2. D. Goodman, P. Krishnan and B. Sugla, "Minimizing queuing delays and number of messages in mobile phone location," *Mobile Networks and Applications*, **1**, 39-48, 1996.
3. C. Rose and R. Yates, "Minimizing the average cost of paging under delay constraints," *Wireless Networks*, **1**, 211-219, 1995.

This Side Up!

Leah Epstein[1],[*] and Rob van Stee[2],[**]

[1] School of Computer Science, The Interdisciplinary Center, Herzliya, Israel
lea@idc.ac.il.
[2] Centre for Mathematics and Computer Science (CWI), Amsterdam,
The Netherlands
Rob.van.Stee@cwi.nl.

Abstract. We consider two- and three-dimensional bin packing problems where 90° rotations are allowed. We improve all known asymptotic performance bounds for these problems. In particular, we show how to combine ideas from strip packing and two-dimensional bin packing to give a new algorithm for the three-dimensional strip packing problem where boxes can only be rotated sideways. We propose to call this problem 'This side up'. Our algorithm has an asymptotic performance bound of 9/4.

1 Introduction

The study of multi-dimensional packing problems gained an increasing interest in the last few years [1, 2, 4, 7]. A main trend was the study of offline and online packing algorithms for oriented items which are rectangles or boxes. Given a large supply of bins which are squares, or cubes, or a strip of infinite height, the goal is to pack items efficiently, without rotation. This problem clearly has applications; however, in many applications, there is no reason to exclude the option of changing the orientation of items before assignment. Some applications may allow rotation only in certain directions.

In this paper we study several rotatable packing problems. The same problem is also known as "packing of non-oriented items" [5]. All packing problems involve an input which is a set of items. In "strip packing" problems, the goal is to pack the items into a strip of unlimited height, so as to minimize the maximum height ever used. In "bin packing" problems, the goal is to use a minimum number of bins for the packing. The items always need to be packed without overlap. The exact structure of the strip, bins and items depends on the specific problem.

The two-dimensional bin packing problem with rotations is defined as follows. The bins are unit squares and the items are rectangles that may be rotated by 90°. In the strip packing version, the strip is two-dimensional, with a base of

---

[*] Research supported by Israel Science Foundation (grant no. 250/01).
[**] Research supported by the Netherlands Organization for Scientific Research (NWO), project number SION 612-30-002.

G. Persiano and R. Solis-Oba (Eds.): WAOA 2004, LNCS 3351, pp. 48–60, 2005.

width one and infinite height. In the three-dimensional bin packing problem with rotations, the bins are three-dimensional cubes, and the items are non-oriented three-dimensional boxes, rotatable by 90° in all possible directions. In the strip packing version we pack these items into a three-dimensional strip with a base which is a unit square, and again, infinite height.

In three dimensions, we can also consider the case of items that may be rotated so that the width and length are interchanged, however the height is fixed. We call this problem "This Side up", as it has applications in packing of fragile objects, where a certain face of the box must be placed on top. This three-dimensional problem, as the rotatable problems has two versions: packing into a three-dimensional strip (also called "the z-oriented 3-D packing problem" [10]) and packing into three-dimensional bins.

The standard measure of algorithm quality for box packing is the *asymptotic performance ratio*, which we now define. For a given input sequence σ, let $\mathcal{A}(\sigma)$ be the number of bins used by algorithm $\mathcal{A}$ on σ. Let $\mathrm{OPT}(\sigma)$ be the minimum possible number of bins used to pack items in σ. The *asymptotic performance ratio* for an algorithm $\mathcal{A}$ is defined to be

$$\mathcal{R}_{\mathcal{A}}^{\infty} = \limsup_{n \to \infty} \sup_{\sigma} \left\{ \frac{\mathcal{A}(\sigma)}{\mathrm{OPT}(\sigma)} \middle| \mathrm{OPT}(\sigma) = n \right\}.$$

Previous Results: The oriented packing problems have been widely studied. The best result for two-dimensional packing into bins is 1.691, due to Caprara [2]. See references in [1, 2, 4, 7] for further results on oriented packing problems. The square and cube packing (into bins) problems are special cases of rotatable packing. An APTAS for these problems was given in [1] and independently in [4].

Although the possibility of allowing rotations was already mentioned in [3], there has been relatively little research into this subject from a worst-case perspective until recently. Fujita and Hada [8] considered the two-dimensional bin packing problem with rotations. They presented two online algorithms and claimed asymptotic performance ratios of at most 2.6112 and 2.56411. Epstein [6] showed that the first algorithm instead has an asymptotic performance ratio of at most 2.63889 and questioned the validity of the second algorithm. She also presented an online algorithm with asymptotic performance ratio slightly below 2.45.

Recently, Miyazawa and Wakabayashi [11] presented an *offline* approximation algorithm for two-dimensional bin packing with rotations with an asymptotic performance bound of 2.64. It is most likely that the extended abstract [6] as well as the earlier paper [8] were unknown to those authors.

The paper [11] also considers several other problems with rotatable items and gives an upper bound of 2.64 for the This Side Up problem which was also considered in [9, 10]. The paper [10] demonstrates a reduction from the general three-dimensional strip packing problem with rotations to the This Side Up problem in a strip, but this reduction does not hold for the case considered in this paper, where the three-dimensional strip always has a square base of side 1.

Our Results. We improve upon the best known results for all the above problems.

- An algorithm of asymptotic performance ratio $3/2 = 1.5$ for two-dimensional rotatable strip packing. This improves on the bound 1.613 in [11].
- An algorithm of asymptotic performance ratio $9/4 = 2.25$ for two-dimensional rotatable packing into bins. This improves on the on-line algorithm in [6] that has an asymptotic performance ratio of slightly less than 2.45. Although this algorithm basically consists of many (easy) cases, it has the advantage that it can easily be adapted to the more complex problems listed below.
- An algorithm which combines methods of the two above algorithms and has asymptotic performance ratio $9/4 = 2.25$ for the "This side up" problem in a strip. This is the main result of the paper. This improves the bound of [11] which is 2.64.
- An adaptation of the previous algorithm to packing of rotatable items in a three-dimensional strip, with the same asymptotic performance ratio $9/4 = 2.25$. This improves the bound of [11] which is 2.76.
- A simple adaptation of the two previous algorithms for the bin packing versions of these problems, with asymptotic performance ratio $9/2 = 4.5$. This improves the bound of [11] for three-dimensional bin packing of rotatable items, which is 4.89.

2   Two-Dimensional Strip Packing

We assume that all items have height and width at most 1. The strip has width 1 and unbounded height. As a subroutine for our algorithm, we use the well-known algorithm First Fit Decreasing Height (FFDH). The following theorem was proved by Coffman et al. in 1980 [3].

Theorem 1. *Let L be any list of rectangles ordered by non-increasing height such that no rectangle in L has width exceeding $1/m$ for some $m \geq 2$. Then $FFDH(L) \leq (1 + 1/m)A(L) + 1$, where $A(L)$ is the total area of the items in L.*

Items that have width and height greater than $1/2$ are called *big*. These items are rotated so that their width is not smaller than their height. For items that have only one dimension greater than $1/2$, we rotate them so that this dimension is the height.

1. The big items are stacked at the bottom of the strip, in order of decreasing width and aligned with the left side of the strip. Denote the height needed for this packing by h_1.
2. Denote the height at which the first item of width at most $2/3$ is packed by h'_1 ($0 \leq h'_1 \leq h_1$). If $h'_1 < h_1$, define a substrip of width $1/3$ that starts at height h'_1, at the right side of the strip. Pack items that have widths in $(0, 1/6]$ inside this strip using FFDH, until all these items are packed or until the next item to be packed would be placed (partially) above height h_1.

3. If all items that have width in $(0, 1/6]$ have now been packed:
   (a) Stack items of widths in $(1/6, 1/2]$ at the right side of the bin, on top of the substrip from step 2. Place these items in order of *increasing width*. Each item is placed as low as possible, at the extreme right of the bin, under the constraint that it does not overlap with previously placed items. Do this until all such items are packed or the next item to be packed would be placed (partially) above height h_1.
   (b) Place the unpacked items of width in $(1/3, 1/2]$ in two stacks starting at height h_1 by each time adding an item to the shortest stack. Pack the unpacked items of width in $(1/6, 1/3]$ using FFDH.
4. Else, place all remaining items above height h_1 using the algorithm FFDH.

Theorem 2. *For this algorithm, we have* $\text{ALG}_1(L) \leq \frac{3}{2}\text{OPT}(L) + 3$ *for any input list L.*

Proof. The proof consists of showing that at all heights of the packing apart from a total height of at least 3, a width of $2/3$ is covered by items. This implies immediately that the height of the optimal packing is at least $2/3$ of the height of the packing of ALG_1. Details are omitted in this extended abstract. □

It is straightforward to see that the complexity of our algorithm is $O(n \log n)$, where n is the number of items to be packed, since apart from sorting by width or height all the steps in the algorithm take linear time. We conjecture that the complexity of any algorithm with an approximation ratio strictly below $3/2$ is substantially higher than $O(n \log n)$.

3   Two-Dimensional Bin Packing

We apply a first partition of items to types in the following way. We rotate all items such that the length is at least as large as the width. We call this the *standard* orientation, and the other one the *reversed* orientation. The one-dimensional intervals we use in the initial partition are

- $(2/3, 1]$ (type 0)
- $(1/2, 2/3]$ (type 1)
- $(1/(i+1), 1/i]$ for $i = 2, \ldots, 8$ (type i)
- $(0, 1/9]$ (type 9)

A two-dimensional item is of type (i, j) if its width is of type i and its length is of type j. Clearly $i \geq j$ due to the orientation we defined. In some cases, we will use a finer classification for type 1. We let type $1a = (1/2, 11/20]$, type $1b = (11/20, 3/5]$ and type $1c = (3/5, 2/3]$.

There are four types which we will call *large*. We will begin by defining their packing. Each such type is packed independently of the other ones. We pack items of types $(1,1), (1,0)$ and $(0,0)$ one per bin, always in the left bottom corner of the bin and in standard orientation. The items of type $(2, 1)$ are further classified according to their length (largest dimension). Items of type $(2, 1a)$ are packed two per bin, both in reverse orientation, touching the same edge of the bin and each other, with one of them in the left bottom corner of the bin. The same holds for items of type $(2, 1b)$ and $(2, 1c)$ (but the items of these three subtypes are not packed together in any bin).

There are also four *medium* types. Items of type $(2, 2)$ are packed four per bin (the bin is first partitioned into four identical sub-bins). Items of type $(i, 0)$ are packed i per bin for $i = 2, 3, 4$.

After the packing of the large and medium types is completed, smaller items are added. They are first added into bins which contain large items. If some items remain unpacked after those bins are considered, they are packed into empty bins.

Bins containing items of the types $(0, 0), (2, 2), (2, 0), (3, 0)$ and $(4, 0)$ do not receive smaller items. We note that all these bins are packed so that a fraction of at least $4/9$ of their area is occupied, except possibly the last bin for each of the last four types. This follows from the types and the amounts of items per bin.

The performance bound of $9/4$ follows from one of the two following reasons.

1. If no new bins are opened for smaller items, we use a weighting function for the analysis. Those functions are usually useful in analyzing on-line algorithms. Here we use it to analyze an offline algorithm.
2. If at least one bin was opened for smaller items, we use an area based analysis. We show that all bins except a constant number have items of total area of at least $4/9$. Then we get $OPT \geq W \geq (4/9)(ALG_2 - c)$ which implies the performance ratio.

Case 1. The weighting function is defined in the following way. Small items get weight 0.

Type	$(0,0)$	$(1,0)$	$(1,1)$	$(2,1)$	$(2,0)$	$(2,2)$	$(3,0)$	$(4,0)$
Weight	1	1	1	1/2	1/2	1/4	1/3	1/4

The following claim is immediate from the definitions.

Claim. All bins packed by our algorithm with large items, except possibly the last one for each (sub)type, contain a weight of 1.

Claim. A bin can contain at most nine items of both width and length larger than $1/4$.

Proof. See [7]. □

It follows from the same result that a bin can contain at most twenty-five items of both width and length larger than $1/6$.

Claim. A bin can contain at most $9/4$ of weight.

Proof. Consider a bin with a certain amount of weight. We may assume there is no item of type $(0,0)$ or $(1,0)$, because the smaller type $(1,1)$ also has weight 1, and also no item of type $(2,0)$ because $(2,1)$ gives the same weight.

We will use Claim 3 to determine the highest possible weight in a bin by expressing all items as multiples of items of width and length just larger than $1/4$ or $1/6$. For instance, by cutting a $(1,1)$ item halfway both horizontally and vertically, it can be seen that other items of 'worth' at most 5 items of width and length just larger than $1/4$ can be placed with it in one bin (otherwise this cutting would create a packing with more than 9 such items, contradicting Claim 3).

An overview can be found in the following table. Here the heading 'items $> 1/4$' means 'number of items of length and width more than $1/4$ that items of this type contain', etc.

Type	items $> 1/4$	items $> 1/6$	weight	weight per item $> 1/6$
$(1,1)$	4	9	1	$1/9$
$(2,1)$	2	6	$1/2$	$1/12$
$(2,2)$	1	4	$1/4$	$1/16$
$(3,0)$	2	4	$1/3$	$1/12$
$(4,0)$	0	4	$1/4$	$1/16$

If there is no item of type $(1,1)$, then by the last column, the weight per 'virtual' item of width and length larger than $1/6$ is at most $1/12$ which gives total weight of at most $25/12 < 9/4$.

Otherwise, by the second column and Claim 3, at most 2 items of type $(2,1)$ or $(3,0)$ can be in the bin together with the item of type $(1,1)$. To get maximum weight, we should maximize the number of virtual items that have weight $1/12$ per item. We can have at most 12 such virtual items because there can be at most 2 items that cover 6 of them. This leaves at most 4 virtual items with weight per item $1/16$, giving additional weight of $1/4$. The total weight therefore is at most 1 (from the largest item) $+1$ (from the $(2,1)$ items) $+1/4 = 9/4$.   □

Case 2. It is left to show how small items are packed to keep a $4/9$ fraction of each bin occupied (except for a constant number of bins). Each bin will contain items of a given small type or set of types. For each type or set of types, we need to show how they are packed in the following three cases.

A. A bin that already contains a $(1,0)$ item, or two $(2,1)$ items.
B. A bin that already contains a $(1,1)$ item.
C. An empty bin.

Consider the area left for further packing in the three cases. See Figure 1. For many small types, summarized in Table 1, the packing of the small items does not depend on the exact size of the large items that they are packed with.

In type A bins, there is a strip of width $1/3$ and length 1 that does not contain any items. Such a bin already contains an area of at least $1/3$.

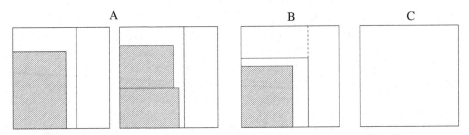

Fig. 1. Unused areas in bins of types A, B and C. Small items are packed here

Table 1. All types that are combined on a single line of the table are packed together, except the types (i,j) for $3 \le j \le i \le 5$ in empty bins (type C bins) and the $(9,i)$ types. For the $(9,i)$ types, shelves of length $1/i$ and width $1/3$ are created in type A and B bins. They are all filled to a length of $1/(i+1)$ and a width of $2/9$. In bins of type B, one extra shelf is created in the smaller part of the L-shape. In bins of type C, shelves of width 1 are created; they are filled to a width of $8/9$

Type	A items	area = 1/3+	B items	area = 1/4+	C items	area
$(3,1)$	1	1/8	$1+1$	1/4	4	1/2
$(3,2)$	2	1/6	$2+1$	1/4	6	1/2
$(4,2)$	2	2/15	$2+1$	1/5	8	8/15
$(i,j)\ 3 \le j \le i \le 4$	3	3/25	$3+2$	1/5	ij	9/16
$(5,j)\ j = 3,4,5$	5	5/36	$5+3$	2/9	$5j$	5/8
$(i,1)\ i = 6,7,8$	2	1/9	$2+2$	2/9	8	4/9
$(i,2)\ i = 6,7,8$	4	4/27	$4+2$	2/9	12	12/27
$(i,3)\ i = 6,7,8$	6	1/6	$6+4$	5/18	18	1/2
$(i,j)\ i = 6,7,8, j = 4,5$	8	4/27	$8+4$	2/9	24	4/9
$(i,j)\ 6 \le j \le i \le 8$	12	12/81	$12+8$	20/81	36	4/9

Type	shelves	area = 1/3+	shelves	area = 1/4+	shelves	area
$(9,i)\ i = 2,\ldots,8$	i	$2/3 \cdot 2/9$	$i+1$	2/9	i	$8/9 \cdot 2/3$

In type B bins, the area outside of a square of $2/3$ by $2/3$ in the left bottom corner does not contain any items. We partition this L-shaped area in two rectangles, one of dimensions 1 and $1/3$ and the other of dimensions $2/3$ and $1/3$. The orientation is not important since rotations are allowed. We pack some number of small items in the larger rectangle and some number in the smaller rectangle; the numbers are written as a sum in the 'items' column for type B. These bins already contain an area of at least $1/4$.

In type C bins, we use the so-called side-by-side packing [6] to pack items. I.e., for type (i,j) items, we place ij of these items in an i by j grid at the bottom of the bin, and then (when possible) add some extra items in reverse orientation at the top of the bin.

Table 2. The variable w in the Condition column refers to the width of the big item(s) in the current bin of type A or B. Recall that the width is the *smallest* size of an item. The type $(4,1)_a$ contains items of width in the interval $(9/40, 1/4]$. The type $(i,0)_a$ $(i = 6,7,8)$ contains items of width in $(2/3, \frac{i-1}{i}]$, the type $(i,0)_b$ contains items of width in $(\frac{i-1}{i}, 1]$. The types $(i,0)$ $(i = 4, \ldots, 8)$ are packed separately (in type B bins: both subtypes separately), the types $(9,0)$ and $(9,1)$ are packed together in type A and C bins. For the $(9,1)$ items in type B bins, we use two shelves of length 1 and 2/3, both of width 1/3. In both shelves, at least a width of 2/9 and length of 1/2 will be occupied

Type	Bins	Condition	items	area
$(4,1)$	A	subtype $(4,1)_a$	1	$1/3 + 9/80$
		$w > 11/20$	1	$11/30 + 1/10$
		$w \le 11/20$	2	$1/3 + 2/10$
	B		$1+1$	$1/4 + 1/5$
	C		5	$1/2$
$(5,0)$	A		1	$1/3 + 1/9$
	B	$w > 3/5$	$1 + 0$	$9/25 + 1/9$
		$w \le 3/5$	$2 + 0$	$1/4 + 2/9$
	C		5	$5/9$
$(5,1)$	A	$w > 3/5$	1	$2/5 + 1/12$
		$w \le 3/5$	2	$1/3 + 1/6$
	B	$w > 3/5$	$1 + 1$	$9/25 + 1/6$
		$w \le 3/5$	$2 + 1$	$1/4 + 1/4$
	C		6	$1/2$
$(5,2)$	A		2	$1/3 + 1/9$
	B	$w > 3/5$	$2 + 1$	$9/25 + 3/18$
		$w \le 3/5$	$4 + 1$	$1/4 + 5/18$
	C		10	$5/9$
$(i,0), i = 6,7,8$	A		2	$1/3 + 4/(3i+3)$
	B	subtype $(i,0)_a$	$2 + 1$	$1/4 + 2/(i+1)$
		subtype $(i,0)_b$	$2 + 0$	$1/4 + 2\frac{i-1}{i}\frac{1}{i+1}$
	C		i	$2/3 \cdot i/(i+1)$

Type	Bins	Condition	shelves	area
$(9,i), i = 0,1$	A		1	$1/3 + 2/9 \cdot 1/2$
$(9,0)$	B	$w > 11/20$	1	$\frac{121}{400} + \frac{2}{3}\frac{2}{9}$
		$w \le 11/20$	1	$1/4 + (\frac{9}{20} - \frac{1}{9})$
$(9,1)$	B		2	$1/4 + 4/9 \cdot 1/2$
$(9,i), i = 0,1$	C		1	$1/2 \cdot 8/9$

Table 2 contains the items that are slightly more complicated to pack, at least in Type A and Type B bins. Here it is usually important what the exact width of the large items is. This is the reason that we classified type (2,1) items further: we can now be assured that if one of them has e.g. width $w \le 3/5$, the other one has this as well. (Note: to keep the analysis uniform, for these items we let the width be the *largest* size. The width of a pair can be taken arbitrarily as either the width of the first or the second item, due to our classification.)

In type A bins, we will now sometimes (where possible) use a strip of width $2/5$ or $9/20$ to pack small items, e.g. for type $(4, 1)$. In type B bins, the L-shaped free area will sometimes also be partitioned such that one strip of width $2/5$ or $9/20$ (and length 1) is created. For items of type $(6, 0)$, $(7, 0)$ and $(8, 0)$, the width of the largest strip depends on the width of the *small* items packed in there. The area that is already in a type A or B bin is of course different if we put restrictions on the width of the large items; it is given by the first term of the sum in the 'area' column.

Finally, there is one type that still remains to be packed. This type, $(9, 9)$, is described below.

Type $(9, 9)$. We show how to pack these items into square sub-bins of width and length $1/3$ so that inside each such sub-bin at least $4/9$ of its area is occupied. We begin by showing that this is a dense enough packing in all three cases.

A. We can create three sub-bins. We get a total area of $1/3 + 12/81 = 13/27 = 4/9$.

B. The item already packed in this bin has length and width no larger than $2/3$. Therefore we can create five sub-bins. The total occupied area would be $1/4 + 5 \cdot 4/81 = 161/324 > 4/9$.

C. We create nine sub-bins and get a total area of $4/9$.

Next we explain the packing into sub-bins. We use Next Fit Decreasing Length (NFDL) to pack items into these sub-bins. All items are rotated such that their length is no smaller than their width, and then sorted by decreasing length. Then, the items are packed into levels using Next Fit, where the length of each level is the length of the first item placed in it. When the next item does not fit in the current level anymore, a new level is started, if necessary in a new sub-bin.

Since all items have width at most $1/9$, we find that each shelf is filled to a width of at least $2/9$. Denote the length of shelf i by H_i. Let k be the number of shelves in the current bin. The first item that does not fit has length H_{k+1}. All items in shelf i have length at least H_{i+1}.

Then for each sub-bin except the last, the packed area is at least $2/9(H_2 + \ldots + H_{k+1}) > 2/9(1/3 - H_1) > 4/81$. This is a $4/9$ fraction of the area of the sub-bin which is $1/9$.

Theorem 3. *For any input list L, we have* $\text{ALG}_2(L) \leq 9/4 \cdot \text{OPT}(L) + 41$.

Proof. If no new bins are opened for small items, we have from Claim 3 that there are at most 7 bins with weight less than 1 (this cannot occur for the types $(0, 0), (1, 0)$ and $(1, 1)$). Combining this with Claim 3 gives $\text{ALG}_2(L) \leq 9/4 \cdot \text{OPT}(L) + 7$.

If there are new bins opened for small items, then almost all bins contain an area of at least $4/9$. By the above analysis, this holds in this case for all bins that contain small items, except possibly the last such bin for each type that is packed separately. Note that it is also possible that we run out of a certain small type while we are packing a bin of type A, in this case this bin is not used further and has a bad area guarantee. Counting the number of types packed separately, there are 21 such types in Table 1 and 12 in Table 2. (Note that the type $(4, 1)$ can only cause a single bin with a low area guarantee, because this cannot happen for subtype $(4, 1)_a$ or for a bin of type A with $w > 11/20$.) Finally there is the type $(9, 9)$.

Moreover, there can be at most 7 bins with large items that have no small items and area less than $4/9$: these are the bins that had weight less than 1 after packing the large items. Since the total area of the items is a lower bound on the optimal number of bins required to pack them, we find in this case $\text{ALG}_2(L) \leq 9/4 \cdot \text{OPT}(L) + 41$.

4   This Side Up

We now show how to use the algorithm in the previous section to get a 9/4-approximation algorithm for the This Side Up problem. Naively, one might think that one could simply group items of similar height and pack each group using the algorithm from the previous section (ignoring the height of the items). However, the problem with this is that some groups might contain only large items and other groups only small items. In this case, the groups with large items will have poor volume guarantees, and we will not get a good approximation ratio.

We therefore have to be more careful. Our algorithm works as follows. All items are classified into types as in the previous section, where the height of the items is (for now) ignored. We then begin by packing the large items. The $(0, 0)$ items are stacked in some way, nothing is placed next to this stack.

For the $(1, 0)$, $(1, 1)$ and $(2, 1)$ items, we classify them further along a dimension that has type 1, using the types $1a$, $1b$ and $1c$ that were defined at the start of section 2. Thus we have in total nine subtypes (the $(1, 1)$ items are only classified further along their width (smallest size)). We sort the items by subtype, items of subtypes of $(2, 1)$ are further sorted by decreasing height.

For each subtype, the items are stacked in the strip so that one corner of each item is directly above a designated corner of the base of the strip, and all items are oriented in the same way. Pairs of $(2, 1)$ items are considered as a single $(1, 0)$ item in this step, where the width is one of the lengths (largest sizes) of the pair. Thus we have six stacks of items on top of each other: three for the $(1, 1)$ items and one each for items of types $(1x, 0)$ and $(2, 1x)$ for $x = a, b, c$. Here we rotate the $(2, 1)$ items such that their length is oriented along the width of the $(1, 0)$ items, and a free strip is left next to these items along one side of the main strip.

If we view any one of these stacks from above, it leaves either an L-shaped area or one strip. We can now start using this extra volume for the small items, using the six stacks one by one. The small items are also packed per type. Within each type, the items are sorted in order of decreasing height. Then the items are packed in levels, where each level is packed as in section 3 next to the current stack. The height of a level is the height of the first item packed in it.

Because this stack contains items of only one large subtype (by considering pairs of $(2,1)$ items as $(1,0)$ items), and the levels contain items from one small type, the packing uses the same unique method on all levels that are used for this small type. It is for this reason that we can ignore the single large items in the stack and only care about the height of the stack. If we did not use this distinction into subtypes for the large items, we might need to change the packing method many times, and we would leave much vertical space unused.

We continue creating levels until we run out of items for this small type or the next small item does not fit next to the current stack (its height would be higher than the height of the stack). In this last case, the remaining items of this type are packed next to the next stack of large items. I.e., the next level for this type is not created immediately above the previous level, but instead at the height where the next stack starts. Also, the packing method might be changed at this point.

Finally, if all six stacks are used in this way, or the small items are all packed, we pack the remaining large and small items according to the methods for packing items into empty bins. That is, for each (large or small) type, the items are sorted according to height and then packed in order of decreasing height using the methods from section 3, using as many levels as necessary.

Analysis. We begin by making a general remark. Whenever items are packed into levels in order of decreasing height, some height in each level is lost because the first item on the level determines the height of the level, and the next items might have smaller height. However, if we denote the heights of the levels by $H_1, \ldots, H_k$, we have that all items on level i have height at least H_{i+1}. To see how much area is occupied, we can move all items from each level i to level $i+1$. Then level $i+1$ is completely covered by items for each i, and only level 1 is left empty. This implies that when we consider the height of the entire packing for these items, at most a height of H_1 does not contain any items.

We have two cases in our analysis. First of all, it can be seen that if all the small items fit next to the six created stacks, we can ignore the small items in the analysis because they do not add to the total height used. In this case, we can use the weighting technique from section 3.

The weight of an item is now defined as the vertical size divided by the number of times that the 'horizontal item' fits in a square. To give a bound on the performance ratio, we introduce a new concept which is the *weight density*. This is the weight of an item *per unit of vertical dimension*, i.e. it is the weight of the two-dimensional item that we get when we ignore the vertical dimension of an item. We will examine the weight densities at arbitrary horizontal planes through the packings of our algorithm and of the optimal packing.

We find that for each large (sub)type, if it is packed in levels between heights h_1 and h_2, the weight density is 1 at all heights $h \in [h_1, h_2]$ apart from a total height of at most 1. For $(0,0)$ and the subtypes of $(1,1)$ and $(1,0)$, there is even a weight density of 1 at the entire height of their stacks, because all these items are placed directly on top of each other. In the optimal solution, according to Claim 3, there can not be a weight density of more than $9/4$ at any height. Since we use seven types that do not have a weight of 1 everywhere (four medium types and the three subtypes of $(2,1)$), we find that $\mathrm{ALG}_3(L) \leq 9/4 \cdot \mathrm{OPT}(L) + 7$.

Now suppose that some small items need to be packed above the large items. Consider some large (sub)type (one stack) and a single small type t. Suppose all items from this type are placed next to this large subtype, between heights h_t and h'_t. Since the small items are sorted by decreasing height, and the large items are all stacked on top of each other, we have for each height $h_t \leq h \leq h'_t$ an area guarantee of $4/9$ using the proof from section 3, apart from a total height of at most 1.

A small type may also be split among two large stacks, or among one stack and levels of its own (not next to any stack). In this case, some height at the top of the first stack might not contain small items. We can assign this additional height loss to the large (sub)type of that stack. We then find that for each large and small (sub)type, there is a height of at most 1 at which we do not have an area guarantee of $4/9$. In total we have 10 large (sub)types in separate stacks and $21 + 13 + 1 = 35$ small (sub)types and we find $\mathrm{ALG}_3(L) \leq 9/4 \cdot \mathrm{OPT}(L) + 45$.

5   Three-Dimensional Strip and Bin Packing

To pack items in a three-dimensional strip, we place each item such that its weight, defined as in the previous section, is minimized. Note that this does not mean simply placing it with its smallest dimension vertical, because the number of times that the implied horizontal item fits in a square might depend on the orientation.

We omit a detailed description of this algorithm in this extended abstract. We show in the full version that the performance ratio of this algorithm (as well as the additive constants) are identical to those of ALG_3. Moreover, we show there that the packing generated by this algorithm can be transformed into a packing for the three-dimensional bin packing problem, which loses a further factor of 2 in the asymptotic performance ratio.

6   Conclusion

In this paper we design offline algorithms for six packing problems. Most of these problems were not studied in on-line environments, which can be interesting as well. It might be the case that some of the bounds in this paper are possible to improve. Specifically we are interested in improving the constant for packing in three dimensional bins (both for rotatable items and for the "This Side Up" problem). This can be done by designing algorithms for these problems directly instead of adaptation of algorithms for other problems.

References

1. Nikhil Bansal and Maxim Sviridenko. New approximability and inapproximability results for 2-dimensional packing. In *Proceedings of the 15th Annual Symposium on Discrete Algorithms*, pages 189–196. ACM/SIAM, 2004.
2. Alberto Caprara. Packing 2-dimensional bins in harmony. In *Proc. 43th IEEE Symp. on Found. of Comp. Science*, pages 490–499, 2002.
3. Edward G. Coffman, Michael R. Garey, David S. Johnson, and Robert E. Tarjan. Performance bounds for level oriented two-dimensional packing algorithms. *SIAM Journal on Computing*, 9:808–826, 1980.
4. Jose Correa and Claire Kenyon. Approximation schemes for multidimensional packing. In *Proceedings of the 15th ACM/SIAM Symposium on Discrete Algorithms*, pages 179–188. ACM/SIAM, 2004.
5. Mauro Dell'Amico, Silvano Martello, and Daniele Vigo. A lower bound for the non-oriented two-dimensional bin packing problem. *Discrete Applied Mathematics*, 118:13–24, 2002.
6. Leah Epstein. Two dimensional packing: the power of rotation. In *Proc. of the 28th International Symposium on Mathematical Foundations of Computer Science (MFCS'2003)*, pages 398–407, 2003.
7. Leah Epstein and Rob van Stee. Optimal online bounded space multidimensional packing. In *Proc. of 15th Annual ACM-SIAM Symposium on Discrete Algorithms (SODA'04)*, pages 207–216. ACM/SIAM, 2004.
8. Satoshi Fujita and Takeshi Hada. Two-dimensional on-line bin packing problem with rotatable items. *Theoretical Computer Science*, 289(2):939–952, 2002.
9. Keqin Li and Kam-Hoi Cheng. Static job scheduling in partitionable mesh connected systems. *Journal on Parallel and Distributed Computing*, 10:152–159, 1990.
10. Flavio Keidi Miyazawa and Yoshiko Wakabayashi. Approximation algorithms for the orthogonal z-oriented 3-d packing problem. *SIAM Journal on Computing*, 29(3):1008–1029, 2000.
11. Flavio Keidi Miyazawa and Yoshiko Wakabayashi. Packing problems with orthogonal rotations. In M. Farach-Colton, editor, *Theoretical Informatics, 6th Latin American Symposium*, number 2976 in Lecture Notes in Computer Science, pages 359–368, 2004.

Approximation Algorithm for Directed Multicuts

Yana Kortsarts[1], Guy Kortsarz[2], and Zeev Nutov[3]

[1] Department of computer science, Widener University Chester, PA, USA
yanako@cs.widener.edu
[2] Department of Computer Science, Rutgers, Camden, USA
guyk@crab.rutgers.edu.
[3] Computer Science department, The Open University, Tel Aviv, Israel
nutov@shaked.openu.ac.il

Abstract. The Directed Multicut (DM) problem is: given a simple directed graph $G = (V, E)$ with positive capacities u_e on the edges, and a set $K \subseteq V \times V$ of ordered pairs of nodes of G, find a minimum capacity K-multicut; $C \subseteq E$ is a K-multicut if in $G - C$ there is no (s, t)-path for every $(s, t) \in K$. In the uncapacitated case (UDM) the goal is to find a minimum size K-multicut. The best approximation ratio known for DM is $\min\{O(\sqrt{n}), opt\}$ by Anupam Gupta [5], where $n = |V|$, and opt is the optimal solution value. All known non-trivial approximation algorithms for the problem solve large linear programs. We give the first combinatorial approximation algorithms for the problem. Our main result is a $\tilde{O}(n^{2/3}/opt^{1/3})$-approximation algorithm for UDM, which improves the $\sqrt{n}$-approximation for $opt = \Omega(n^{1/2+\varepsilon})$. Combined with the paper of Gupta [5], we get that UDM can be approximated within better than $O(\sqrt{n})$, unless $opt = \tilde{\Theta}(\sqrt{n})$. We also give a simple and fast $O(n^{2/3})$-approximation algorithm for DM.

1 Introduction and Preliminaries

Problem formulation: An instance to the *Directed Multicut* (DM) problem consists of a *simple* directed graph $G = (V, E)$ with integral capacities u_e on the edges and a set $K \subseteq V \times V$ of ordered pairs of nodes of G. The goal is to find a minimum K-*multicut*, that is, a minimum capacity edge set C so that in $G - C$ there is no (s, t)-paths for every $(s, t) \in K$. In the *uncapacitated case* (UDM), all edges have capacities 1.

Related work: The case $|K| = 1$ is polynomially solvable based on the fundamental Max-Flow Min-Cut Theorem. For $|K| > 1$ the min-cut max-flow equality breaks down even on undirected graphs. In fact, the undirected multicut problem is MAXSNP-hard even on stars [6]. [6] gives a 2-approximation algorithm for the undirected multicut problem on trees. The best approximation ratio for the minimum multicut problem on general undirected graphs is $O(\log |K|)$ [7].

G. Persiano and R. Solis-Oba (Eds.): WAOA 2004, LNCS 3351, pp. 61–67, 2005.

In [8], a related problem is studied. The input is as in the DM problem, except that the pairs in K are unordered. The goal is to remove a min-cost edge set C so that in $G - C$ no cycle contains a pair from K. This problem seems easier than the DM problem. In particular, divide and conquer methods similar to the ones in [3, 7, 9] give an $O(\log^2 |K|)$-approximation for this variant [8]. In [3] a relatively general scheme is presented handling many problems that are "decomposable", but DM does not seem to lend itself in any way to the divide and conquer approach. Given this fact, it may be that the directed multi-cut problem is harder to approximate than the undirected one. In particular, a (poly)logarithmic approximation is not known for DM, nor for UDM. However, so far, an exact proof separating the approximability of the undirected and directed problems does not exist. In fact, the only approximation threshold known for the directed case is the one derived from the undirected case: namely, that the problem is MAXSNP-hard.

The first nontrivial approximation ratio $O(\sqrt{n \lg n})$ for DM is due to Cheriyan, Karloff, and Rabani [1]. This was slightly improved by Anupam Gupta [5] to $O(\sqrt{n})$. Gupta's analysis also gives an $O(opt^2)$ cost solution with opt the optimal multicut capacity. This can be considered as an opt-approximation algorithm and is useful in the case the value of opt is "small". Both algorithms [1] and [5] require solving large linear programs.

Our results: We design *combinatorial* approximation algorithms for DM. Let n and m be the number of nodes and of edges, respectively, in the input graph. Our main result is:

Theorem 1. *For UDM there exists an algorithm with running time $\tilde{O}(n^2 m)$ that finds a multicut C of size $O\left((n \lg n \cdot opt)^{2/3}\right) = \tilde{O}\left((n \cdot opt)^{2/3}\right)$.*

The approximation ratio is $\tilde{O}(n^{2/3}/opt^{1/3})$. Therefore, Theorem 1 implies that for UDM the $\sqrt{n}$-approximation can be improved if opt is large (e.g., $opt = \Omega(n^{1/2+\varepsilon})$ for some $\varepsilon > 0$). This is the first algorithm whose approximation ratio *improves* as opt gets larger. Combined with the results of [5] that provides an $O(opt)$-approximation, we get approximation ratio better than $\tilde{O}(\sqrt{n})$, unless $opt = \tilde{\Theta}(\sqrt{n})$.

Our additional result is:

Theorem 2. *DM admits an $O(n^{2/3})$-approximation algorithm with running time $\tilde{O}(nm^2)$.*

The approximation ratio in [5, 1] is better than the one in Theorem 2. However, our algorithm is very simple and runs faster than the algorithms in [5, 1]; the later can be implemented in $O(n^2 m^2)$ time using the algorithm of Fleisher [4] for finding an approximate solution of multicommodity-flow type linear programs.

We prove Theorems 1 and 1 in Sections 2 and 3, respectively.

Notation: Let $G = (V, E)$ be a directed graph. For $s, t \in V$ the *distance* $d_G(s, t)$ from s to t in G is the minimum number of edges in an (s, t)-path; $d_G(s, t) = \infty$ if no (s, t)-path exists in G. For disjoint subsets $S, T \subseteq V$ of V let $\delta_G(S, T) = \{st \in E : s \in S, t \in T\}$.

We often omit the subscript G if it is clear from the context. An edge set $C \subseteq E$ is an (s, t)-*cut* if $C = \delta(S)$ for some $S \subseteq V - t$ with $s \in S$. Let $u(C) = \sum\{u_e : e \in C\}$ be the *capacity* of C; $u(C) = |C|$ if no capacities are given. For simplicity of the exposition, we ignore that some numbers are not integral. The adaptation using floors and ceilings is immediate.

Preliminaries: Our algorithms run with certain parameters, which should get appropriate values that depend on n and opt to achieve the claimed approximation ratios. Specifically, for UDM we show an algorithm that for any integer ℓ computes a multicut of size $\ell \cdot opt + O((n \lg n)^2 / \ell^2)$. Setting $\ell = (n \lg n)^{2/3} / opt^{1/3}$ gives the claimed approximation ratio. Since opt is not known, we execute the algorithm for $\ell = 1, \ldots, (n \lg n)^{2/3}$, and among the multicuts computed output one of minimum size. For DM we show an algorithm that for any integers ℓ, μ with $1 \le \ell \le n - 1$ and $\mu \ge opt$ computes a K-multicut of capacity $\le \mu \cdot (2\ell + n^2 / \ell^2)$. Setting $\ell = n^{2/3}$ and $\mu = opt$ gives the claimed approximation ratio. Since opt is not known, we apply binary search to find the minimum integer μ so that a multicut of capacity $\le \mu \cdot (2\ell + n^2 / \ell^2)$ is returned. Note that if $\mu \ge opt$, a multicut C of capacity $\le \mu(2\ell + n^2 / \ell^2)$ is returned. If $\mu < opt$, then either the returned multicut C is of capacity $\le \mu(2\ell + n^2 / \ell^2) < 3opt\,n^{2/3}$ which is fine or we know that $\mu < opt$ as the above inequality fails.

Remark: Recently we became aware of the [10] paper, which gives an $\tilde{O}(n^{2/3})$-approximation algorithm for the related Edge-Disjoint Paths problem. Our result for UDM, which was derived independently, and the main result in [10] rely on the same combinatorial statement (Corollary 1 in our paper, Theorem 1.1 in [10]), but the proofs are different.

2   The Uncapacitated Case

Definition 1. *For $X, Y \subseteq V$, let $R_G(X, Y) = |\{(x, y) \subseteq X \times Y : x \ne y, d_G(x, y) < \infty\}|$ denote the number of pairs $(x, y) \subseteq X \times Y$, so that an (x, y)-path exists; let $R(G) = R_G(V, V)$.*

We say that $G = (V, E)$ is a *p-layered graph* if V can be partitioned into p *layers* $L_1, \ldots, L_p$ so that every $e \in E$ belongs to $\delta_G(L_i, L_{i+1})$ for some $i \in \{1, \ldots, p - 1\}$.

Lemma 1. *Let $G = (V, E)$ be a 4-layered graph containing k edge-disjoint (L_1, L_4)-paths such that $G - \delta_G(L_2, L_3)$ is a simple graph. Then $R(L_1, L_3) + R(L_2, L_4) \ge k$.*

Remark: Observe that the graph induced by $L_2 \cup L_3$ may contain parallel edges.

Proof. We will prove the statement by induction on k. The case $k = 0$ is obvious. Assume $k \geq 1$, and that E is a union of k edge-disjoint paths. Let $st \in \delta_G(L_2, L_3)$, let $G' = G - \{s, t\}$, and let $S = \{v \in L_1 : vs \in E\}$, $T = \{v \in L_4 : tv \in E\}$. Then G' contains at least $k - (|S| + |T|)$ edge-disjoint (L_1, L_4)-paths. Also, $R_{G'}(L_1, L_3) \leq R_G(L_1, L_3) - |S|$ and $R_{G'}(L_2, L_4) \leq R_G(L_2, L_4) - |T|$. This follows because of the removal of $\{s, t\}$. By the induction hypothesis, $R_{G'}(L_1, L_3) + R_{G'}(L_2, L_4) \geq k - (|S| + |T|)$. Combining, we get the statement.

Lemma 2. *Let G be a simple ℓ-layered graph containing k-edge disjoint paths from the first layer to the last layer, and let S and T be the union of $p_S \geq 2$ first and $p_T \geq 2$ last layers, respectively, so that $S \cap T = \emptyset$. Then $R(S, T) = \Omega(k p_S p_T)$.*

Proof. By Lemma 1, $R(L_i, L_j) + R(L_{i+1}, L_{j+1}) \geq k$ for every two pairs $L_i, L_{i+1} \subseteq S$ and $L_j, L_{j+1} \subseteq T$. The statement follows by summing the contribution of all such pairs.

Lemma 3. *Let s, t be a pair of nodes in a simple graph G with $d_G(s, t) \geq 2p \lg n$. Then there exists an (s, t)-cut C so that $R(G) - R(G - C) = \Omega(|C| p^2)$.*

Proof. Consider the corresponding $d_G(s, t)$ BFS layers from s to t, where nodes that cannot reach t are deleted. Let X_i be the layer at distance i from s, and let Y_i be the layer at distance i to t. Let k_j be the maximum number of edge-disjoint $(X_{j \cdot p}, Y_{j \cdot p})$-paths in the graph G_j induced by all the layers starting with $X_{j \cdot p}$ and ending at $Y_{j \cdot \ell}$, $j = 1, \ldots, 2 \lg n$.

We claim that there exists an index j with $k_j \leq 2 \cdot k_{j-1}$. Otherwise, since $k_0 \geq 1$, we have $k_j \geq 2^j$. For $j = \log n$ we get $k_j \geq n^2$, which is not possible in a simple graph.

Let j be such an index with $k_j \leq 2 \cdot k_{j-1}$, and let C be a minimum $(X_{j \cdot p}, Y_{j \cdot p})$-cut, so $|C| = k_j$. We now apply Lemma 2 on the graph G_{j-1}. Note that G_{j-1} contains $|C|/2$ edge-disjoint paths between its first layer $X_{(j-1) \cdot \ell}$ and its last layer $Y_{(j-1) \cdot \ell}$; this is since $k_j = |C|$ by Menger's Theorem, and $k_{j-1} \geq k_j / 2$ by the choice of j. Since C separates the first and the last p layers of G_{j-1}, the statement follows from Lemma 2.

Corollary 1. *For UDM there exists an algorithm that for any integer ℓ finds in $\tilde{O}(mn^2/\ell^2)$ time a K-multicut B with $|B| = O\left((n \lg n)^2 / \ell^2\right)$, where $K = \{(u, v) : d(u, v) \geq \ell\}$.*

Proof. Let $p = \ell/(2 \lg n)$. The algorithm starts with $B = \emptyset$. While there is an (s, t)-path for some $(s, t) \in K$ it computes an (s, t)-cut $C = C_{st}$ as in Lemma 3, and sets $B \leftarrow B \cup C$, $G \leftarrow G - C$. We claim that at the end of the algorithm $|B| = O(R(G)/p^2) = O(n^2/p^2)$; we get that $|B| = O\left((n \lg n)^2 / \ell^2\right)$ by substituting $p = \ell/(2 \lg n)$. Lemma 2 implies that there exists a constant $\alpha > 0$ so that each time C_{st} is deleted, $R(G)$ is reduced by at least $\alpha |C_{st}| p^2$. Thus we get:

$$\alpha p^2 \cdot |B| \leq \alpha p^2 \cdot \sum_{(s,t) \in K} |C_{st}| \leq R(G) \leq n^2.$$

The dominating time at each iteration is spent for computing a cut as in Lemma 3. This can be done using $O(\lg n)$ max-flow computations, thus in $\tilde{O}(m|C_{st}|)$ time using the Ford-Fulkerson algorithm. Thus the total time required is $\tilde{O}(m|B|) = \tilde{O}(mn^2/\ell^2)$.

Given an integer ℓ, apply the following algorithm starting with $A, B = \emptyset$:
While there is an (s,t)-path P with $|P| \le \ell$ for some $(s,t) \in K$ do:
$$A \leftarrow A + P, \ G \leftarrow G - P.$$
End While
Find in $G - A$ a K-multicut B as in Corollary 1.

For any integer ℓ, the algorithm computes a K-multicut $C = A \cup B$ of size $\ell \cdot opt + O((n \lg n)^2/\ell^2)$; $|A| \le \ell \cdot opt$ since any K-multicut contains at least one edge of each path removed, and $|B| = O((n \lg n)^2/\ell^2)$ by Corollary 1. As was explained in the introduction, we execute the algorithm for $\ell = 1, \ldots, (n \lg n)^{2/3}$, and among the multicuts computed output one of minimum size. For $\ell = (n \lg n)^{2/3}/opt^{1/3}$ we get the claimed approximation ratio.

Let us now discuss the implementation of the algorithm. After executing Procedure 1 at iteration ℓ, the graph $G - A$ is used as an input for iteration $\ell + 1$. As the total length of the paths removed is at most n^2, the total time of Phase 1 executions is $O(mn^2)$. The total time of Phase 2 executions is $\tilde{O}\left(\sum_{\ell=1}^{n^{2/3}} mn^2/\ell^2\right) = \tilde{O}(mn^2)$. Thus the time complexity is as claimed, and the proof of Theorem 1 is complete.

3 An $O(n^{2/3})$-Approximation Algorithm for DM

The algorithm: Consider the following algorithm:
Input: An instance (G, u, K) of DM, and integers ℓ, μ.
Initialization: $C \leftarrow \emptyset$.
While in G there is an (s,t)-path P for some $(s,t) \in K$ do:
    (a) Let P' is the union of the first and the last ℓ edges of P ($P' = P$ if $|P| < 2\ell$);
    (b) Among the (s,t)-cuts in G disjoint to P' compute one C' of minimum capacity $\quad (u(C') = \infty$ if $P' = P$);
    (c) *If* $u(C') > \mu$ *then:* $u_e \leftarrow u_e - \min\{u_e : e \in P'\}$ for every $e \in P'$;
$$C \leftarrow C \cup P_0', \ G \leftarrow G - P_0', \text{ where } P_0' = \{e \in P' :$$
$u_e = 0\}$.
        Else $(u(C') \le \mu)$ $C \leftarrow C \cup C', \ G \leftarrow G - C'$.
End While

Theorem 3. *At the end of the algorithm C is a K-multicut. If $\mu \ge opt$ then* $u(C) \le \mu \cdot (2\ell + n^2/\ell^2)$.

Proof. Assume that $\mu \ge opt$. Consider a specific iteration of the main loop, and the edge sets P', C' found. There are two possible cases.

If $u(C') > \mu$ then $u(C') > \mu \ge opt$. This implies that *any* minimum K-multicut contains at least one edge from P'. Hence, after setting $u_e \leftarrow u_e -$

$\min\{u_e : e \in P'\}$ for every $e \in P'$ the optimum decreased by at least $\min\{u_e : e \in P'\}$. Since $|P'| = 2\ell$, the total capacity of the edges in all sets P'_0 added into C during the algorithm is at most $2\ell opt \leq 2\ell\mu$.

Otherwise, if $u(C') \leq \mu$ then $R(G) - R(G - C') \geq \ell^2$. Thus the total *number* of cuts C' removed during the algorithm $\leq n^2/\ell^2$, and their total capacity $\leq \mu n^2/\ell^2$.

To see that $R(G) - R(G - C') \geq \ell^2$, let P'_F and P'_L be the first and the last ℓ nodes in P, respectively. We claim that $R_G(P'_F, P'_L) = |P'_F| \cdot |P'_L| = \ell^2$ and $R_{G-C'}(P'_F, P'_L) = 0$. The first statement follows from the simple observation that P'_F, P'_L belong to the same path P of G, and thus $d_G(u, v) < \infty$ for every pair u, v with $u \in P'_F, v \in P'_L$. To see the second statement, note that in $d_{G-C'}(u, v) = \infty$ for every such pair u, v, as otherwise there would be an (s,t)-path in $G - C'$, contradicting that C' is an (s,t)-cut in G.

As was mentioned in the introduction, for $\ell = n^{2/3}$ we use binary search to find the minimum integer μ so that a multicut of capacity $\leq \mu \cdot (2\ell + n^2/\ell^2)$ is returned. Theorem 3 implies that $\mu \leq opt$, and the required ratio follows.

Implementation: We can assume that $u_e \in \{1, \ldots, n^4\}$ or $u_e = \infty$ for every $e \in E$. In this case binary search for appropriate μ requires $O(\lg(n^6)) = O(\lg n)$ iterations. Indeed, let c be the least integer so that $\{e \in E : u_e \leq c\}$ is a K-multicut. Edges of capacity $\geq cn^2$ do not belong to any optimal solution, and their capacity is set to ∞. Edges of capacity $\leq c/n^2$ are removed, as adding all of them to the solution affects only the constant in the approximation ratio. This gives an instance with $u_{\max}/u_{\min} \leq n^4$, where $u_{\max}$ and $u_{\min}$ denote the maximum finite and the minimum nonzero capacity of an edge in E, respectively. Further, for every $e \in E$ set $u_e \leftarrow \lceil u_e/u_{\min} \rceil$. It is easy to see that the loss incurred in the approximation ratio is only a constant, which is negligible in our context.

The dominating time is spent for computing $O(m)$ minimum cuts at step (b); each such computation leads to a removal of an edge, since reducing the capacities along P' by $\min\{u_e : e \in P'\}$ guarantees that at least one edge gets capacity zero. As a max-flow/min-cut computation can be done in $\tilde{O}(nm)$ time (c.f., [2]), the total running time is $\tilde{O}(nm^2)$. This finishes the proof of Theorem 2.

Acknowledgment. The second author thanks Joseph Cheriyan for suggesting the problem and for helpful discussions, and Howard Karloff and Aravind Srinivasan for useful discussions.

References

1. J. Cheriyan, H. Karloff, and Y. Rabani. Approximating directed multicuts *Combinatorica, to appear.*
2. T. H. Cormen, C E. Leiserson, R. L. Rivest and C. Stein, Introduction to Algorithms, Second Edition, MIT press, 1998.

3. G. Even, S. Naor, B. Schieber, and S. Rao, Divide-and-Conquer Approximation Algorithms Via Spreading Metrics, Journal of the ACM 47(4):585 - 616, 2000.
4. L. Fleisher Approximating Fractional Multicommodity Flows Independent of the Number of Commodities , SIAM J. Discrete Math. 13 (2000), no. 4, 505-520
5. Anupam Gupta, Improved approximation algorithm for directed multicut In *SODA 2003*, pages 454-455.
6. N. Garg, V. Vazirani and M. Yannakakis Primal-Dual Approximation Algorithms for Integral Flow and Multicut in Trees. Algorithmica, 18(1):3-20, 1997.
7. N. Garg, V. Vazirani and M. Yannakakis Approximate max-flow min-(multi)cut theorems and their applications, SIAM Journal on Computing, 25(2):235-251, 1996.
8. P. N. .Klein, S. A. Plotkin, S. Rao, and E. Tardos", Approximation Algorithms for Steiner and Directed Multicuts, J. Algorithms, 22(2):241-269, 1997.
9. P. D. Seymour, Packing Directed Circuits Fractionally, Combinatorica 15(2): 281-288 1995.
10. K. Varadarajan and G. Venkataraman, Graph Decomposition and a Greedy Algorithm for Edge-disjoint Paths, Symposium on Discrete Algorithms (SODA) 2004, Pages 379-380 .

Improved Bounds for Sum Multicoloring and Scheduling Dependent Jobs with Minsum Criteria

Rajiv Gandhi[1], Magnús M. Halldórsson[2], Guy Kortsarz[1],
and Hadas Shachnai[3*]

[1] Department of Computer Science, Rutgers University, Camden, NJ 08102
{rajivg, guyk}@camden.rutgers.edu.
[2] Department of Computer Science, University of Iceland, IS-107 Reykjavik, Iceland
mmh@hi.is.
[3] Department of Computer Science, The Technion, Haifa 32000, Israel
hadas@cs.technion.ac.il.

Abstract. We consider a general class of scheduling problems where
a set of *dependent* jobs needs to be scheduled (preemptively or non-
preemptively) on a set of machines so as to minimize the weighted sum
of completion times. The dependencies among the jobs are formed as an
arbitrary conflict graph. An input to our problems can be modeled as
an instance of the *sum multicoloring* (SMC) problem: Given a graph and
the number of colors required by each vertex, find a proper multicoloring
which minimizes the sum over all vertices of the largest color assigned
to each vertex. In the *preemptive* case (pSMC), each vertex can receive
an arbitrary subset of colors; in the *non-preemptive* case (npSMC), the
colors assigned to each vertex need to be contiguous. SMC is known to
be no easier than classic graph coloring, even in the case of unit color
requirements.

Building on the framework of Queyranne and Sviridenko (*J. of Scheduling, 5:287-305, 2002*), we present a general technique for reducing the
sum multicoloring problem to classical graph multicoloring. Using the
technique, we improve the best known results for pSMC and npSMC on
several fundamental classes of graphs, including line graphs, $(k+1)$-claw
free graphs and perfect graphs. In particular, we obtain the first constant factor approximation ratio for npSMC on interval graphs, on which
our problems have numerous applications. We also improve the results
of Kim (*SODA 2003, 97–98*) for npSMC of line graphs and for resource-
constrained scheduling.

1   Introduction

We consider a general class of problems in which jobs that utilize non-sharable
resources need to be scheduled (preemptively or non-preemptively) on multiple

---

* Part of this work was done while the author was on leave at Bell Laboratories, Lucent
Technologies, 600 Mountain Ave., Murray Hill, NJ 07974.

G. Persiano and R. Solis-Oba (Eds.): WAOA 2004, LNCS 3351, pp. 68–82, 2005.

machines. Scheduling any job j depends on whether another job sharing resources with j is being scheduled. The dependencies among the jobs are modeled by an *arbitrary* conflict graph, in which the vertices represent the jobs, and an edge between two vertices means that the corresponding jobs cannot be scheduled simultaneously. Then the problem of scheduling dependent jobs can be formulated as a coloring problem: a proper coloring of the conflict graph partitions the set of jobs to subsets of non-conflicting jobs. Thus, when all jobs have the same (unit) execution time, we get a graph *coloring* problem, and when the execution times are arbitrary, we get a graph *multicoloring* problem.

In this work, we focus on the *sum of completion times* measure. For unit-length jobs, this is known as the *chromatic sum* or *sum coloring* (SC) of the conflict graph. Let $G = (V, E)$ be the conflict graph. Given a coloring ψ of G, the sum coloring of ψ is given by $\text{SC}(G, \psi) = \sum_v \psi(v)$. The minimum chromatic sum of G is given by $\text{SC}(G) = \min_\psi \text{SC}(G, \psi)$. In the weighted case, each vertex v has a weight, w_v, and we need to minimize $\sum_v w_v \psi(v)$ over all proper colorings.

An instance of a multicoloring problem is a pair (G, x), where $G = (V, E)$ is a graph, and x is a vector of *color requirements* (or *lengths*) of the vertices. A *multicoloring* of G is an assignment $\psi : V \to 2^{\mathbf{N}}$, such that each vertex $v \in V$ is assigned a set of x_v distinct colors, and adjacent vertices receive non-intersecting sets of colors. Denote by $f_v(\psi) = \max_{i \in \psi(v)} i$ the largest color assigned to v by a multicoloring ψ. The *sum multicoloring* (SMC) of ψ on G is $\text{SMC}(G, \psi) = \sum_{v \in V} f_v(\psi)$. The SMC problem is to find a multicoloring ψ, such that $\text{SMC}(G, \psi)$ is minimized. In the weighted case, we want to minimize $\sum_{v \in V} w_v f_v(\psi)$, over all proper multicolorings ψ. When all the color requirements are equal to 1, the problem reduces to SC. A multicoloring, ψ, is called *non-preemptive* if the colors assigned to each vertex v are contiguous, i.e., if for any $v \in V$, $(\max_{i \in \psi(v)} i) - (\min_{i \in \psi(v)} i) + 1 = x_v$. We denote this version of the problem by npSMC; the preemptive problem, where each vertex v can receive *any* set of x_v colors, is denoted by pSMC.

Scheduling dependent jobs, and the resulting variants of the sum (multi) coloring problem, have numerous applications, in particular on interval graphs. The following practical scenarios yield instances of our problems on this natural subclass of graphs.

Session scheduling on a path: In a path network, pairs of nodes need to communicate, for which they need use of the intervening path. If two paths intersect, the corresponding sessions cannot be held simultaneously. In this case, it would be natural to expect the sessions (i.e., "jobs") to be of different lengths, leading to the sum multicoloring problem on interval graphs.

Storage allocation: Storage allocation in a warehouse involves minimizing the total distance traveled by a robot [W97]. Goods are checked in and out at known times; thus, goods that are not in the warehouse at the same time can share the same location. We represent each of the goods by an interval on the line, which gives the time interval in which it is available at the warehouse. Numbering the storage locations by their distance from the counter, the total distance corresponds to sum coloring the intervals formed by the goods.

VLSI design: In the wire-minimization problem [NSS99], terminals lie on a single vertical line (each terminal is represented by an interval on this line), and with unit spacings are vertical bus lanes. Pairs of terminals are to be connected via horizontal wires on each side to a vertical lane, with non-overlapping pair utilizing the same lane. With the vertical segments fixed, the wire cost corresponds to the total length of horizontal segments. Numbering the lanes in increasing order of distance from the terminal line, lane assignment to a terminal corresponds to coloring the terminal's interval by an integer. The wire-minimization problem then corresponds to sum coloring an interval graph.

Other applications of sum (multi)coloring include traffic intersection control, session scheduling in local-area networks and compiler design (a comprehensive survey appears in [BHK+00]). Instances of SMC on line graphs and, more generally, on $(k + 1)$-claw free graphs, are derived mainly from applications that involve resource constrained scheduling. Our results apply also to permutation graphs, which model, e.g., train scheduling problems.

1.1   Our Results

We present (in Section 2.1) a general technique for reducing SMC to the classic graph (multi)coloring problem. Using the technique, we improve the best known results for pSMC and npSMC on several fundamental classes of graphs, including line graphs, $(k+1)$-claw free graphs and perfect graphs. In particular, we obtain the first constant factor approximation ratio for npSMC on interval graphs. Our improved bound of 7.682 for npSMC of line graphs is achieved by a simple greedy algorithm (see in Section 3.1). The previous best ratio of 10, achieved by an algorithm of Kim [K-03], involved solving an LP with an exponential number of constraints.

While our main focus is on minimizing the sum of completion times of the *jobs*, our technique can be applied to other minsum optimization problems, such as *resource constrained scheduling (RCS)*. In RCS, we have a set of jobs, each requesting up to k resources; jobs that need to utilize the same resource cannot be processed simultaneously. We say that a resource has completion time i if the last job utilizing this resource completes at time i. Our goal is to find a non-preemptive schedule that minimizes the sum of completion times of all the *resources*. We show (in Section 4) that our technique yields an approximation ratio of $2e \cdot k \approx 5.437k$. This improves the best ratio known of $8k - 7$ given in [K-03], for any $k \geq 3$.

For simplicity, in formulating our results it is implicitly assumed that the number of machines is "unbounded". The technique can, however, be applied in a system with any given number of machines, with slightly weaker performance ratios (see in [?]). Also, we formulate our results for the unweighted case, and show (in Section 4) how to generalize the results for the weighted versions of the problems.

Relation to Min-sum Set Cover: Our results include an approximation ratio of 3.591 for sum coloring of perfect graphs. This improvement upon the previous ratio of 4 (of [BBH⁺98]) is of particular interest, due to the relation of SC to the *min-sum set cover* problem. The input to min-sum set cover consists of a universe $\mathcal{U}$ and a collection of subsets $\mathcal{S} = \{S_i\}$, $S_i \subseteq \mathcal{U}$. A feasible solution is an ordered sub-collection of subsets $\mathcal{S}' = \{S'_1, S'_2, \ldots\}$, such that $\bigcup_i S'_i = \mathcal{U}$. We say that $u \in \mathcal{U}$ has *cover time i* if S'_i is the first subset in the order of $\mathcal{S}'$ to include u. The goal is to minimize the sum of cover times over all the elements of $\mathcal{U}$. Feige *et al.* [FLT-02] showed that min-sum set cover admits a 4-approximation and that, unless P=NP, for any constant $\epsilon > 0$, there is no $(4 - \epsilon)$-approximation. Observe that SC is a special case of min-sum set cover, in which $\mathcal{S}$ is the collection of all independent sets in G. Hence, our 3.591-approximation implies that the min-sum set cover problem in its full generality is provably harder to approximate than SC on perfect graphs.

Techniques: Our general approximation technique builds on the framework of Queyranne and Sviridenko [QS-02] for scheduling jobs with release times on parallel machines. As in [QS-02], we divide the time line into intervals of geometrically increasing size (see also [HSW-96, HSSW-97]), using randomized starting points (as introduced in [CP⁺96]), and approximate the classic makespan problem on each block. Note, however, that the results in [QS-02] do not apply to *arbitrary* conflict graphs. The class of problems studied in [QS-01, QS-02] include shop scheduling (open shop and job shop) and entail a different optimization criteria than SMC. (As shown in [GHKS04], open shop scheduling is in fact a special case of the data migration problem [K-03].)

1.2   Related Work

The SC problem was introduced in [K89] and the SMC problems in [BHK⁺00]. Table 1 summarizes the known results for SC, pSMC and npSMC in various classes of graphs. New bounds given in this paper are shown in boldface. In each of these entries, we give in parenthesis the previous best known bound for the problem. Entries marked with · follow by inference, either by using containment of graph classes (interval graphs are perfect), or by SC being a special case of SMC. When omitted, [BBH⁺98] is the references for SC and [BHK⁺00] for SMC. Also, in the table below, c represents some constant.

There is a wide literature on parallel machine scheduling with the objective of minimizing the sum of completion times. These works generally deal with scheduling *independent* jobs, or allow for *precedence constraints* which are directed dependencies. The undirected conflict graphs considered here require quite different treatment.

Some work has been done on resource-constrained scheduling. Kubale [K-96] studied the complexity of scheduling biprocessor tasks. They also investigate special classes of graphs, and showed that npSMC of line graphs of trees is NP-hard in the weak sense. Afrati et al. [AB⁺00] gave a polynomial time approximation scheme for the problem that we consider, minimizing sum of completion times of dedicated tasks. However, their method applies only to the case where the total number of processors is a fixed constant. Coffman et al. [CG⁺85] analyzed the

Table 1. Known results for sum (multi-)coloring problems

	SC			SMC
	u.b.	l.b.	pSMC	npSMC
General graphs	·	$n^{1-\epsilon}$	$n/\log^2 n$	$n/\log n$
Perfect graphs	**3.591** (4)	$c > 1$ [BK98]	**5.436** (16)	$O(\log n)$
Interval graphs	1.796 [HKS03]	$c > 1$ [G01]	**5.436** (7.184)	**11.273** $(O(\log n))$
Bipartite graphs	27/26 [G$^+$02]	$c > 1$ [BK98]	1.5	2.8
Planar graphs	PTAS [HK02]	NPC [HK02]	PTAS [HK02]	PTAS [HK02]
Trees	1 [K89]		PTAS [HKP$^+$03]	1 [HKP$^+$03]
$k + 1$-claw free	k		k	**1.796k^2+.5** ($4k^2-2k$) [HKS03]
k-sets	k		k	**3.591k+.5** ($6k-2$) [K-03]
Line graphs	2	NPC	2	**7.682** (10) [K-03]

makespan version of npSMC of line graphs, which arises in the *file transfer problem*. They showed that a class of greedy algorithms yields a 2-approximation and gave a $(2+\epsilon)$-approximation for a version with more general resource constraints. Kim [K-03] gave an LP formulation of the npSMC problem on line graphs and intersection graphs of k-sets,[1] improving the earlier bounds of [HKS03]. The paper presents also a ratio of $8k - 7$ for the RCS problem with k resources.

2 Sum Multicoloring via Makespan Approximations

In this section we describe and analyze our main approximation technique. Later, we show how to obtain our results by applying the general technique to specific classes of graphs, and to the different variants of the sum multicoloring problem that we consider here.

2.1 Algorithms and Implementation

Our technique uses two components: (i) a lower bound, f_v^*, on the completion time of the vertex v in an optimal solution, for any $v \in V$; a parameter $d \geq 1$, which indicates how well the lower bound captures the optimal value; (ii) a (makespan) multicoloring algorithm $\mathcal{A}$ with performance ratio ρ, for some $\rho \geq 1$.

Given the f_v^* values, the algorithm schema, ALG, breaks the time line (or the color sequence $1, 2, \ldots$) into intervals. We use in the partition two parameters: α, chosen uniformly at random from $[0, 1)$, and a constant $\beta > 1$ (to be optimized). Let $c_k = \beta^{\alpha+k}$, for $k = 0, 1, \ldots, L$, where $c_L \geq \max_v x_v$. The intervals induce a partition of the graph into *blocks* $V_\ell = \{v \in V : c_{\ell-1} < f_v^* \leq c_\ell\}$, $\ell = 1, \ldots, L$, of vertices whose completion times (f_v^*) fall in the respective interval. We then apply the makespan multicoloring algorithm on each block in sequence. We show that when this is possible, our algorithm attains a ratio of $d \cdot e\rho \approx 2.718d\rho$ for pSMC, $1.796d\rho + 0.5$ for npSMC, and $1.796d\rho$ for SC.

---

[1] We give the precise definition in Section 3.1.

The lower bounds, f_v^*, can be obtained either by solving a linear program, or by using an approximation algorithm for the preemptive sum multicoloring problem. This results in two algorithms described below. As shown in Section 2.2, we can unify the analyses of the two algorithms, once we guarantee that each satisfies certain properties.

LP-based Algorithm: One way to obtain the f_v^* values is by solving the LP relaxation of an integer programming formulation of the problem. (Such LP relaxations have been used in the past in scheduling *independent* jobs; see, e.g., [W-85, Q-93, S-96].) Before we describe our LP-based algorithm, we give some underlying properties of this algorithm. Let OPT be the cost of an optimal solution, and $OPT^* = \sum_v f_v^*$ the total of the lower bounds f_v^*. Also, we denote by $\omega(H, x)$ the maximum weight of any clique in a subgraph H, i.e., largest sum of color requirements. For a subset U of vertices, let $x(U) = \sum_{u \in U} x_u$.

We require that the following properties be satisfied:

(P1) $OPT^* \leq OPT$.
(P2a) $\max_{v \in V_\ell} f_v^* \geq \omega(V_\ell, x)/d$, for some $d \geq 1$, for all $1 \leq \ell \leq L$.
(P2b) There is a multicoloring algorithm, $\mathcal{A}$, that approximates the makespan of any graph in the given graph class within a ρ factor of the weighted clique size, and in particular,

$$\mathcal{A}(V_\ell, x) \leq \rho \cdot \omega(V_\ell, x), \text{ for } \ell = 1, 2, \ldots, L. \tag{1}$$

We formulate sum multicoloring with an integer program that uses *linear ordering* variables (see, e.g., [P-80, HSSW-97]). For any edge $uv \in E$, there is a variable $\delta_{uv} \in \{0, 1\}$, such that $\delta_{uv} = 1$ if u precedes v in the schedule, and 0 otherwise. Let $N(v)$ denote the set of neighbors of v in G, and $C_1, \ldots, C_{N_v}$ denote the maximal cliques in $N(v)$. The constraints (2) follow from the requirement that the vertices in any clique C are assigned *disjoint* sets of colors; thus the completion time f_v of a vertex v in a clique C is at least the sum of the color requirements of the vertices in C that completed before v plus that of v itself.

$$(LP) \quad \text{minimize} \sum_{v \in V} f_v$$

$$\text{subject to: } \forall v \in V, \ 1 \leq r \leq N_v : \ f_v \geq x_v + \sum_{u \in C_r} x_u \delta_{uv} \tag{2}$$

$$\forall uv \in E \qquad : \ \delta_{uv} + \delta_{vu} = 1$$

In the linear relaxation of LP, we allow f_v to take non-integral values ≥ 1. We denote by f_v^* the value of f_v in an optimal LP solution. Note that the program is equally valid for the preemptive and non-preemptive variants.

The next lemma shows that the above LP formulation satisfies property (P2a) with $d = 2$. It is based on a result of [K-03] (Lemma 2.3), attributed to [HSSW-97].

Lemma 1. *For any $1 \le \ell \le L$, $\max_{v \in V_\ell} f_v^* \ge \dfrac{\omega(V_\ell, x)}{2}$.*

In particular, since $\max_{v \in V_\ell} f_v^* \le c_\ell$, this implies that $c_\ell \ge \omega(V_\ell, x)/2$ for $\ell = 1, \ldots, L$.

Proof. Let C be a clique in G. Let f_v be the completion time of $v \in C$ in the solution for LP. Indeed, $C \setminus \{v\} \subseteq N(v)$. From LP, we get that

$$\sum_{v \in C} x_v f_v \ge \sum_{v \in C} x_v \left(x_v + \sum_{u \in C, u \ne v} x_u \delta_{uv} \right)$$

$$= \sum_{v \in C} x_v^2 + \sum_{u, v \in C} (x_v x_u \delta_{uv} + x_v x_u \delta_{vu})$$

$$\ge \frac{\sum_{v \in C} x_v^2 + \left(\sum_{u \in C} x_u \right)^2}{2} \tag{3}$$

Now, let C_ℓ be a maximum weight clique in V_ℓ, and let v_ℓ be the vertex in C_ℓ with the largest completion time in V_ℓ, $f_{v_\ell}^*$. From (3), we have that $\sum_{u \in C_\ell} x_u f_u \ge x(C_\ell)^2 / 2 = \omega(V_\ell, x)^2 / 2$. We also have that $\sum_{u \in C_\ell} x_u f_u \le f_{v_\ell}^* \sum_{u \in C_\ell} x_u = f_{v_\ell}^* x(C_\ell) = f_{v_\ell}^* \omega(V_\ell, x)$. ∎

We now summarize the steps of the LP-based algorithm with parameters $\beta, \alpha > 1$.

Algorithm ALG$_{\text{LP}}$
(i) Solve the linear program LP to obtain the f_v^* values.
(ii) Partition the vertices in the graph to the blocks $V_1, V_2, \ldots$ by their f_v^* values.
(iii) Color the blocks in sequence using a non-preemptive multicoloring algorithm $\mathcal{A}$ which satisfies Property (P2b); that is, suppose that the last color used for the block V_ℓ is col_ℓ, then $\mathcal{A}$ starts coloring the block $V_{\ell+1}$ with $col_\ell + 1$.

Applying an Approximation Algorithm for pSMC: An alternative way of obtaining the infeasible solution, f_v^*, is to use the preemptive solution when solving the non-preemptive problem. In this case, we replace (P2a) and (P2b) by the following properties.

(**P2a'**) There is a d-approximation algorithm for pSMC, for some $d \ge 1$.
(**P2b'**) There is non-preemptive multicoloring algorithm, $\mathcal{A}$, that approximates the makespan of any graph in the given graph class within a ρ factor of the number of colors used by a preemptive multicoloring, and in particular,

$$\mathcal{A}(V_\ell, x) \le \rho \cdot \text{pMC}(V_\ell, x), \text{ for } \ell = 1, 2, \ldots, L. \tag{4}$$

We now summarize the steps of the algorithm based on the approximation for pSMC. The algorithm gets as parameters the values $\beta, \alpha > 1$.

Algorithm ALG$_{\text{PRE}}$
(i) Apply to G a d-approximation algorithm for pSMC. Let f_v^{pre} be the completion time of $v \in V$. Set for any $v \in V$, $f_v^* = f_v^{pre} / d$,
(ii) Partition the vertices in the graph to the blocks $V_1, V_2, \ldots$ by their f_v^* values.
(iii) Color the blocks in sequence using a non-preemptive multicoloring algorithm $\mathcal{A}$ which satisfies Property (P2b').

2.2 Analysis

We use in the analysis the following notation. Recall that the *(multi)chromatic number* $\chi(G)$ of a graph G is the minimal number of colors required for (multi) coloring the vertices in G properly. In scheduling terms, this is the minimal total length (or *makespan*) of any legal schedule. We use the notation pMC, npMC for the preemptive and non-preemptive versions of this problem, respectively. Let ℓ_v denote the block into which v falls (ℓ_v is a function of α). Let t_ℓ denote the number of colors used by the multicoloring algorithm $\mathcal{A}$ on block ℓ. If we apply algorithm $\mathsf{ALG}_{\mathrm{LP}}$, then by properties (P2a) and (P2b),

$$t_\ell \leq \rho\omega(V_\ell, x) \leq \rho dc_\ell. \tag{5}$$

Similarly, if we use $\mathsf{ALG}_{\mathrm{PRE}}$, we have that $t_l \leq \rho \cdot \mathsf{pMC}(V_\ell, x) \leq \rho \max_{v \in V_\ell} f_v^{pre} = \rho d \max_{v \in V_\ell} f_v^* \leq \rho dc_\ell$. We proceed to analyze our algorithm schema, ALG, without making any assumptions on the algorithm used (i.e., the analysis applies for both $\mathsf{ALG}_{\mathrm{LP}}$ and $\mathsf{ALG}_{\mathrm{PRE}}$).

Denote by $\tilde{f}_v$ the last color (completion time) of a vertex v by our algorithm schema ALG. This color is the sum of the makespans of the colorings of the previous blocks, plus the completion time f_v' of v within the current block, i.e. $\tilde{f}_v = \sum_{r=1}^{\ell-1} t_r + f_v'$.

Bound for pSMC: We first consider a general scenario, that captures, e.g., the preemptive case. We trivially bound the last color of $v \in V_\ell$ under $\mathcal{A}$ by the total number of colors used, i.e., $f_v' \leq t_\ell$. Hence, we get for each vertex independently that

$$\tilde{f}_v \leq \sum_{r=1}^{\ell} t_r \leq \frac{d \cdot \rho\beta^{\alpha+\ell+1}}{\beta - 1}, \tag{6}$$

and

$$\mathsf{ALG}(V, x) = \sum_{v \in V} \tilde{f}_v \leq d \cdot \rho \sum_{v \in V} \frac{\beta^{\alpha+\ell_v+1}}{\beta - 1} = d \cdot \rho \cdot \frac{\beta}{\beta - 1} \sum_{v \in V} c_{\ell_v}, \tag{7}$$

where ℓ_v is the block in which v is colored and c_ℓ is the largest color in block ℓ.

We now select α uniformly at random from $[0, 1)$. Then ℓ_v and c_ℓ are also random variables.

Lemma 2. *For any $\beta > 1$ and $v \in V$, $\mathbf{E}[c_{\ell_v}] = \frac{\beta-1}{\ln \beta} f_v^*$, where the expectation is over the random choices of α.*

Proof. By the definition of ℓ_v, $c_{\ell_v - 1} = \beta^{\alpha+\ell_v-1} < f_v^* \leq \beta^{\alpha+\ell_v} = c_{\ell_v}$. Let us write $f_v^* = \beta^x$, i.e. $x = \log_\beta f_v^*$. Let $y_v = \ell_v + \alpha - x$ and note that y_v is in the range $[0, 1)$. We may write $y_v = (\alpha - x) \bmod 1$. The values f_v^* and x are fixed and independent of α. Thus, when α is chosen uniformly at random from $[0, 1)$, y_v is also uniformly distributed in $[0, 1)$. The random variable β^{y_v} then has expected value

$$\mathbf{E}[\beta^{y_v}] = \int_0^1 \beta^t dt = \frac{\beta - 1}{\ln \beta}.$$

Hence,

$$\mathbf{E}[c_{\ell_v}] = \mathbf{E}[\beta^{\ell_v+\alpha}] = \mathbf{E}[\beta^{\ell_v+\alpha-x}] \cdot \beta^x = \frac{\beta-1}{\ln\beta} f_v^*. \tag{8}$$

∎

Recall that $OPT^* = \sum_v f_v^*$. Combining (7) with Lemma 2 we get that

$$\mathbf{E}[\mathsf{ALG}(V,x)] \le d\rho\frac{\beta}{\beta-1} \sum_{v \in V} \frac{\beta-1}{\ln\beta} f_v^* \le d\rho\frac{\beta}{\ln\beta} OPT^*.$$

The function $f(\beta) = \beta/\ln\beta$ is minimized when $\beta = e \approx 2.718$. This gives the following.

Theorem 1. *There is a $(d \cdot e\rho)$-approximation algorithm for* pSMC.

Bound for npSMC**:** In the non-preemptive case, we may use the schedule output by algorithm $\mathcal{A}$ for V_ℓ either directly or reversed. In the reverse order, any vertex v, whose last color is f_v, is colored with $(t_\ell - f_v + 1), (t_\ell - f_v + 2), \dots, (t_\ell - f_v + x_v)$. By selecting the order that yields the better weighted average completion time, we may assume that on average, each job is at least half-way through completion at the half-way mark for V_ℓ. That is, on average, for any vertex $v \in V_\ell$, $f_v' \le (t_\ell + x_v)/2$. Thus, we have

$$\tilde{f}_v \le \sum_{r=1}^{\ell-1} t_r + \frac{t_\ell}{2} + \frac{x_v}{2}$$

$$\le d \cdot \rho \left(\frac{\beta^{\alpha+\ell}}{2} + \sum_{r=0}^{\ell-1} \beta^{\alpha+r} \right) + \frac{x_v}{2} \tag{9}$$

$$\le d \cdot \rho\beta^{\alpha+\ell} \left(\frac{1}{2} + \frac{1}{\beta-1} \right) + \frac{x_v}{2}$$

$$= d \cdot \rho \cdot c_\ell \left(\frac{\beta+1}{2(\beta-1)} \right) + \frac{x_v}{2} \tag{10}$$

Combining (10) with Lemma 2 we have

$$\mathbf{E}[\mathsf{ALG}(V,x)] = \sum_{v \in V} \mathbf{E}[\tilde{f}_v] \le d \cdot \rho\frac{\beta+1}{2(\beta-1)} \sum_v \mathbf{E}[c_{\ell_v}] + \frac{x(V)}{2}$$

$$= d \cdot \rho\frac{\beta+1}{2\ln\beta} OPT^* + \frac{x(V)}{2}$$

The function $f(\beta) = (\beta+1)/\ln\beta$ is minimized when $\beta = \gamma \approx 3.59112$, for a ratio of $d\gamma\rho/2 + 0.5$.

Note that the above schema can be derandomized, by partitioning the interval $(0,1]$ to smaller intervals; we can then search for the best value for α in these intervals, to within desired precision. We summarize in the next result.

Theorem 2. *There is a $(d\gamma\rho/2+0.5)$-approximation algorithm for* npSMC, *where $\gamma \approx 3.59112$.*

Deterministic and simultaneous approximation: If we make do without randomization, we can still obtain reasonable bounds that translate to simultaneous approximations of makespan and weighted completion time.

By the definition of V_ℓ, $f_v^* > \beta^{\alpha+\ell-1}$. Then, from (6) we obtain, for each vertex v, a bound of

$$\frac{\tilde{f}_v}{f_v^*} \leq d \cdot \rho \frac{\beta^2}{\beta - 1} \ .$$

This is optimized when $\beta = 2$,

Theorem 3. *There is an algorithm that approximates simultaneously* pSMC *(*npSMC*) and* pMC *(*npMC*), to within factor $4d\rho$.*

Sum coloring approximation: When the graph has unit color requirements, we get the SC problem. For this case, we obtain a slight improvement.

Theorem 4. *There is a $(d\gamma\rho/2)$-approximation algorithm for* SC, *where $\gamma \approx 3.59112$.*

Proof. Continuing from (9), we have

$$\sum_{v \in V_\ell} \tilde{f}_v \leq d\rho |V_\ell| (\frac{\beta^{\alpha+\ell}}{2} + \sum_{r=0}^{\ell-1} \beta^{\alpha+r}) + \frac{1}{2} \sum_{v \in V_\ell} x_v$$

$$= d\rho |V_\ell| (\frac{\beta^{\alpha+\ell}}{2} + \beta^\alpha \frac{\beta^\ell - 1}{\beta - 1}) + \frac{1}{2} \sum_{v \in V_\ell} x_v$$

$$= d\rho |V_\ell| (\beta^{\alpha+\ell} \frac{\beta + 1}{2(\beta - 1)} - \frac{\beta^\alpha}{\beta - 1}) + \frac{1}{2} \sum_{v \in V_\ell} x_v.$$

Thus,

$$\mathsf{ALG}(V, x) = \sum_{v \in V} \tilde{f}_v \leq d\rho \sum_{\ell \geq 1} |V_\ell| \left[\frac{\beta + 1}{2(\beta - 1)} \beta^{\alpha+\ell} - \frac{\beta^\alpha}{\alpha - 1} \right] + \frac{|V|}{2} \ .$$

Hence, applying Lemma 1, we have

$$\mathbf{E}[\mathsf{ALG}(V, x)] = \sum_{v \in V} \mathbf{E}[\tilde{f}_v] \leq d\rho \sum_v \left[\mathbf{E}[c_{\ell_v}] \cdot \frac{\beta + 1}{2(\beta - 1)} - \frac{1}{\beta - 1} \right] + \frac{|V|}{2}$$

$$= d\rho \cdot \frac{\beta + 1}{2 \ln \beta} OPT^* - \frac{d\rho |V|}{\beta - 1} + \frac{|V|}{2} \leq d\rho \cdot \frac{\beta + 1}{2 \ln \beta} OPT^*.$$

The last inequality follows from the fact that $\frac{\rho d}{\beta - 1} > 1/2$, since $\rho \geq 1$, $\beta < 5$, and in the cases we have studied, $d \geq 2$. ∎

3   Approximating Sum Multicoloring

We now apply our technique to the npSMC problem on several classes of graphs. We use both the preemptive approximation and the LP-based algorithm.

3.1   Approximating npSMC

Line graphs: Here we can apply both the LP and the preemptive relaxations with equal performance ratio, but the latter is both combinatorial and more efficient. A greedy 2-approximation algorithm for pSMC on line graphs is presented in [BHK+00] (that holds also in the weighted case). Thus, we can apply algorithm $\mathsf{ALG}_{\mathrm{PRE}}$, with $d = 2$.

For non-preemptive multicoloring line graphs, we use the greedy algorithm of [CG+85] that schedules each job as early as possible (i.e. colors each vertex with the smallest possible colors), breaking ties arbitrarily. This ensures that each vertex is always waiting for a neighbor until it is scheduled to completion. The completion time of a vertex is then at most the sum of the lengths of its neighbors, which is at most twice the length of the larger clique involving the vertex (see [CG+85]). Thus, in this case, we have $\rho = 2$. Now, using Theorem 2, we get a performance bound for line graphs.

Theorem 5. *There is a 7.683-approximation algorithm for npSMC on line graphs.*

This improves on the recent factor of 10 by Kim [K-03] and the factor of 12 obtained by a combinatorial (greedy) algorithm in [HKS03]. Observe that the non-preemptive algorithms are all measured in terms of the preemptive optimum.

Intersection graphs of k-sets: Resource-bounded scheduling when each job uses at most k resources is modeled by graphs that are intersection graphs of sets of size at most k. For each resource r, the vertices using that resource form a clique C_r. Then, for any $v \in V$, $N(v)$ can be partitioned into at most k maximal cliques.

We can extend the LP-based strategy for line graphs to intersection graphs of k sets. In this case, the non-preemptive greedy multicoloring algorithm of [CG+85] uses at most $k\omega$ colors, where ω is the maximal size of any of the resource cliques. Thus, it suffices to consider only cliques induced by individual resource, and not those cliques formed by interplay of a collection of resources. In other words, the clique constraints in LP need only involve the resource-cliques, therefore the number of constraints in polynomial. Hence, we obtain a non-preemptive solution with $d = 2$ and $\rho = k$, and by Theorem 2, we get

Theorem 6. *There is a $(3.591k + 0.5)$-approximation for npSMC on intersection graphs of k-sets.*

This improves on the ratio of $6k - 2$ of [K-03].

$(k + 1)$-claw free graphs: The combinatorial strategy for line graphs can be generalized for $(k + 1)$-claw free graphs, albeit with a worse ratio function than for LP-based algorithm for intersection graphs of k-sets. The sorted greedy algorithm of [BHK+00] yields a ratio of k for pSMC in $(k + 1)$-claw free graphs, resulting in a preemptive relaxation with $d = k$ in our schema. Also, as above, the makespan algorithm has performance ratio $\rho = k$. Thus, we get

Theorem 7. *There is a combinatorial* $(1.796k^2 + 0.5)$-*approximation for* npSMC *on* $(k + 1)$-*claw free graphs.*

Interval graphs: The npMC problem on interval graphs is better known as *dynamic storage allocation*. Gergov gave an algorithm that uses at most $3\omega(G)$ colors [G-99]. The number of maximal cliques in an interval graph is at most n. Thus, LP has a polynomial number of constraints and we can use it to obtain a multicoloring satisfying (P1) and (P2a), with $d = 2$. We can also use the approximation of the preemptive solution of [HKS03] as a relaxation with $d = 7.184$. Applying Theorem 2, we obtain the first constant approximation factor for this problem.

Theorem 8. *There is an* 11.273-*approximation and a combinatorial* 38.7-*approximation for* npSMC *on interval graphs.*

3.2 Approximating pSMC

Perfect graphs: On perfect graphs, LP can be solved in polynomial time, even though the number of constraints may be exponential, because there is a polynomial time separation algorithm: given a solution $\mathbf{f}$ for LP, we can test in polynomial time whether all the constraints are satisfied. For a vertex $v \in V$, we set, for each neighbor $u \in N(v)$, $x'_u = x_u \delta_{uv}$. We can now find the maximum weight clique in $N(v)$ with respect to $\mathbf{x}'$, since any subgraph of G is perfect. Then, we can test in polynomial time whether f_v satisfies the constraint (2) by checking whether the inequality holds for this maximum weight clique. (For more details, see e.g., [Q-93].) The solution for LP yields a multicoloring ψ^* that satisfies (P1) and (P2), with $d = 2$. The multicoloring problem pMC on perfect graphs is solvable in polynomial time, within arbitrary desired precision, as shown in [GLS-93], yielding our $\rho = 1 + O(1/n)$. Applying Theorems 1 and 4, we improve on the previous best factors of 16 for pSMC [BHK$^+$00] and 4 for SC [BBH$^+$98].

Theorem 9. *There is a* $2e \approx 5.436$-*approximation for* pSMC *and a* 3.592-*approximation for* SC *on perfect graphs.*

4 Extensions

Weights: Note that vertex weights can be added in our LP formulation, to get the fractional values f^*_v that satisfy (P1) and (P2) for the weighted minsum objective. We then apply as before for each block ℓ the makespan algorithm $\mathcal{A}$.

Release times: Our technique can be applied also in the case where each job J_j has a release time, r_j. In this case, in the LP formulation we add for any vertex v the constraint $f_v \geq r_v + x_v$. This ensures that, for any $v \in V_\ell$, $r_v \leq c_\ell$. Hence, when applying the makespan algorithm, $\mathcal{A}$, we start scheduling the vertices in V_ℓ at $\max(\sum_{r=1}^{\ell-1} t_r, \beta^{\alpha+\ell})$. This is attained by taking $\beta = 2$, which slightly increases the performance bounds that we obtained for ALG, both in the preemptive and the non-preemptive case.

Theorem 10. ALG *attains a ratio of* $d\rho 1.5/\ln 2 \approx 2.16 d\rho$ *for* npSMC *and* $d\rho 2/\ln 2 \approx 2.89 d\rho$ *for* pSMC *instances with release times.*

Resource Constrained Scheduling: Recall that in RCS, the resources are represented as cliques in our conflict graph G. Let $\mathcal{C}$ denote the set of maximal cliques in G, then RCS can be formulated as the following linear program.

$$(LP - RCS) \quad \text{minimize} \quad \sum_{\hat{C} \in \mathcal{C}} f_{\hat{C}}$$

$$\text{subject to:} \quad \forall \hat{C} \in \mathcal{C}, \ \forall v \in \hat{C} \quad : \quad f_v \geq x_v + \sum_{u \in \hat{C}} x_u \delta_{uv}$$

$$\forall \hat{C} \in \mathcal{C}, \ \forall v \in \hat{C} \quad : \quad f_C \geq f_v \tag{11}$$

$$\forall uv \in E \quad : \quad \delta_{uv} + \delta_{vu} = 1 \tag{12}$$

This corresponds to only the last vertex of each clique contributing to the objective function in the npSMC problem. Our analysis in the preemptive case was separate for each vertex, bounding the cost for the vertex only by the last color used in that block. Thus, we obtain an approximation ratio of $2e \cdot k$ for RCS. This improves on the previous ratio of $8k - 7$ presented by Kim [K-03], for any $k \geq 3$. For $k = 2$, the ratio of 10.45 is worse than the best known approximation ratio of 5.055 [GHKS04], but is achieved by a polynomial-size linear program.

Acknowledgments. We thank Moses Charikar and Chandra Chekuri for helpful comments and suggestions.

References

[AB+00]   F. Afrati, E. Bampis, A. Fishkin, K. Jansen, and C. Kenyon. Scheduling to minimize the average completion time of dedicated tasks. *FSTTCS '00.*

[BBH+98]   A. Bar-Noy, M. Bellare, M. M. Halldórsson, H. Shachnai, T. Tamir. On chromatic sums and distributed resource allocation. *Inf. Comp.* **140**:183–202, 1998.

[BHK+00]   A. Bar-Noy, M. M. Halldórsson, G. Kortsarz, H. Shachnai, and R. Salman. Sum multicoloring of graphs. *J. Algorithms* **37**(2):422–450, 2000.

[BK98]   A. Bar-Noy and G. Kortsarz. The minimum color-sum of bipartite graphs. *J. Algorithms,* **28**:339–365, 1998.

[CP+96]   S. Chakrabarti, C. A. Phillips, A. S. Schulz, D. B. Shmoys, C. Stein and J. Wein. Improved scheduling algorithms for minsum criteria. *ICALP '96,* 875–886.

[CG+85]   E. G. Coffman, Jr., M. R. Garey, D. S. Johnson and A. S. LaPaugh. Scheduling file transfers. *SIAM J. Comput.* **14**:744–780, 1985.

[FLT-02]   U. Feige, L. Lovász, P. Tetali. Approximating min-sum set cover. *AP-PROX'02,* 94–107.

[G-99]   J. Gergov. Algorithms for compile-time memory allocation. *SODA'99*.

[G+02]   K. Giaro, R. Janczewski, M. Kubale and M. Malafiejski.   A 27/26-approximation algorithm for the chromatic sum coloring of bipartite graphs. *APPROX '02*, 131–145.

[G01]   M. Gonen. Coloring Problems on Interval Graphs and Trees. M.Sc. thesis, The Open Univ., Tel-Aviv, 2001.

[G+04]   R. Gandhi, M. M. Halldórsson, G. Kortsarz and H. Shachnai, Improved Bounds for Sum Multicoloring and Scheduling Dependent Jobs with Minsum Criteria, full version. http://www.cs.technion.ac.il/~hadas/PUB/smc-waoa04.ps.

[GHKS04]   R. Gandhi, M. M. Halldórsson, G. Kortsarz and H. Shachnai, Improved Results for Data Migration and Open Shop Scheduling. *ICALP '04*.

[GLS-93]   M. Grötschel, L. Lovász and A. Schrijver. Geometric Algorithms and Combinatorial Optimization. Springer-Verlag, 1993.

[HSW-96]   L. A. Hall, D. B. Shmoys, and J. Wein. Scheduling to minimize average completion time: Off-line and on-line algorithms. *SODA'96*, 142–151. Jan 1996.

[HSSW-97] L. A. Hall, A. Schulz, D. B. Shmoys, and J. Wein. Scheduling to minimize average completion time: Off-line and on-line approximation algorithms. *Math. Operations Research* 22:513–544, 1997.

[HK02]   M. M. Halldórsson and G. Kortsarz. Tools for multicoloring with applications to planar graphs and partial k-trees. *J. Algorithms* 42(2), 334–366, 2002.

[HKP+03]   M. M. Halldórsson, G. Kortsarz, A. Proskurowski, R. Salman, H. Shachnai, and J. A. Telle. Multicoloring trees. *Inf. Computation* 180(2):113–129, 2003.

[HKS03]   M. M. Halldórsson, G. Kortsarz, H. Shachnai. Sum coloring interval and *k*-claw free graphs with application to scheduling dependent jobs. *Algorithmica* 37:187–209, 2003.

[J-97]   K. Jansen. The optimum cost chromatic partition problem. *CIAC '97*, 25–36.

[K-03]   Y. A. Kim. Data migration to minimize the average completion time, *SODA'03*.

[K-96]   M. Kubale. Preemptive versus non preemptive scheduling of biprocessor tasks on dedicated processors. *European J. Operational Research* **94**:242–251, 1996.

[K89]   E. Kubicka. The chromatic sum of a graph. PhD thesis, Western Michigan, 1989.

[NSS99]   S. Nicoloso, M. Sarrafzadeh and X. Song. On the sum coloring problem on interval graphs. *Algorithmica* **23**:109–126,1999.

[P-80]   C. N. Potts. An algorithm for the single machine sequencing problem with precedence constraints, *Math. Prog. Stud.* 13, 78–87, 1980.

[Q-93]   M. Queyranne. Structure of a simple scheduling polyhedron. *Math. Prog.* 58:263–285, 1993.

[QS-01]   M. Queyranne, M. Sviridenko. A $2+\epsilon$-approximation algorithm for generalized preemptive open shop problem with minsum objective. *J.Alg.* 45:202–212, 2002.

[QS-02]   M. Queyranne, M. Sviridenko. Approximation algorithms for shop scheduling problems with minsum objective. *J. Scheduling* 5:287-305, 2002.

[S-96]   A. S. Schulz. Scheduling to minimize total weighted completion time: Performance guarantees of LP-based heuristics and lower bounds. *IPCO '96*, 301–315.

[W97]   G. Woeginger. Private communication, 1997.

[W-85]   L. Wolsey. Mixed Integer Programming Formulations for Production Planning and Scheduling Problems. Invited talk at *12th ISMP*, MIT, 1985.

Approximation Algorithms for Spreading Points*

Sergio Cabello

Institute for Mathematics, Physics and Mechanics,
Department of Mathematics, Ljubljana, Slovenia
sergio.cabello@imfm.uni-lj.si

Abstract. We consider the problem of placing n points, each one inside its own, prespecified disk, with the objective of maximizing the distance between the closest pair of them. The disks can overlap and have different sizes. The problem is NP-hard and does not admit a PTAS. In the L_∞ metric, we give a 2-approximation algorithm running in $O(n\sqrt{n}\log^2 n)$ time. In the L_2 metric, similar ideas yield a quadratic time algorithm that gives an $\frac{8}{3}$-approximation in general, and a ~ 2.2393-approximation when all the disks are congruent.

1 Introduction

The problem of distant representatives was recently introduced by Fiala et al. [11, 12]: given a collection of subsets of a metric space and a value $\delta > 0$, we want a representative of each subset such that any two representatives are at least δ apart. They introduced this problem as a variation of the problem of systems of disjoint representatives in hypergraphs [1]. It generalizes the problem of systems of distinct representatives, and it has applications in areas such as scheduling or radio frequency (or channel) assignment to avoid interferences.

As shown by Fiala et al. [11,12], and independently by Baur and Fekete [3], the problem of deciding the existence of distant representatives is NP-hard even in the plane under natural metrics. This problem naturally embeds within the context of packing and map labelling problems, which has a much longer history; see the discussion in [3] and references therein.

However, in most applications, rather than systems of representatives at a given distance, we would be more interested in systems of representatives whose closest pairs are as separated as possible. Therefore, the design of approximation algorithms for the latter problem seems a suitable alternative. Here, we consider the problem of maximizing the distance of the closest pair in systems of representatives in the plane with either the L_∞ or the Euclidean L_2 metric. The subsets that we consider are (possibly intersecting) disks.

The geometric optimization problem under consideration finds applications in cartography [7], graph drawing [8], and more generally in data visualization,

---

* Extended abstract. A full version is available as [4]. This research was done as PhD student at the Institute of Information and Computing Sciences, Utrecht University, partially supported by Cornelis Lely Stichting, NWO, and DIMACS.

where the readability of the displayed data is a basic requirement, and often a difficult task. In many cases, there are some restrictions on how and where each object has to be drawn, as well as some freedom. For example, cartographers improve the readability of a map by displacing some features with respect to their real position. The displacement has to be small to preserve correctness, and the problem can be abstracted as follows. We want to place a fixed number of points (0-dimensional cartographic features) in the plane, but with the restriction that each point has to lie inside a prespecified region. The regions may overlap, and we want the placement that maximizes the distance between the closest pair. The region where each point has to be placed is application dependent. We will assume that they are given, and that they are disks.

Formulation of the problem. Given a distance d in the plane, consider the function $D : (\mathbb{R}^2)^n \to \mathbb{R}$ that gives the distance between a closest pair of n points

$$D(p_1, \ldots, p_n) = \min_{i \neq j} d(p_i, p_j).$$

Let $\mathcal{B} = \{B_1, \ldots, B_n\}$ be a collection of (possibly intersecting) disks in $\mathbb{R}^2$ under the metric d. A *feasible* solution is a placement of points $p_1, \ldots, p_n$ with $p_i \in B_i$. We are interested in a feasible placement $p_1^*, \ldots, p_n^*$ that maximizes D

$$D(p_1^*, \ldots, p_n^*) = \max_{(p_1, \ldots, p_n) \in B_1 \times \cdots \times B_n} D(p_1, \ldots, p_n).$$

We use $D(\mathcal{B})$ to denote this optimal value.

A t-approximation, with $t \geq 1$, is a feasible placement $p_1, \ldots, p_n$, with $t \cdot D(p_1, \ldots, p_n) \geq D(\mathcal{B})$. We will use $B(p, r)$ to denote the disk of radius r centered at p. Recall that under the L_∞ metric, $B(p, r)$ is an axis-aligned square centered at p and side length $2r$. We assume that the disk B_i is centered at c_i and has radius r_i, so $B_i = B(c_i, r_i)$.

Related work. The decision problem associated to our optimization one is the original distant representatives problem: for a given value δ, is $D(\mathcal{B}) \geq \delta$? Fiala et al. [11, 12] showed that this problem is NP-hard in the Euclidean and Manhattan metrics. Furthermore, by repeating at regular intervals their construction [11– Figures 1 and 2], it follows from the slackness of the construction that, unless $NP = P$, there is a certain constant $T > 1$ such that no T-approximation is possible. See [13] for a similar argument related to the slackness. They also notice that the one dimensional problem can be solved using the scheduling algorithm by Simons [17].

Closely related are *geometric dispersion* problems: we are given a polygonal region of the plane and we want to place n points on it such that the closest pair is as far as possible. This problem has been considered by Baur and Fekete [3] (see also [6, 10]), where both inapproximability results and approximation algorithms are presented. Their NP-hardness proof and inapproximability results can easily be adapted to show inapproximability results for our problem, showing also that no polynomial time approximation scheme is possible, unless $P = NP$.

Dispersion problems have also been considered in arbitrary metric spaces and with various optimization functions; see [6, 14] and references therein.

In a more general setting, we can consider the following problem: given a collection $S_1, \ldots, S_n$ of regions in $\mathbb{R}^2$, and a function $f : S_1 \times \cdots \times S_n \to \mathbb{R}$ that describes the quality of a feasible placement $(p_1, \ldots, p_n) \in S_1 \times \cdots \times S_n$, we want to find a feasible placement $p_1^*, \ldots, p_n^*$ such that

$$f(p_1^*, \ldots, p_n^*) = \max_{(p_1, \ldots, p_n) \in S_1 \times \cdots \times S_n} f(p_1, \ldots, p_n).$$

Geometric dispersion problems are a particular instance of this type where we want to maximize the function D over k copies of the same polygonal region. Minimum diameter covering problems try to minimize the diameter of the placement [2]. In [5], given a graph on the vertices $p_1, \ldots, p_n$, placements that maximize the number of straight-line edges in a given set of orientations are considered.

Our results. The main idea in our approach is to consider an "approximate-placement" problem in the L_∞ metric: given a value δ that satisfies $2\delta \leq D(\mathcal{B})$, we can provide a feasible placement $p_1, \ldots, p_n$ such that $D(p_1, \ldots, p_n) \geq \delta$. The proof can be seen as a suitable packing argument. This placement can be computed in $O(n\sqrt{n} \log n)$ time using the data structure by Mortensen [16] and the technique by Efrat et al. [9] for computing a matching in geometric settings. See Sections 2 and 3 for details.

We then combine the "approximate-placement" algorithm with the geometric features of our problem to get a 2-approximation in the L_∞ metric. This is done by a two-stage binary search on some special values by paying an extra logarithmic factor; see Section 4.

Section 5 *summarizes* in L_2 the results that are equivalent to those described in L_∞. In particular, the same idea of "approximate-placement" can be used in the L_2 metric, but the approximation ratio becomes 8/3 and the running time increases to $O(n^2)$. Using binary search leads to an (8/3)-approximation algorithm for the L_2 metric.

However, when we restrict ourselves to congruent disks in L_2, a trivial adaptation of the techniques gives an approximation ratio of ~ 2.2393. This is explained in Section 6.

2   Placement Algorithm in L_∞

Consider an instance $\mathcal{B} = \{B_1, \ldots, B_n\}$ of the problem in the L_∞ metric, and let $\delta^* = D(\mathcal{B})$ be the maximum value that a feasible placement can attain. We will consider the "approximate-placement" problem that follows: given a value δ, we provide a feasible placement $p_1, \ldots, p_n$ such that, if $\delta \leq \frac{1}{2}\delta^*$ then $D(p_1, \ldots, p_n) \geq \delta$, and otherwise there is no guarantee on the placement. In this section we present an algorithm and discuss its approximation performance, while in next section we discuss a more efficient version of it.

Let $\Lambda = \mathbb{Z}^2$, that is, the lattice $\Lambda = \{(a,b) \mid a, b \in \mathbb{Z}\}$. For any $\delta \in \mathbb{R}$ and any point $p = (p_x, p_y) \in \mathbb{R}^2$, we define $\delta p = (\delta p_x, \delta p_y)$ and $\delta\Lambda = \{\delta p \mid p \in \Lambda\}$. Observe that $\delta\Lambda$ is also a lattice. The reason to use this notation is that we can use $p \in \Lambda$ to refer to $\delta p \in \delta\Lambda$ for different values of δ. An *edge* of the lattice $\delta\Lambda$ is a horizontal or vertical segment joining two points of $\delta\Lambda$ at distance δ. The edges of $\delta\Lambda$ divide the plane into squares of side length δ, which we call the *cells* of $\delta\Lambda$.

The idea is that whenever $2\delta \leq \delta^*$, the lattice points $\delta\Lambda$ almost provide a solution. However, we have to treat as a special case the disks with no lattice point inside. More precisely, let $Q \subset \delta\Lambda$ be the set of points that cannot be considered as a possible placement because there is another already placed point too near by. Initially, we have $Q = \emptyset$. If a disk B_i does not contain any point from the lattice, there are two possibilities:

- B_i is contained in a cell C of $\delta\Lambda$; see Fig. 1 left. In this case, place $p_i := c_i$ in the center of B_i, and remove the vertices of the cell C from the set of possible placements for the other disks, that is, add them to Q.
- B_i intersects an edge E of $\delta\Lambda$; see Fig. 1 right. In this case, choose p_i on $E \cap B_i$, and remove the vertices of the edge E from the set of possible placements for the other disks, that is, add them to Q.

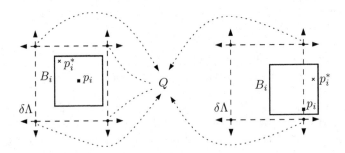

Fig. 1. Special cases where the disk B_i does not contain any lattice point. Left: B_i is fully contained in a cell of $\delta\Lambda$. Right: B_i intersects an edge of $\delta\Lambda$

We are left with disks, say $B_1, \ldots, B_k$, that have some lattice points inside. Consider for each such disk B_i the set of points $P_i := B_i \cap (\delta\Lambda \setminus Q)$ as candidates for the placement corresponding to B_i. Observe that P_i may be empty if $(B_i \cap \delta\Lambda) \subset Q$. We want to make sure that each disk B_i gets a point from P_i, and that each point gets assigned to at most one disk B_i. We deal with this by constructing a bipartite graph G_δ with $B := \{B_1, \ldots, B_k\}$ as one class of nodes and $P := P_1 \cup \cdots \cup P_k$ as the other class, and with an edge between $B_i \in B$ and $p \in P$ whenever $p \in P_i$.

It is clear that a (perfect) matching in G_δ provides a feasible placement. When a matching is not possible, the algorithm reports a feasible placement by

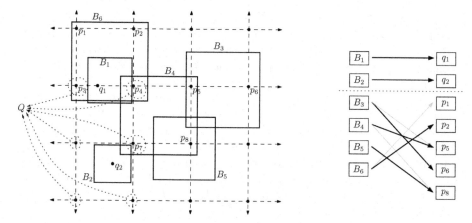

Fig. 2. Example showing the main features of the placement algorithm in L_∞

placing each point in the center of its disk. We call this algorithm PLACEMENT.
See Figure 2 for an example.

In any case, PLACEMENT always gives a feasible placement $p_1, \ldots, p_n$, and we
can then compute the value $D(p_1, \ldots, p_n)$ finding a closest pair in the placement.
Below we show that, whenever $2\delta \leq \delta^*$, PLACEMENT(δ) gives a placement whose
closest pair is at distance at least δ. In particular, this implies that if $B_i \cap \delta\Lambda \neq \emptyset$
but $P_i = B_i \cap (\delta\Lambda \setminus Q) = \emptyset$, then there is no matching in G_δ because the node
B_i has no edges, and so we can conclude that $2\delta > \delta^*$.

Definition 1. *In the L_∞ metric, PLACEMENT(δ) succeeds if the computed placement $p_1, \ldots, p_n$ satisfies $D(p_1, \ldots, p_n) \geq \delta$. Otherwise, PLACEMENT($\delta$) fails.*

Lemma 1. *If $2\delta \leq \delta^*$, then PLACEMENT(δ) succeeds.*

Proof. The proof is divided in two steps. Firstly, we will show that if $2\delta \leq \delta^*$ then
the graph G_δ has a matching. Secondly, we will see that if $p_1, \ldots, p_n$ is a placement computed by PLACEMENT(δ) when $2\delta \leq \delta^*$, then indeed $D(p_1, \ldots, p_n) \geq \delta$.

Consider an optimal placement $p_1^*, \ldots, p_n^*$. The points that we added to Q due
to a disk B_i are in the interior of $B(p_i^*, \delta^*/2)$ because of the following analysis:

- If $B_i \cap \delta\Lambda = \emptyset$ and B_i is completely contained in a cell C of $\delta\Lambda$, then p_i^* is
  in C, and $C \subset B(p_i^*, \delta) \subset B(p_i^*, \delta^*/2)$; see Figure 1 left.
- If $B_i \cap \delta\Lambda = \emptyset$ and there is an edge E of $\delta\Lambda$ such that $B_i \cap E \neq \emptyset$, then
  $E \subset B(p_i^*, \delta) \subset B(p_i^*, \delta^*/2)$; see Figure 1 right.

The interiors of the disks (in L_∞) $B(p_i^*, \delta^*/2)$ are disjoint, and we can
use them to construct a matching in G_δ as follows. If $B_i \cap \delta\Lambda \neq \emptyset$, then
$B(p_i^*, \delta^*/2) \cap B_i$ contains some lattice point $p_i \in \delta\Lambda$. Because the interiors of the
disks $B(p_i^*, \delta^*/2)$ are disjoint, we have $p_i \notin Q$ and $p_i \in P_i$. We cannot directly
add the edge (B_i, p_i) to the matching that we are constructing because it may
happen that p_i is on the boundary of $B(p_i^*, \delta^*/2) \cap B_i$, but also on the boundary

of $B(p_j^*, \delta^*/2) \cap B_j$. However, in this case, $B(p_i^*, \delta^*/2) \cap B_i \cap \delta\Lambda$ contains an edge of $\delta\Lambda$ inside. If we match each B_i to the lexicographically smallest point in $B(p_i^*, \delta^*/2) \cap B_i \cap \delta\Lambda$, then, because the interiors of disks $B(p_i^*, \delta^*/2)$ are disjoint, each point is claimed by at most one disk. This proves the existence of a matching in G_δ provided that $2\delta \leq \delta^*$.

For the second part of the proof, let p_i, p_j be a pair of points computed by PLACEMENT(δ). We want to show that p_i, p_j are at distance at least δ. If both were computed by the matching in G_δ, they both are different points in $\delta\Lambda$, and so they are at distance at least δ. If p_i was not placed on a point of $\delta\Lambda$ (at c_i or on an edge of $\delta\Lambda$), then the lattice points closer than δ to p_i were included in Q, and so the distance to any p_j placed during the matching of G_δ is at least δ. If both p_i, p_j were not placed on a point of $\delta\Lambda$, then B_i, B_j do not contain any point from $\delta\Lambda$, and therefore $r_i, r_j < \delta/2$. Two cases arise:

- If both B_i, B_j do not intersect an edge of $\delta\Lambda$, by the triangle inequality we have $d(p_i, p_j) \geq d(p_i^*, p_j^*) - d(p_i, p_i^*) - d(p_j, p_j^*) > \delta^* - \delta/2 - \delta/2 \geq \delta$, provided that $2\delta \leq \delta^*$.
- If one of the disks, say B_i, intersects an edge E of $\delta\Lambda$, then B_i is contained in the two cells of $\delta\Lambda$ that have E as an edge. Let C be the six cells of $\delta\Lambda$ that share a vertex with E. If B_j does not intersect any edge of $\delta\Lambda$, then $B_j \cap C = \emptyset$ because otherwise $d(p_i^*, p_j^*) < 2\delta$, and so $d(p_i, p_j) \geq \delta$. If B_j intersects an edge E' of $\delta\Lambda$, we have $E \cap E' = \emptyset$ because otherwise $d(p_i^*, p_j^*) < 2\delta$. It follows that $d(p_i, p_j) \geq \delta$.

$\square$

Notice that, in particular, if r_{min} is the radius of the smallest disk and we set $\delta = (r_{min}/\sqrt{n})$, then the nodes of type B_i in G_δ have degree n, and there is always a matching. This implies that $\delta^* = \Omega(r_{min}/\sqrt{n})$.

Observe also that whether PLACEMENT fails or succeeds is not a monotone property. That is, there may be values $\delta_1 < \delta_2 < \delta_3$ such that both PLACEMENT(δ_1) and PLACEMENT(δ_3) succeed, but PLACEMENT(δ_2) fails. This happens because for values $\delta \in (\frac{\delta^*}{2}, \delta^*]$, we do not have any guarantee on what PLACEMENT(δ) does.

The algorithm can be adapted to compute PLACEMENT($\delta+\epsilon$) for an infinitesimal $\epsilon > 0$ because only the points of $\delta\Lambda$ lying on the boundaries of $B_1, \ldots, B_n$ are affected. Therefore, for an infinitesimal $\epsilon > 0$, we can decide if PLACEMENT($\delta+\epsilon$) succeeds or fails.

Observation 2. *If PLACEMENT(δ) succeeds for $\mathcal{B}$, but PLACEMENT(δ) fails for a translation of $\mathcal{B}$, then $\delta \leq \delta^* < 2\delta$ and we have a 2-approximation.*

If for some $\delta > \delta'$, PLACEMENT(δ) succeeds, but PLACEMENT(δ') fails, then $\delta^* < 2\delta' < 2\delta$ and we have a 2-approximation.

If PLACEMENT(δ) succeeds, but PLACEMENT($\delta + \epsilon$) fails for an infinitesimal $\epsilon > 0$, then $\delta^* \leq 2\delta$ and we have a 2-approximation.

3   Efficiency of the Placement Algorithm in L_∞

The algorithm PLACEMENT, as stated so far, is not strongly polynomial because the sets $P_i = B_i \cap (\delta\Lambda \setminus Q)$ can have arbitrarily many points, depending on the value δ. However, when P_i has more than n points, we can just take any n of them. This is so because a node B_i with degree at least n is never a problem for the matching: if $G_\delta \setminus B_i$ does not have a matching, then G_δ does not have it either; if $G_\delta \setminus B_i$ has a matching M, then at most $n-1$ nodes from the class P participate in M, and one of the n edges leaving B_i has to go to a node in P that is not in M, and this edge can be added to M to get a matching in G_δ.

For a disk B_i we can decide in constant time if it contains some point from the lattice $\delta\Lambda$: we round its center c_i to the closest point p of the lattice, and depending on whether p belongs to B_i or not, we decide. Each disk B_i adds at most 4 points to Q, and so $|Q| \leq 4n$. We can construct Q and remove repetitions in $O(n \log n)$ time.

If a disk B_i has radius bigger than $3\delta\sqrt{n}$, then it contains more than $5n$ lattice points, that is, $|B_i \cap \delta\Lambda| > 5n$. Because Q contains at most $4n$ points, P_i has more than n points. Therefore, we can shrink the disks with radius bigger than $3\delta\sqrt{n}$ to disks of radius exactly $3\delta\sqrt{n}$, and this does not affect to the construction of the matching. We can then assume that each disk $B_i \in \mathcal{B}$ has radius $O(\delta\sqrt{n})$. In this case, each B_i contains at most $O(n)$ points of $\delta\Lambda$, and so the set $P = \bigcup_i P_i$ has $O(n^2)$ elements.

In fact, we only need to consider a set P with $O(n\sqrt{n})$ points. The idea is to divide the disks $\mathcal{B}$ into two groups: the disks that intersect more than $\sqrt{n}$ other disks, and the ones that intersect less than $\sqrt{n}$ other disks. For the former group, we can see that they bring $O(n\sqrt{n})$ points in total to P. As for the latter group, we only need to consider $O(\sqrt{n})$ points per disk.

Lemma 3. *It is sufficient to consider a set P with $O(n\sqrt{n})$ points. Moreover, we can construct such a set P in $O(n\sqrt{n}\log n)$ time.*

We are left with the following problem: given a set P of $O(n\sqrt{n})$ points, and a set $\mathcal{B}$ of n disks, find a maximum matching between P and $\mathcal{B}$ such that a point is matched to a disk that contains it. However, the graph G_δ does not need to be constructed explicitly because its edges are implicitly represented by the disk-point containment. This type of matching problem, when both sets have the same cardinality, has been considered by Efrat et al. [9]. Although in our setting one of the sets may be much larger than the other one, we can make minor modifications to the algorithm in [9] and use the data structure designed by Mortensen [16] to get the following result.

Lemma 4. *In L_∞, PLACEMENT can be adapted to run in $O(n\sqrt{n}\log n)$ time.*

4   Approximation Algorithms for L_∞

We want a value $\tilde{\delta}$ such that PLACEMENT($\tilde{\delta}$) succeeds, but PLACEMENT($\tilde{\delta} + \epsilon$) fails for an infinitesimally small $\epsilon > 0$; then $\tilde{\delta}$ is a 2-approximation because of Observation 2. General techniques based on Megiddo's parametric search [15] are not very fruitful in this case because the known parallel algorithms for computing maximum matchings do not have an appropriate tradeoff between the number of processors and the running time.

Instead, we can use the geometric characteristics of our problem to find a 2-approximation $\tilde{\delta}$ in $O(n\sqrt{n}\log^2 n)$ time. The idea is to consider for which values δ the algorithm changes its behavior, and use it to narrow down the interval where $\tilde{\delta}$ can lie. First, we state how to solve a special type of instances, and then present the main result of this section.

Lemma 5. *Let $\mathcal{B}$ be an instance consisting of m disks such that each disk $B_i \in \mathcal{B}$ has radius $O(r\sqrt{k})$, and assume that there is a disk B of radius $R = O(mr\sqrt{k})$ enclosing all the disks in $\mathcal{B}$. If PLACEMENT($\frac{r}{3\sqrt{k}}$) succeeds, then we can compute in $O(m\sqrt{m}\log^2 mk)$ time a placement $p_1, \ldots, p_m$ with $p_i \in B_i$ that yields a 2-approximation of $D(\mathcal{B})$.*

Proof. (Sketch) The proof is divided into three parts. Firstly, we show that we can assume that the origin is placed at the center of the enclosing disk B. Secondly, we narrow down our search space to an interval $[\delta_1, \delta_2]$ such that PLACEMENT(δ_1) succeeds but PLACEMENT(δ_2) fails. Finally, we consider all the critical values $\delta \in [\delta_1, \delta_2]$ for which the flow of control of PLACEMENT(δ) is different than for PLACEMENT($\delta + \epsilon$) or PLACEMENT($\delta - \epsilon$). The important observation is that the values δ_1, δ_2 are such that not many critical values are in the interval $[\delta_1, \delta_2]$.

Let $\mathcal{B}'$ be a translation of $\mathcal{B}$ such that the center of the enclosing disk B is at the origin. By hypothesis, PLACEMENT($\frac{r}{3\sqrt{k}}$) for $\mathcal{B}$ succeeds. If PLACEMENT($\frac{r}{3\sqrt{k}}$) for $\mathcal{B}'$ fails, then PLACEMENT($\frac{r}{3\sqrt{k}}$) for $\mathcal{B}$ gives a 2-approximation due to Observation 2, and we are done. This finishes the first part of the proof.

As for the second part, consider the horizontal axis h. Because the enclosing disk B has radius $R = O(mr\sqrt{k})$, the lattice $(\frac{r}{3\sqrt{k}})\Lambda$ has $O(mk)$ points in $B \cap h$. Equivalently, we have $t = \max\{z \in \mathbb{Z} \text{ s.t.}(\frac{r}{3\sqrt{k}})(z,0) \in B\} = \lfloor \frac{3R\sqrt{k}}{r} \rfloor = O(mk)$. In particular, $\frac{R}{t+1} \le \frac{r}{3\sqrt{k}}$.

If PLACEMENT($\frac{R}{t+1}$) fails, then PLACEMENT($\frac{r}{3\sqrt{k}}$) is a 2-approximation due to Observation 2. So we can assume that PLACEMENT($\frac{R}{t+1}$) succeeds. We can also assume that PLACEMENT($\frac{R}{1}$) fails, as otherwise $\mathcal{B}$ consists of only one disk.

We perform a binary search in $\mathbb{Z} \cap [1, t+1]$ to find a value $t' \in \mathbb{Z}$ such that PLACEMENT($\frac{R}{t'}$) succeeds but PLACEMENT($\frac{R}{t'-1}$) fails. We can do this with $O(\log t) = O(\log mk)$ calls to PLACEMENT, each taking $O(m\sqrt{m}\log m)$ time due to Lemma 4, and we have spent $O(m\sqrt{m}\log^2 mk)$ time in total. Let $\delta_1 := \frac{R}{t'}$ and $\delta_2 := \frac{R}{t'-1}$. This finishes the second part of the proof.

Before we start the third part, let us state, without proof, the property of δ_1, δ_2 that we will use later. If $p \in \Lambda$ is such that $\delta_1 p$ is in the interior of B, and C_p is the union of all four cells of $\delta_1 \Lambda$ having $\delta_1 p$ as a vertex, then $\delta_2 p \in C_p$, and more generally, $\delta p \in C_p$ for any $\delta \in [\delta_1, \delta_2]$. Therefore, if for a point $p \in \Lambda$ there is a $\delta \in [\delta_1, \delta_2]$ such that $\delta p \in \partial B_i$, then ∂B_i must intersect C_p.

We are ready for the third part of the proof. Consider the critical values $\delta \in [\delta_1, \delta_2]$ for which the flow of control of the PLACEMENT changes. They are the following:

- A point $p \in \Lambda$ such that $\delta p \in B_i$ but $(\delta + \epsilon)p \notin B_i$ or $(\delta - \epsilon)p \notin B_i$ for an infinitesimal $\epsilon > 0$. That is, $\delta p \in \partial B_i$.
- B_i intersects an edge of $\delta \Lambda$, but not of $(\delta + \epsilon)\Lambda$ $(\delta - \epsilon)\Lambda$ for an infinitesimal $\epsilon > 0$.

Because of the property of δ_1, δ_2 stated above, only the vertices V of cells of $\delta_1 \Lambda$ that intersect ∂B_i can change the flow of control of PLACEMENT. In the L_∞ metric, because the disks are axis-aligned squares, the vertices V are distributed along two axis-aligned rectangles, and each disk B_i induces $O(1)$ such critical values Δ_i changing the flow of control of PLACEMENT.

We can compute all the critical values $\Delta = \bigcup_{i=1}^{m} \Delta_i$ and sort them in $O(m \log m)$ time. Using a binary search on Δ, we find $\delta_3, \delta_4 \in \Delta$, with $\delta_3 < \delta_4$, such that PLACEMENT(δ_3) succeeds but PLACEMENT(δ_4) fails. Because $|\Delta| = O(m)$, this can be done in $O(m\sqrt{m} \log^2 m)$ time with $O(\log m)$ calls to PLACEMENT. The flow of control of PLACEMENT(δ_4) and of PLACEMENT$(\delta_3 + \epsilon)$ are the same. Therefore, we know that PLACEMENT$(\delta_3 + \epsilon)$ also fails, and conclude that PLACEMENT(δ_3) yields a 2-approximation because of Observation 2. $\qquad\square$

Theorem 1. *Let $\mathcal{B} = \{B_1, \ldots, B_n\}$ be a collection of disks in the plane with the L_∞ metric. We can compute in $O(n\sqrt{n} \log^2 n)$ time a placement $p_1, \ldots, p_n$ with $p_i \in B_i$ that yields a 2-approximation of $D(\mathcal{B})$.*

Proof. Let us assume that $r_1 \leq \cdots \leq r_n$, that is, B_i is smaller than B_{i+1}. Consider the values $\Delta = \{\frac{r_1}{3\sqrt{n}}, \ldots, \frac{r_n}{3\sqrt{n}}, 4r_n\}$. We know that PLACEMENT$(\frac{r_1}{3\sqrt{n}})$ succeeds, and we can assume that PLACEMENT$(4r_n)$ fails; if it would succeed, then the disks in $\mathcal{B}$ would be disjoint, and placing each point $p_i := c_i$ would give a 2-approximation.

We use PLACEMENT to make a binary search on the values Δ and find a value r_{max} such that PLACEMENT$(\frac{r_{max}}{3\sqrt{n}})$ succeeds, but either PLACEMENT$(\frac{r_{max}+1}{3\sqrt{n}})$ fails or $r_{max} = r_n$. This takes $O(n\sqrt{n} \log n)$ time, and two cases arise:

- If PLACEMENT$(4r_{max})$ succeeds, then $r_{max} \neq r_n$. In the case that $4r_{max} > \frac{r_{max}+1}{3\sqrt{n}}$, we have a 2-approximation due to Observation 2. In the case that $4r_{max} \leq \frac{r_{max}+1}{3\sqrt{n}}$, consider any value $\delta \in [4r_{max}, \frac{r_{max}+1}{3\sqrt{n}}]$. On the one hand, the balls $B_{max+1}, \ldots, B_n$ are not problematic because they have degree n in G_δ. On the other hand, the balls $B_1, \ldots, B_{max}$ have to be disjoint because $\delta^* \geq 4r_{max}$, and they determine the closest pair in PLACEMENT(δ).

In this case, placing the points $p_1, \ldots, p_{max}$ at the centers of their corresponding disks, computing the distance $\tilde{\delta}$ of their closest pair, and using PLACEMENT($\tilde{\delta}$) for the disks $B_{max+1}, \ldots, B_n$ provides a 2-approximation.

- If PLACEMENT($4r_{max}$) fails, then we know that for any $\delta \in [\frac{r_{max}}{3\sqrt{n}}, 4r_{max}]$ the disks B_j with $\frac{r_j}{3\sqrt{n}} \geq 4r_{max}$ have degree at least n in G_δ. We shrink them to have radius $12r_{max}\sqrt{n}$, and then they keep having degree at least n in G_δ, so they are not problematic for the matching. We also use $\mathcal{B}$ for the new instance (with shrunk disks), and we can assume that all the disks have radius $O(12r_{max}\sqrt{n}) = O(r_{max}\sqrt{n})$.

We group the disks $\mathcal{B}$ into clusters $\mathcal{B}_1, \ldots, \mathcal{B}_t$ as follows: a *cluster* is a connected component of the intersection graph of the disks $B(c_1, r_1+4r_{max}), \ldots,$ $B(c_n, r_n + 4r_{max})$. This implies that the distance between different clusters is at least $4r_{max}$, and that each cluster $\mathcal{B}_j$ can be enclosed in a disk of radius $O(r_{max}|\mathcal{B}_j|\sqrt{n})$.

For each subinstance $\mathcal{B}_j$, we can use Lemma 5, where $m = |\mathcal{B}_j|$ and $k = n$, and compute in $O(|\mathcal{B}_j|\sqrt{|\mathcal{B}_j|}\log^2(|\mathcal{B}_j|n))$ time a placement yielding a 2-approximation of $D(\mathcal{B}_j)$. Joining all the placements we get a 2-approximation of $D(\mathcal{B})$, and we have used $\sum_{j=1}^{t} O\left(|\mathcal{B}_j|\sqrt{|\mathcal{B}_j|}\log^2(|\mathcal{B}_j|n)\right) = O(n\sqrt{n}\log^2 n)$ time overall. $\qquad\square$

5   Analogous Results in L_2

The rest of the paper studies how the L_2 metric changes the approximation ratio and running time of the algorithms studied for the L_∞ metric. We just give the main observations, and refer to the full version for details.

5.1   Placement Algorithm in L_2

For the L_∞ metric, we used the optimal packing of disks that is provided by an orthogonal grid. For the Euclidean L_2 metric we will consider the regular hexagonal packing of disks, given by $\Lambda := \{(a + \frac{b}{2}, \frac{b\sqrt{3}}{2}) \mid a, b \in \mathbb{Z}\}$. For disks of radius $\delta/2$, the hexagonal packing is provided by placing the disks centered at $\delta\Lambda$. The *edges* of $\delta\Lambda$ are the segments connecting each pair of points in $\delta\Lambda$ at distance δ. They decompose the plane into equilateral triangles of side length δ, which are the *cells* of $\delta\Lambda$.

Consider a version of PLACEMENT using the new lattice $\delta\Lambda$ and modifying it slightly for the cases when B_i contains no lattice point:

- If B_i is contained in a cell C, place $p_i := c_i$ and add the vertices of C to Q; see Figure 3a.
- If B_i intersects some edges of $\delta\Lambda$, let E be the edge that is closest to c_i. Then place p_i at the projection of c_i onto E, and add the vertices of E to Q; see Figure 3b.

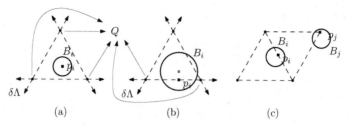

Fig. 3. Cases and properties of PLACEMENT for the L_2 metric. (a) Placement when B_i is fully contained in a cell. (b) Placement when B_i intersects an edge: we project the center c_i onto the closest edge. (c) A case showing that the closest pair in PLACEMENT(δ) may be at distance $\frac{\delta\sqrt{3}}{2}$

Observe that, in this case, the distance between a point placed on an edge and a point in $\delta\Lambda \setminus Q$ may be $\frac{\delta\sqrt{3}}{2}$; see Figure 3c. We modify accordingly the criteria of Definition 1 regarding when PLACEMENT succeeds, and then we state the result corresponding to Lemma 1.

Definition 2. *In the L_2 metric,* PLACEMENT(δ) *succeeds if the computed placement $p_1,\ldots,p_n$ satisfies $D(p_1,\ldots,p_n) \geq \frac{\delta\sqrt{3}}{2}$. Otherwise,* PLACEMENT($\delta$) *fails.*

Lemma 6. *If $\frac{4\delta}{\sqrt{3}} \leq \delta^*$, then* PLACEMENT($\delta$) *succeeds.*

Observe that now, if PLACEMENT(δ) succeeds, but PLACEMENT($\delta + \epsilon$) fails for an infinitesimal $\epsilon > 0$, then we are getting an approximation ratio of 8/3: we have $\delta \geq \frac{\delta^*\sqrt{3}}{4}$, and PLACEMENT($\delta$) gives a placement $p_1,\ldots,p_n$ that satisfies $D(p_1,\ldots,p_n) \geq \frac{\delta\sqrt{3}}{2} \geq \frac{3\delta^*}{8}$.

5.2   Approximation Algorithms in L_2

Lemma 3 also applies to the L_2 metric. However, the proof of Lemma 4 relies on some data structures that do not carry over to L_2. Instead we can show the following result, whose running time depends on whether the original disks are congruent or not.

Lemma 7. *The Algorithm* PLACEMENT *can be adapted to run in $O(n^{1.5+\epsilon})$ time. When all the disks are congruent, it can be adapted to run in $O(n\sqrt{n}\log n)$ time.*

The proof of Lemma 5 is not valid for the L_2 metric because it relies on the fact that disks in the L_∞ metric are squares. Instead, we can solve in quadratic time the type of instances that are considered in Lemma 5. This leads to the following result, where we spend roughly $O(\sqrt{n})$ times more time than in the L_∞ case (Theorem 1).

Theorem 2. *Let $\mathcal{B} = \{B_1,\ldots,B_n\}$ be a collection of disks in the plane with the L_2 metric. We can compute in $O(n^2)$ time a placement $p_1,\ldots,p_n$ with $p_i \in B_i$ that yields an $\frac{8}{3}$-approximation of $D(\mathcal{B})$.*

6   Congruent Disks in L_2

When the disks $B_1, \ldots, B_n$ are all congruent, say, of diameter one, we can improve the approximation ratio in Theorem 2. For general disks, the problematic cases are those balls that do not contain any lattice point. But when all the disks are of diameter one, we can rule out those cases.

Assume $1 \leq \delta^* \leq 2$ and take δ such that $\delta \leq \frac{-\sqrt{3}+\sqrt{3}\delta^*+\sqrt{3+2\delta^*-\delta^{*2}}}{4}$. When running PLACEMENT(δ) under this hypothesis, it is possible to show that $Q = \emptyset$ and that the graph G_δ has a matching. This requires some non-trivial geometric considerations; see the full version of the paper.

In this case, if $p_1, \ldots, p_n$ is the placement computed by PLACEMENT(δ), we have $D(p_1, \ldots, p_n) \geq \delta$ because $Q = \emptyset$ and so all the points $p_i \in \delta\Lambda$. Therefore, for $1 \leq \delta^* \leq 2$, we can get an approximation ratio of

$$\frac{\delta^*}{\delta} \geq \frac{4\delta^*}{-\sqrt{3} + \sqrt{3}\delta^* + \sqrt{3 + 2\delta^* - \delta^{*2}}}.$$

For any $\delta^* \leq 1$, also some geometric considerations imply that PLACEMENT gives a 2-approximation.

On the other hand, we have the trivial approximation algorithm CENTERS consisting of placing each point $p_i := c_i$, which gives a $\frac{\delta^*}{\delta^*-1}$-approximation when $\delta^* > 1$. In particular, CENTERS gives a 2-approximation when $\delta^* \geq 2$.

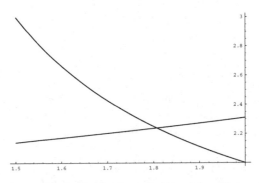

Fig. 4. Approximation ratios for both approximation algorithms as a function of the optimum δ^*

The idea is that the performances of PLACEMENT and CENTERS are reversed for different values δ^* in the interval $[1, 2]$. For example, when $\delta^* = 2$, the algorithm PLACEMENT gives a $\frac{4}{\sqrt{3}}$-approximation, while CENTERS gives a 2-approximation because the disks need to have disjoint interiors to achieve $\delta^* = 2$. But for $\delta^* = 1$, the performances are reversed: PLACEMENT gives a 2-approximation, while CENTERS does not give any constant factor approximation.

The approximation ratios of both algorithms are plotted in Figure 4. Applying both algorithms and taking the best of both solutions, we get an approximation ratio that is the minimum of both approximation ratios, which attains a maximum of

$$\alpha := 1 + \frac{13}{\sqrt{65 + 26\sqrt{3}}} \sim 2.2393.$$

Theorem 3. *Let $\mathcal{B} = \{B_1, \ldots, B_n\}$ be a collection of congruent disks in the plane with the L_2 metric. We can compute in $O(n^2)$ time a placement $p_1, \ldots, p_n$ with $p_i \in B_i$ that yields a ~ 2.2393-approximation of $D(\mathcal{B})$.*

Acknowledgements

The author wishes to thank Pankaj Agarwal, Refael Hassin, Marc van Kreveld, Mark Overmars, and Günter Rote for several fruitful comments that improved the content and shape of the paper. Thanks again to Marc for correcting the draft version and pointing out Lemma 3, which improved the previous running time. This research was initiated when Sergio was visiting DIMACS and Duke University, which provided good atmosphere for it.

References

1. R. Aharoni and P. Haxell. Hall's theorem for hypergraphs. *Journal of Graph Theory*, 35:83–88, 2000.
2. E. Arkin and R. Hassin. Minimum diameter covering problems. *Networks*, 36:147–155, 2000.
3. C. Baur and S. Fekete. Approximation of geometric dispersion problems. *Algorithmica*, 30:451–470, 2001. A preliminary version appeared in *APPROX'98*.
4. S. Cabello. Approximation algorithms for spreading points. Technical report UU-CS-2003-040, Available at http://www.cs.uu.nl/research/techreps/UU-CS-2003-040.html, 2003.
5. S. Cabello and M. van Kreveld. Approximation algorithms for aligning points. *Algorithmica*, 37:211–232, 2003. A preliminary version appeared in *SoCG'03*.
6. B. Chandra and M. M. Halldórsson. Approximation algorithms for dispersion problems. *J. Algorithms*, 38:438–465, 2001.
7. B. Dent. *Cartography: Thematic Map Design*. McGraw-Hill, 5th edition, 1999.
8. G. Di Battista, P. Eades, R. Tamassia, and I. G. Tollis. *Graph Drawing*. Prentice Hall, Upper Saddle River, NJ, 1999.
9. A. Efrat, A. Itai, and M. J. Katz. Geometry helps in bottleneck matching and related problems. *Algorithmica*, 31(1):1–28, 2001.
10. S. Fekete and H. Meijer. Maximum dispersion and geometric maximum weight cliques. To appear in *Algorithmica* 38(3), 2004.
11. J. Fiala, J. Kratochvíl, and A. Proskurowski. Geometric systems of disjoint representatives. In *Graph Drawing, 10th GD'02, Irvine, California*, number 2528 in Lecture Notes in Computer Science, pages 110–117. Springer Verlag, 2002.

12. J. Fiala, J. Kratochvíl, and A. Proskurowski. Systems of sets and their representatives. Technical Report 2002-573, KAM-DIMATIA, 2002. Available at http://dimatia.mff.cuni.cz/.

13. M. Formann and F. Wagner. A packing problem with applications to lettering of maps. In *Proc. 7th Annu. ACM Sympos. Comput. Geom.*, pages 281–288, 1991.

14. R. Hassin, S. Rubinstein, and A. Tamir. Approximation algorithms for maximum dispersion. *Operations Research Letters*, 21:133–137, 1997.

15. N. Megiddo. Applying parallel computation algorithms in the design of serial algorithms. *J. ACM*, 30(4):852–865, 1983.

16. C. W. Mortensen. Fully-dynamic two dimensional orthogonal range and line segment intersection reporting in logarithmic time. In *Proceedings of the fourteenth annual ACM-SIAM symposium on Discrete algorithms*, pages 618–627. Society for Industrial and Applied Mathematics, 2003.

17. B. Simons. A fast algorithm for single processor scheduling. In *Proc. 19th Annu. IEEE Sympos. Found. Comput. Sci.*, pages 246–252, 1978.

More Powerful and Simpler Cost-Sharing Methods*

(When Cross-Monotonicity Is the Wrong Way)

Paolo Penna and Carmine Ventre

Dipartimento di Informatica ed Applicazioni "R.M. Capocelli"
Università di Salerno, via **S. Allende** 2,
I-84081 Baronissi (SA), Italy
{penna, ventre}@dia.unisa.it

Abstract. We provide a new technique to derive *group strategyproof* mechanisms for the cost-sharing problem. Our technique is simpler and provably more powerful than the existing one based on so called *cross-monotonic* cost-sharing methods given by Moulin and Shenker [1997]. Indeed, our method yields the first *polynomial-time* mechanism for the Steiner tree game which is group strategyproof, *budget balance* and also meets other standard requirements (No Positive Transfer, Voluntary Participation and Consumer Sovereignty). A known result by Megiddo [1978] implies that this result cannot be achieved with cross-monotonic cost-sharing methods, even if using exponential-time mechanisms.

1 Introduction

Consider a service providing company P with a set of possible customers, also called *users*, U. For each subset $S \subseteq U$ of users, $C_P(S)$ denotes the cost incurred by the company P to *jointly* service the users in S. The function $C_P(\cdot)$ is usually termed the *cost function*. A typical scenario is that of company P broadcasting some kind of transmission (e.g., movies, sport events, news, etc) over a given network: in this case, $C_P(S)$ is the cost of implementing a multicast tree connecting a source node s to all users is S. Each user i valuates the transmission an amount v_i: this value quantifies how much user i likes the transmission (or how much he/she would pay for it). A key point is that v_i is a property of user i (and not of the network) and, thus, this value is known to i only. If user i is required to pay p_i for receiving the transmission, then her *utility* is equal to $v_i - p_i$. The utility is naturally what each user i tries to maximize. Users may act *selfishly* and, thus, a user i may misreport her valuation at some other number b_i. (Consider a simple mechanism which charges to every user i an amount equal to her *reported* valuation b_i).

---

* Work supported by the European Project IST-2001-33135, Critical Resource Sharing for Cooperation in Complex Systems (CRESCCO).

G. Persiano and R. Solis-Oba (Eds.): WAOA 2004, LNCS 3351, pp. 97–110, 2005.

A *cost-sharing mechanism* should decide which user S should receive the transmission and at which price. The mechanism is said *strategyproof* if, for each user i, revealing the true value v_i is a dominant strategy: that is, reporting any $b_i \neq v_i$ cannot improve the utility of i (see Sect. 2 for a formal definition). The mechanism is *group strategyproof* if this holds also for coalitions of users. The mechanism is *budget balance* if the total amount of money payed by the users equals the servicing cost $C_\mathsf{P}(S)$. Finally, a mechanism is *efficient* if it maximizes, over all subsets S, the sum of the valuations of users in S minus the cost $C_\mathsf{P}(S)$.

A fundamental result by Moulin and Shenker [8, 7] shows that the existence of a so called *cross-monotonic cost-sharing method* (see Sect. 2 for a formal definition) for $C_\mathsf{P}(\cdot)$ gives rise to a group strategyproof and budget balance mechanism. Moreover, if $C_\mathsf{P}(\cdot)$ is submodular (see the definition in Sect. 2.1), the converse holds as well. These results point all in the direction of cross-monotonic cost-sharing methods: on the one hand, no other technique is known to derive such mechanisms; on the other hand, the "converse" part of Moulin and Shenker's theorem says that, for submodular functions, these type of mechanisms capture all possible ones.

Unfortunately, meeting the cross-monotonicity requirement is often far from trivial: some (optimal) cost functions do not admit such methods [6]; others require a rather involved use of primal-dual algorithms [3, 9, 5, 1]; for others, only some sort of approximation of the budget balance condition is guaranteed (e.g., the mechanism may create some surplus or recover only a fraction of the cost) [9, 5, 1].

In this work we provide a more powerful method to derive such mechanisms by introducing the concept of *self cross-monotonic* cost-sharing method (see Sect. 3). Our main result is that, given any such cost-sharing method, it is possible to obtain group strategyproof mechanisms. The resulting technique extends the one by Moulin and Shenker and is provably more powerful: it indeed applies to some optimal cost functions for which the method by Moulin and Shenker cannot be used and/or gives simpler constructions of the mechanisms (see Sect.s 2.1 and 2.2 for a more detailed discussion of previous and our results).

2   Model

We are given a set U of n users. Depending on the problem instance at hand, for every $Q \subseteq U$, and for every feasible solution T_Q which allows to provide the service to users in Q (e.g., a multicast tree connecting a source node s to all users in Q), we denote by $\mathsf{COST}(T_Q)$ the cost of this solution.[1] Hence, for a service providing company P that decides to service Q by implementing solution T_Q, we have a cost $C_\mathsf{P}(Q) = \mathsf{COST}(T_Q)$.

---

[1] A formal definition should be $\mathsf{COST}(Q, G)$ since the cost depends on the instance. However, for the sake of clarity, we will omit 'G' whenever this will be clear from the context.

Each user is a *selfish agent* reporting some (not necessarily true) valuation b_i; the true value v_i is *privately known* to agent i. Based on the *reported* values $b = (b_1, b_2, \ldots, b_n)$ a *mechanism* $M = (A, P)$ uses algorithm A to compute the following:

- A subset $Q(b) \subseteq U$ of users to be serviced;
- A feasible solution $T_{Q(b)}$ to be implemented in order to provide the service to the set $Q(b)$; solution $T_{Q(b)}$ does *not* provide the service to any $j \notin Q(b)$.

For the sake of convenience, one can imagine that an algorithm $\mathsf{A}(\cdot)$ is used by M in order to compute $T_{Q(b)}$ once a set $Q(b)$ has been selected, that is, $T_{Q(b)} = \mathsf{A}(Q(b))$. (For instance, $\mathsf{A}(\cdot)$ may be a multicast algorithm computing a tree connecting a source node s to a subset Q of the nodes of a network.) In this case, we let $C_{\mathsf{A}}(Q(b)) := \mathsf{COST}(\mathsf{A}(Q(b)))$.

In addition, for every user $i \in Q(b)$, the mechanism computes the cost $P^i(b)$ that user i must pay for getting the service, with $P = (P^1, P^2, \ldots, P^n)$. Hence, the *utility* of agent i when she reports b_i, and the other agents report $b_{-i} := (b_1, \ldots, b_{-i}, b_{i+1}, \ldots, b_n)$, is equal to

$$u_i(b_i, b_{-i}) := v_i \cdot \sigma_i(Q(b_i, b_{-i})) - P^i(b_i, b_{-i}),$$

where $(x, b_{-i}) = (b_1, \ldots, b_{i-1}, x, b_{i+1}, \ldots, b_n)$ and $\sigma_i(X)$ equals 1 if $i \in X$, and 0 otherwise. In the sequel, for every $C \subseteq U$ and any two vectors x and y of length n, (x_C, y_{-C}) denotes the vector $z = (z_1, \ldots, z_n)$ such that $z_i = x_i$ if $i \in C$ and $z_i = y_i$ if $i \notin C$.

There is a number of natural constraints/goals that, for every problem instance G, a mechanism $M = (A, P)$ should satisfy/meet:

1. **Cost Optimality (CO).** Let $C_{\mathsf{opt}}(Q)$ denote the minimum cost required to service all users in Q, for every $Q \subseteq U$. We require that the computed solution $T_{Q(b)}$ is optimal w.r.t. the set $Q(b)$, that is, $C_{\mathsf{A}}(Q(b)) = C_{\mathsf{opt}}(Q(b))$.
2. **No Positive Transfer (NPT).** No user receives money from the mechanism, i.e., $P^i(\cdot) \geq 0$.
3. **Voluntary Participation (VP).** We never charge an user an amount of money grater than her *reported* valuation, that is, $\forall b_i, \forall b_{-i} \ \ b_i \geq P^i(b_i, b_{-i})$. In particular, a user has always the option to not paying for a service for which she is not interested. Morever, $P^i(b) = 0$, for all $i \notin Q(b)$, i.e., only the users getting the service will pay.
4. **Consumer Sovereignty (CS).** Every user is guaranteed to get the service if she reports a high enough valuation.
5. **Budget Balance (BB).**
   (a) **Cost recovery.** $\sum_{i \in Q(b)} P^i(b) \geq C_{\mathsf{A}}(Q(b))$, i.e., the cost of the computed solution is recovered from all the users being serviced;
   (b) **Competitiveness.** $\sum_{i \in Q(b)} P^i(b) \leq C_{\mathsf{A}}(Q(b))$, i.e., no surplus is created. If some surplus were created, then a competitor may offer the same service at a better price.

6. **Group Strategyproofness.** We require that a user $i \in U$ that misreport her valuation (i.e., $b_i \neq v_i$) cannot improve her utility (strategyproofness or truthfulness) nor improve the utility of other users without worsening her own utility (otherwise, a coalition C containing i would secede). Consider a coalition $C \subseteq U$ of users. Let $b_i = v_j$ for all $j \notin C$. The group strategyproofness requires that if the inequality

$$v_i \cdot \sigma_i(Q(b_C, v_{-C})) - P^i(b_C, v_{-C}) \geq v_i \cdot \sigma_i(Q(v_C, v_{-C})) - P^i(v_C, v_{-C}) \quad (1)$$

holds for all $i \in C$ then it must hold with equality for all $i \in C$ as well.

A *cost-sharing method* is a function $\xi(\cdot)$ which distributes the cost $C_A(\cdot)$ to the users that get the service. Intuitively speaking, we will use the function $\xi(\cdot)$ in order to define the payments $P^i(\cdot)$. More formally, $\xi(\cdot)$ takes two arguments: a set of users Q and a user i and returns a *nonnegative* real number satisfying the following:

$$\text{if } i \notin Q \text{ then } \xi(Q, i) = 0 \text{ and} \quad (2)$$

$$\sum_{i \in Q} \xi(Q, i) = C_A(Q). \quad (3)$$

Observe that, if we take $P^i(b) := \xi(Q(b), i)$, then the payments recover exactly the cost $C_A(Q(b))$ from all and only users in $Q(b)$. Also the NPT condition holds. The other requirements depend on how the mechanism selects $Q(b)$ and $T_{Q(b)}$.

In the context of *multicast routing*, we are given a weighted undirected graph $G = (U \cup \{s\}, E, c)$, where $s \notin U$ is the source node and c_e is the cost of using link $e \in E$. A feasible solution is a pair $T_Q = (Q, T)$, where T is a tree connecting s to a subset Q of users contained in T. The corresponding cost is the weight of T, i.e., $\sum_{e \in T} c_e$. The optimal cost function $C_{opt}(Q)$ to service Q is the cost of an optimal Steiner tree of G connecting s to Q, thus possibly containing some Steiner nodes in $U \setminus Q$. This is the *Steiner tree game* and we let $\sigma_i(T_Q) = 1$ if and only if $i \in Q$. In the *minimum spanning tree game* the feasible solution is any spanning tree T_Q containing s and the set Q only.

Approximation Concepts. The use of optimal cost functions $C_{opt}(\cdot)$ for the given problem may suffer from the following drawbacks: (1) there may not exists a cross-monotonic cost-sharing method, and (2) computing a solution having that cost may be NP-hard. Therefore, one considers the effects of using approximation algorithms on the CO and the BB conditions.

Let $M = (A, P)$ be a mechanism whose cost function is $C_A(\cdot)$. Mechanism M is α-*approximate BB* if it is cost recovery for $C_A(\cdot)$ and α-*competitive*, that is, $\sum_{i \in U} P^i(b) \leq \alpha \cdot C_{opt}(Q(b))$. A β-*surplus* mechanism M satisfies $\sum_{i \in U} P^i(b) \leq (1 + \beta) \cdot C_A(Q(b))$. A ρ-*recovery* mechanism guarantees that $\sum_{i \in U} P^i(b) \geq \rho \cdot C_A(Q(b))$, for some $\rho \leq 1$. Clearly, if A is an α-approximation algorithm and the mechanism is 0-surplus, then it is also α-approximate BB. The converse does not always hold as an α-approximate BB mechanism may not be 0-surplus. A β-*cost-sharing method* $\xi(\cdot)$ satisfies Eq. 2 and the following relaxation of Eq. 3: $C_A(Q) \leq \sum_{i \in Q} \xi(Q, i) \leq \beta \cdot C_A(Q)$.

2.1   Previous Work

A fundamental result by Moulin and Shenker states that, if a cost-sharing *cross-monotonic* method for $C_A(\cdot)$ exists, then it is possible to define a group strategyproof mechanism (see Theorem 3): a cost-sharing method $\xi(\cdot)$ is cross-monotonic if, for every $Q' \subset Q \subseteq U$, $\xi(Q', i) \geq \xi(Q, i)$, for all $i \in Q'$. The converse of their result also holds whenever $C_A(\cdot)$ is *submodular* [8,7], that is, $C_A(\emptyset) = 0$ and, for any two subsets of users Q_1 and Q_2, it holds that

$$C_A(Q_1) + C_A(Q_2) \geq C_A(Q_1 \cup Q_2) + C_A(Q_1 \cap Q_2).$$

The *Shapley value* for multicast routing [11] and the *egalitarian* method due to Dutta and Ray [2] are just two examples of cost-sharing methods which, for functions that are nondecreasing[2] and submodular, are cross monotonic.

The existence of a cross-monotonic method can be related to the *core* concept (see e.g. [3] for a definition): if the core of $C_A(\cdot)$ is empty, then no cross-monotonic cost-sharing method $\xi(\cdot)$ for this cost function exists.

Megiddo proved that the optimal cost function $C_{opt}(\cdot)$ for the Euclidean Steiner tree game has an empty core [6]. Kent and Skorin-Kapov provided the first cross-monotonic cost-sharing method for the minimum spanning tree game [4]. A more general approach has been given by Jain and Vazirani that use primal-dual methods in order to obtain a family of *polynomial-time computable* cross-monotonic methods [3]. These results yield, for the case of metric graphs, a 2-approximate BB, 0-surplus, group strategyproof mechanism for the Steiner tree and for the TSP games. The mechanism also meets NPT, VP and CS.

Biló *et al* [1] considered the muticast routing game in *wireless* networks. They proved that the resulting optimal cost function has an empty core even for d-dimensional Euclidean instances, for $d \geq 2$. Moreover, upon the results for the MST game, they build a $2(3^d - 1)$-approximate BB, group strategyproof mechanism which also meets NPT, VP and CS. This mechanism, however, is not 0-surplus.

2.2   Our Contribution

In this work, we first show how to get around the difficulties of dealing with cross-monotonic cost-sharing methods by providing a new technique for obtaining group strategyproof cost-sharing mechanisms. In particular, we prove that the a weaker property (which we call *self cross-monotonicity*– see Def. 2) suffices (Theorem 2). We prove the following results showing that our method is simpler and more powerful than the one by Moulin and Shenker [8,7]:

- Self cross-monotonic methods for $C_A(\cdot)$ can be trivially obtained whenever the algorithm A is *reasonable* (see Def. 3).
  The resulting mechanism M_A satisfies the NPT, VP, CS, cost recovery, is 0-surplus and, if A is an (polynomial-time) α-approximation algorithm, then M_A is (polynomial-time) α-approximate BB (Theorem 2).

---

[2] A function $C_A(\cdot)$ is nondecreasing if, for every $Q \subset Q' \subseteq U$, $C_A(Q) \leq C_A(Q')$.

- Our method gives the first *polynomial-time* mechanism for the Steiner tree game which is group strategyproof, meets NPT, VP, CS and, more importantly, is BB (Corollary 1). Notice that the latter property implies that we are able to build a multicast tree which is *optimal* for the chosen receivers $Q(b)$, that is, $C_A(Q(b)) = C_{opt}(Q(b))$.

Besides the improvement over the 2-approximate BB mechanism in [3], the fact that our mechanism is BB is somewhat surprising: indeed, the result of Megiddo [6] implies that our result cannot be achieved using cross-monotonic methods; moreover, the NP-hardness of the underlying problem (i.e., given $Q \subset U$, find a minimum cost Steiner tree) seems to require α-approximate BB if we aim at polynomial-time mechanisms (see e.g. [3]). This intuition is wrong! Clearly, our result does *not* imply P = NP since our mechanism is "driven" through a family of sets $Q_0, Q_1, \ldots, Q_n$ for which an optimal Steiner tree does *not* use any Steiner node (thus solvable in polynomial-time). We accomplish this by relating the sets Q_j's to the execution of Prim's MST algorithm (Theorem 4).

These results already prove that focusing (only) on cross-monotonic methods may be the "wrong" thing to do. We continue along this line and consider the wireless multicast game [1], another problem for which our method is provably better. We indeed obtain the following results on it:

- A polynomial-time mechanism which is $(3^d - 1)$-approximate BB, 0-surplus, group strategyproof, and meets NPT, VP, and CS (Theorem 5). This improves over the $2(3^d - 1)$-approximate BB mechanism in [1] which is not 0-surplus.
- A wide class of mechanisms for this game cannot be 0-surplus. This class includes the mechanism by Bilò *et al* and, for certain "bad" instances, the surplus increases exponentially in d (for $d = 1$ and $d = 2$ it cannot be smaller than 1 and 5, respectively).

Mechanism in this class are those which use a multicast algorithm A for which an A-bad instance G exists (see Def. 4). These algorithms are not optimal (Theorem 8) and the cost function $C_A(\cdot)$ is *not* submodular (Theorem 9). Hence, the "inverse" of the result by Moulin and Shenker [8] does not apply to such functions $C_A(\cdot)$. Therefore, it is possible to have BB mechanisms which do *not* use cross-monotonic cost-sharing methods. Finally, we observe that there is no equivalence between bad algorithms A and the non submodularity of $C_A(\cdot)$: indeed, there exists an instance G for which $C_{MST}(\cdot)$ is not submodular, while G is not MST-bad (Theorem 10).

Paper Organization. We briefly recall the result by Moulin and Shenker in Sect. 3, and provide our extension in Sect.s 3.1-3.2; We apply our result to the Steiner tree game in Sect. 3.3; Wireless muticast is considered in Sect. 4.

Due to lack of space some proofs are omitted. The interested reader can find them in [10].

3   A New Method for Cost Sharing

Moulin and Shenker [8, 7] provide an elegant solution by considering the following scheme for obtaining mechanisms:

Mechanism $M(\xi)$

1. Q is initialized to U;
2. If there exists a user i in Q with $v_i < \xi(Q, i)$ then drop i from Q. Keep repeating this step, in arbitrary order, until for every user i in Q, $v_i \geq \xi(Q, i)$;
3. Set $P^i(b) := \xi(Q, i)$, for all $i \in U$.

A sharing method $\xi(\cdot)$ is *cross-monotonic* if, for every two sets Q and Q', with $Q' \subset Q \subseteq U$, it holds that $\xi(Q, i) \leq \xi(Q', i)$, for every $i \in Q'$.

The fundamental result by Moulin and Shenker reduces the problem of designing a mechanism to the problem of finding a *cross-monotonic* sharing method $\xi(\cdot)$ for a cost function $C_A(\cdot)$. The resulting mechanism $M_A(\xi)$ uses the scheme $M(\xi)$ to compute the set $Q = Q(b)$ and the payments $P^i(b) = \xi(Q(b), i)$, and then simply builds a feasible solution $T_{Q(b)} = A(Q(b))$. Then the following holds:

Theorem 1. *[8, 7]* [3] *For any optimal (respectively, α-approximation) algorithm* A *and any cross-monotonic cost-sharing method* $\xi(\cdot)$ *for* $C_A(\cdot)$, *the mechanism* $M_A(\xi)$ *is group strategyproof, BB (respectively, α-approximate BB), 0-surplus and satisfies NPT, VP and CS.*

3.1   Extending Moulin and Shenker Approach

We will show that the cross-monotonicity property can be relaxed so to hold only for certain sets that mechanism $M(\xi)$ can actually output.

Definition 1. *Given any function* $\xi : 2^U \times U \to \mathbf{R}^+ \cup \{0\}$, *we define* $\mathcal{Q}_0^\xi := U$, *and* $\mathcal{Q}_j^\xi := \{Q \setminus \{i\} | \ Q \in \mathcal{Q}_{j-1}^\xi \wedge \xi(Q, i) > 0\}$. *Moreover,* $\mathcal{Q}^\xi := \cup_{j \geq 0} \mathcal{Q}_j^\xi$.

A key point is that mechanisms $M_A(\xi)$ can generate only those subsets of receivers in $\mathcal{Q}^\xi$:

Lemma 1. *At each round of* $M(\xi)$, *the set* Q *considered in Step 2 satisfies* $Q \in \mathcal{Q}^\xi$.

Definition 2. *A function* $\xi : 2^U \times U \to \mathbf{R}^+ \cup \{0\}$ *is self cross-monotonic if, for every* $Q, Q' \in \mathcal{Q}^\xi$ *with* $Q' \subset Q$, *it holds that* $\xi(Q', i) \geq \xi(Q, i)$, *for every* $i \in Q'$.

We next prove the main result of this section. Its proof is similar to the one given in [3].

---

[3] The result presented here is sightly more general then the one in [8, 7]; indeed, as first observed in [3], their result can also deal with α-approximate BB mechanism.

Theorem 2. *For any optimal (respectively, α-approximation) algorithm* A *and any self cross-monotonic β-cost-sharing method $\xi(\cdot)$ for $C_A(\cdot)$, the mechanism $M_A(\xi)$ is group strategyproof, β-approximate BB (respectively, $\alpha\beta$-approximate BB), $(\beta - 1)$-surplus and satisfies NPT, VP and CS. Moreover, $M_A(\xi)$ runs in polynomial time if* A *and $\xi(\cdot)$ are polynomial time.*

Proof. Condition CS follows from the fact that a user i is dropped in Step 2 only if $b_i < \xi(Q, i)$. The NPT and VP conditions thus follow from the properties of $\xi(\cdot)$.

We next prove the group strategyproofness. Consider a coalition $C \subseteq U$ such that

$$j \notin C \Rightarrow b_j = v_j, \tag{4}$$

$$i \in C, b_i \neq v_i \Rightarrow v_i \cdot \sigma_i(Q^{false}) - P^i(b_C, v_{-C}) \geq$$
$$v_i \cdot \sigma_i(Q^{true}) - P^i(v_C, v_{-C}), \tag{5}$$

where Q^{false} and Q^{true} denote the sets of receivers returned by $M_A(\xi)$ on input (b_C, v_{-C}) and (v_C, v_{-C}), respectively. We have to show that the above inequality cannot hold with '>'. Observe that, if $i \notin Q^{false}$, then the NPT and the CS conditions imply that Eq. 5 holds with '='. We thus assume $i \in Q^{false}$ and we consider the following two cases:

$Q^{false} \subseteq Q^{true}$. From Lemma 1, $Q^{false} \in \mathcal{Q}^\xi$ and $Q^{true} \in \mathcal{Q}^\xi$. Since $i \in C$, by self cross-monotonicity and by the definition of $P^i(\cdot)$ in $M(\xi)$,

$$P^i(b_C, v_{-C}) = \xi(Q^{false}, i) \geq \xi(Q^{true}, i) = P^i(v_C, v_{-C}). \tag{6}$$

Since $Q^{false} \subseteq Q^{true}$, $\sigma_i(Q^{false}) \leq \sigma_i(Q^{true})$. By contradiction, if Eq. 5 holds with '>', then we would obtain

$$v_i \cdot \sigma_i(Q^{true}) - P^i(b_C, v_{-C}) > v_i \cdot \sigma_i(Q^{true}) - P^i(v_C, v_{-C}),$$

which contradicts Eq. 6.

$Q^{false} \not\subseteq Q^{true}$. We will show that this case cannot arise. Let $s_1, \ldots, s_k$ be the sequence of users that $M_A(\xi)$ drops on input (v_C, v_{-C}), i.e., $Q^{true} = U \setminus \{s_1, \ldots, s_k\}$. Let s_j be the first user in $s_1, \ldots, s_k$ such that $s_j \in Q^{false}$. Therefore $b_{s_j} \geq \xi(Q^{false}, s_j)$. By definition of $s_1, \ldots, s_{j-1}$, $Q^{false} \subseteq Q_{j-1} := U \setminus \{s_1, \ldots, s_{j-1}\}$. By Lemma 1 and by the self cross-monotonicity of $\xi(\cdot)$, we have $\xi(Q^{false}, s_j) \geq \xi(Q_{j-1}, s_j)$. Since s_j is dropped in Q^{true}, the definition of $M_A(\xi)$ implies that $\xi(Q_{j-1}, s_j) > v_{s_j}$. Putting things together we obtain

$$b_{s_j} \geq \xi(Q^{false}, s_j) \geq \xi(Q_{j-1}, s_j) > v_{s_j}. \tag{7}$$

If $s_j \notin C$, $b_{s_j} = v_{s_j}$, thus contradicting the above inequalities. Otherwise, when $s_j \in C$, Eq. 5 yields $v_{s_j} - P^{s_j}(b_C, v_{-C}) \geq 0$, thus implying $v_{s_j} \geq \xi(Q^{false}, s_j)$, which contradicts Eq. 7.

Finally, since, for every $Q \subseteq U$, $C_A(Q) \leq \sum_{i \in Q} \xi(Q, i) \leq \beta C_A(Q) \leq \alpha\beta \cdot C_{\mathsf{opt}}(Q)$, where α is the approximation ratio of A, $M_A(\xi)$ is $\alpha\beta$-approximate and $(\beta - 1)$-surplus.

3.2 Reasonable Algorithms Is All We Need

In the remaining of this work, given an instance G and a feasible solution T_Q for it, the corresponding set of users that are serviced is denoted to as $\mathsf{Serv}(T_Q, G)$.

Definition 3. *An algorithm* A *is* reasonable *if, for every instance* G, *there exists a sequence* $i_1, i_2, \ldots, i_n$ *of users such that, denoted by* $Q_j := U \setminus \{i_1, i_2, \ldots, i_j\}$, *for* $1 \leq j \leq n$, *it holds that* $\mathsf{Serv}(\mathsf{A}(Q_j), G) = Q_j$, *i.e., algorithm* A *is able to compute a solution which serves all and only the users in* Q_j, *for* $0 \leq j \leq n$. *(We let* $Q_0 := U$.*)*

Theorem 3. *If* A *is reasonable then there exists a self cross-monotonic cost-sharing method* $\xi(\cdot)$ *for* $C_{\mathsf{A}}(\cdot)$.

Proof. Let Q_j be the set defined as in Def. 3. To ensure self cross-monotonicity, we define

$$\xi(Q_j, i) = \begin{cases} C_{\mathsf{A}}(Q_j) & \text{if } i = j+1, \\ 0 & \text{otherwise.} \end{cases} \tag{8}$$

We first show that $\mathcal{Q}_j^\xi = Q_j$. Indeed, at round j of $M_{\mathsf{A}}(\xi)$, the only user which can be dropped is $j+1$, for $0 \leq j \leq n$. Consider $Q, Q' \in \mathcal{Q}^\xi$ with $Q \subset Q'$. Then it must be the case $Q = Q_a$ and $Q' = Q_b$, for some $a > b$. Let $i \in Q$, with $\xi(Q, i) > 0$ (otherwise the theorem holds). Then $i = i_a$, thus implying $\xi(Q_b, i) = 0 = \xi(Q', i) < \xi(Q, i)$. Finally, $\xi(\cdot)$ can be easily extended outside $\mathcal{Q}^\xi$ so to enforce Eq.s 2-3 for every $Q \subseteq U$.

3.3 Steiner Tree Game

Consider a graph $G = (U \cup \{s\}, E, c)$ where the set of terminals coincides with the set of users U. Consider the execution of Prim's algorithm on graph G, starting from node $a_0 := s$. Let a_j be the j-th node added it the j-th iteration: a_j is the closest, among all nodes in $U \setminus \{a_1, \ldots, a_{j-1}\}$, to the connected component built so far, i.e., $S_{j-1} := \{s\} \cup \{a_1, \ldots, a_{j-1}\}$. Let T_j be the tree containing S_j. Then, for every $j \geq 0$, $\mathsf{COST}(T_j) = \mathsf{COST}(\mathsf{MST}(S_j))$.

We next strengthen this result and prove that $\mathsf{COST}(T_j)$ is also the optimal cost for the Steiner tree of S_j:

Theorem 4. *For every* $j \geq 0$, *let* $ST^*(S_j)$ *be an optimal Steiner tree in* G *with terminal set* S_j *and possibly using Steiner points in* $U \setminus S_j$. *Then, it holds that* $\mathsf{COST}(ST^*(S_j)) = \mathsf{COST}(T_j)$.

Proof. The proof is by induction on $r := n - j$, i.e., $S_j = S_{n-r}$ and $0 \leq r \leq n$.

Base step ($r = 0$). For $S_n = U$ there are no Steiner points, thus implying that $ST^*(S_n)$ must be a MST of G.

Inductive step (from $r = n - j - 1$ **to** $r + 1 = n - j$**).** Let $j + 1 = n - r$ and let (a_k, a_{j+1}) be the edge added at step $j + 1$ to connect a_{j+1} to S_j. By contradiction, assume $\mathsf{COST}(T_j) \neq \mathsf{COST}(ST^*(S_j))$. Since $ST^*(S_j)$ is optimal for S_j, it must hold that $\mathsf{COST}(T_j) > \mathsf{COST}(ST^*(S_j))$. If a_{j+1} is not a node of $ST^*(S_j)$, then we let $T'(S_{j+1}) := ST^*(S_j) \cup (a_k, a_{j+1})$; otherwise, we let $T'(S_{j+1}) := ST^*(S_j)$. Since $a_k \in S_j$, then $T'(S_{j+1})$ is a tree spanning S_{j+1}. By definition, $T_{j+1} = T_j \cup (a_k, a_{j+1})$, thus implying

$$\mathsf{COST}(T'(S_{j+1})) \leq \mathsf{COST}(ST^*(S_j)) + c_{(a_k, a_{j+1})}$$
$$< \mathsf{COST}(T_j) + c_{(a_k, a_{j+1})} = \mathsf{COST}(T_{j+1}).$$

By the inductive hypothesis $\mathsf{COST}(ST^*(S_{j+1})) = \mathsf{COST}(T_{j+1})$, and the above inequality contradicts the optimality of $ST^*(S_{j+1})$.

This completes the proof.

Theorem 4 implies that MST is reasonable and optimal for all sets $Q_j := S_{n-j}$, $0 \leq j \leq n$. Theorems 2 and 3 thus yield the following:

Corollary 1. *The Steiner tree game admits a mechanism $M_{\mathsf{MST}}(\xi)$ running in polynomial time which is group strategyproof, budget balance and satisfies NPT, VP and CS.*

4   Wireless Multicast and Limits of Cross-Monotonic Methods

Wireless multicast game. In *wireless* multicast routing, a feasible solution is a *directed tree* T containing a path from s to all of its nodes (i.e., T must be rooted at s and directed downwards). The cost of T is the total energy consumption required to implement all of its edges, which is equal to $\mathsf{COST}(T) := \sum_{i \in U} \max_{(i,j) \in T} c_{(i,j)}$. In the d-dimensional Euclidean version, $c_{(i,j)} = d(i,j)^\gamma$, for some $\gamma > 1$ and $d(i,j)$ being the Euclidean distance between i and j, and the instance G is a complete graph with nodes U. We assume $\gamma \geq d$ as in [1]. Fig. 1 shows a 2-dimensional Euclidean instance G:[4] the cost of the tree $T = \{(s, x_1), (s, x_2), (x_1, q_1), (x_2, q_2)\}$ is equal to $\epsilon + 2$. Interestingly, $T = \mathsf{MST}(G)$, which is not the optimal one:[5] the tree T^* connecting s directly to every other node has cost $(1 + \epsilon)^\gamma$, which is better for sufficiently small ϵ. Observe that, $\mathsf{COST}(T) = C_{\mathsf{MST}}(U) < \sum_{(i,j) \in T} c_{(i,j)}$.

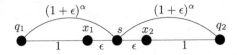

Fig. 1. The "bad" graph B_2

---

[4] For the sake of readability we do not draw all edges of the complete weighted graph G.
[5] For this problem, algorithm MST builds a MST of G and then orients it downwards s.

Theorem 5. *There exists a polynomial-time mechanism for the wireless multicast game which, for d-dimensional Euclidean networks, is group strategyproof, $(3^d - 1)$-approximate BB, 0-surplus and meets NPT, VP and CS.*

We next argue that graph B_2 in Fig. 1 constitutes an example of a "bad" graph for the MST algorithm in that, under certain hypothesis, it forces certain mechanisms $M_{MST}(\xi)$ (the one by Biló *et al* [1] being one of them) to generate some surplus.

The two main ideas can be summarized as follows: (i) the two users $\{q_1, q_2\}$ must always pay at least the marginal cost $C_{MST}(U) - C_{MST}(U \setminus \{q_1, q_2\}) = (\epsilon + 2) - \epsilon = 2$; (ii) the MST algorithm, on input $U \setminus \{x_1, x_2\} = \{q_1, q_2\}$ yields a solution of cost $(1+\epsilon)^\gamma$ which is less than the above mentioned payment provided by $\{q_1, q_2\}$. Hence, some surplus is created if $Q(b) = \{q_1, q_2\}$.

Instead of proving the result for the graph B_2, we first generalized the above example to a wide class of graphs for which it is possible to prove that certain algorithms must create some surplus. Towards this end we first introduce some notation.

Notation. For any tree T, let $c(i, T) := \max_{(i,j) \in T} c_{(i,j)}$. Also let $\mathsf{pay}(T, i)$ be true if and only if $i = \arg\max\{l| \ (j, l) \in T \wedge c_{(j,l)} = c(j, T)\}$. Given an algorithm A, we let $A(Q)$ denote the tree returned by A on input the set of receivers Q.[6] For every $Q \subseteq U$, we define the following two quantities:

$$C_A(Q, i) := \begin{cases} c_{(j,i)} & \text{if } (j, i) \in A(Q) \wedge \mathsf{pay}(A(Q), i), \\ 0 & \text{otherwise.} \end{cases} \tag{9}$$

$$\forall X \subseteq U, \ C_A(Q, X) := \sum_{i \in X} C_A(Q, i). \tag{10}$$

In particular, $C_A(Q) = C_A(Q, Q)$.

For every $i \in Q$, let Q_i^A be the subset of nodes that are reachable through i in $A(Q)$ (i.e., those nodes that have i as an ancestor in $A(Q)$). Let $A_i(Q)$ be set of edges connecting i to Q_i^A in $A(Q)$ (i.e., the edges in the subtree of $A(Q)$ rooted at i). Notice that $A_i(Q)$ does not contain i.

Definition 4. *A communication graph $G = (U \cup \{s\}, E, w)$ is A-bad if there exist $Q \subseteq U$, $X \subset Q$ and $Y \subset Q$ such that the following hold:*

$$A(Q \setminus Q_X) = A(Q) \setminus \bigcup_{i \in X} A_i(Q) \tag{11}$$

$$C_A(Q \setminus Y) < C_A(Q, Q_X) \tag{12}$$

with $Y \cap Q_X = \emptyset$.

---

[6] We assume the algorithm A to return a tree connecting the source s to all and only the nodes in Q.

Theorem 6. *If G is* A-*bad, then there is no cross-monotonic cost-sharing method* $\xi(\cdot)$ *for* $C_A(\cdot)$.

The mechanism by Biló *et al* [1] employs the cross-monotonic methods $\xi_F(\cdot)$ for the MST game by Jain and Vazirani [3]: given a family $F = \{f_1, \ldots, f_n\}$ of functions $f_i : \mathbf{R}^+ \to \mathbf{R}^+$, the function $\xi_F(\cdot)$ is a β-cost-sharing method for the wireless multicast cost function yielded by algorithm MST.

In the sequel, we will show that this kind of approach must always create some surplus. Intuitively, their mechanism $M_{MST}(\xi_F)$ can potentially output every subset $Q \subseteq U$, which requires the method $\xi_F(\cdot)$ to be cross-monotonic. Theorem 6 thus implies that $\beta > 1$.

Definition 5. *A function* $\xi : 2^U \times U \to \mathbf{R}^+ \cup \{0\}$ *is* Y-*critical if, for all* $j \in Y$, $\xi(U, j) > 0$, *where* $Y \subseteq U$.

Theorem 7. *Let* $G = (U \cup \{s\}, E, c)$ *be a* A-*bad graph. If* $\xi(\cdot)$ *is a cross-monotonic* β-*cost-sharing method for* $C_A(\cdot)$ *which is* Y-*critical, where* Y *is the set in Def. 4, then the mechanism* $M_A(\xi)$ *is not* 0-*surplus (on the instance* G).

The above result can be applied to a family of graphs B_k generalizing graph B_2 in Fig. 1:

Definition 6. *For every integer* $k \geq 2$, *the graph* $B_k = (U_k \cup s, E_k, c)$ *is defined as follows:* $U_k := \{q_l\}_{1 \leq l \leq k} \cup \{x_l\}_{1 \leq l \leq k}$, $E_k := \{(s, i) | i \in U_k\} \cup \{(q_l, x_l)\}_{1 \leq l \leq k}$. *Moreover,* $c_{(s,x_l)} = \epsilon$, $c_{(x_l, q_l)} = 1$ *and* $c_{(s,q_l)} = (1 + \epsilon)^\gamma$.

For B_k graphs, we can strengthen Theorem 7 and provide a lower bound on the surplus that all mechanisms using a U-critical function $\xi(\cdot)$ must generate. It is easy to verify that, for every F, $\xi_F(\cdot)$ is U-critical for *every* weighted graph G with non-zero edge weights. We thus obtain the following result on the mechanism $M_{MST}(\xi_F)$ proposed by Biló *et al* [1]:

Corollary 2. *Let* $\xi(\cdot)$ *be cross-monotonic and* U-*critical. Then, for every graph* B_k, *mechanism* $M_{MST}(\xi)$ *cannot be* β-*surplus, for any* $\beta < k - 1$. *Moreover, for d-dimensional Euclidean instances,* $M_{MST}(\xi_F)$ *cannot be less than* $(\tau_d - 1)$-*surplus, with* $\tau_1 = 2$, $\tau_2 = 6$, *and* τ_d *increasing exponentially in* d. *These results apply to* $M_{MST}(\xi_F)$, *for every* F.

The next result states that no A-bad graph exists if A is an optimal algorithm.

Theorem 8. *If* $G = (U \cup \{s\}, E, c)$ *is* A-*bad, then* A *is not optimal.*

Proof. Let $Q \subseteq U$ and $X, Y \subseteq Q$ be the sets as in Def. 4. Then Eq.s 11-12 imply respectively

$$C_A(Q) = C_A(Q \setminus Q_X) + C_A(Q, Q_X) > C_A(Q \setminus Q_X) + C_A(Q \setminus Y). \tag{13}$$

Since $Y \cap Q_X = \emptyset$, $\{Q \setminus Q_X\} \cup \{Q \setminus Y\} = Q$. Hence, the tree $T := A(Q \setminus Q_X) \cup A(Q \setminus Y)$ reaches all nodes in Q and its cost satisfies

$$\mathsf{COST}(T) \leq C_A(Q \setminus Q_X) + C_A(Q \setminus Y) < C_A(Q),$$

thus implying that A was not optimal on input Q.

One could try to prove that no BB mechanism employing algorithm A exists by showing that (i) there exists an A-bad instance and (ii) the function $C_A(\cdot)$ is submodular. Unfortunately, this never happens:

Theorem 9. *If $G = (U \cup \{s\}, E, c)$ is A-bad, then C_A is not submodular.*

Notice that, the above theorem also implies that, if A-bad instances exist, it is still possible to have BB mechanisms which are not based on cross-monotonic cost-sharing functions for $C_A(\cdot)$. In order to prove Theorem 9, we first need the following two intermediate results.

Lemma 2. *For every A-bad graph it holds that*

$$C_A(Q \setminus Q_X) = C_A(Q) - C_A(Q, Q_X),$$

where X is the same as in Def. 4.

Proof. Eq. 11 implies that $A(Q \setminus Q_X) = A(Q) \setminus \{(i,j)| \ (i,j) \in A(Q) \wedge j \in Q_X\}$. Hence, since $A(Q)$ is a tree, we have

$$C_A(Q \setminus Q_X) = \sum_{i \in Q} C_A(Q, i) - \sum_{i \in Q_X} C_A(Q, i) = C_A(Q) - C_A(Q, Q_X).$$

Lemma 3. *If $C_A(\cdot)$ is submodular, then for any $Q', Q, A \subseteq U$, with $Q' \subset Q$ and $A \cap Q' = \emptyset$, it holds that*

$$C_A(Q' \cup A) - C_A(Q') \geq C_A(Q \cup A) - C_A(Q). \tag{14}$$

Proof. Since $C_A(\cdot)$ is submodular, then for any $Q', Q, A \subseteq U$, with $Q' \subset Q$, and any $a \notin Q'$, it holds that

$$C_A(Q) - C_A(Q') \geq C_A(Q \cup \{a\}) - C_A(Q' \cup \{a\}). \tag{15}$$

By contradiction, assume that there exists $A = \{a_1, \ldots .a_k\}$, with $A \cap Q' = \emptyset$ such that $C_A(Q) - C_A(Q') < C_A(Q \cup A) - C_A(Q' \cup A)$. By repeatedly applying Eq. 15, with $a = a_1, a = a_2, \ldots, a = a_k$, we obtain

$$
\begin{aligned}
C_A(Q \cup A) - C_A(Q' \cup A) &> C_A(Q) - C_A(Q') \\
&\geq C_A(Q \cup \{a_1\}) - C_A(Q' \cup \{a_1\}) \\
&\geq C_A(Q \cup \{a_1, a_2\}) - C_A(Q' \cup \{a_1, a_2\}) \\
&\vdots \\
&\geq C_A(Q \cup A) - C_A(Q' \cup A),
\end{aligned}
$$

thus a contradiction.

We are now in a position to prove Theorem 9.

Proof of Theorem 9. From Def. 4 and Lemma 2 there exist $Q \subseteq U$ and $X, Y \subseteq Q$ such that

$$C_A(Q, Q_X) = C_A(Q) - C_A(Q \setminus Q_X) > C_A(Q \setminus Y). \qquad (16)$$

By contradiction, assume that $C_A(\cdot)$ is submodular. The fact that $C_A(\cdot) \geq 0$, Lemma 3 (with $A = Q_X$) and Eq. 16 imply the following inequalities, respectively:

$$C_A(Q \setminus Y) \geq C_A(Q \setminus Y) - C_A(Q \setminus \{Y \cup Q_X\}) \geq C_A(Q) - C_A(Q \setminus Q_X) > C_A(Q \setminus Y).$$

The above contradiction implies that $C_A(\cdot)$ is not submodular.

The following result states that the converse of the above theorem does not hold. Hence, there is no equivalence between "non submodularity" and "badness" of cost functions.

Theorem 10. *There exists a two-dimensional Euclidean instance $G = (U \cup \{s\}, E, c)$ for which G is not MST-bad and $C_{\mathsf{MST}}(\cdot)$, restricted to G, is not submodular.*

Acknowledgements. We wish to thank the authors of [1] for providing us with a copy of their work.

References

1. V. Biló, C. Di Francescomarino, M. Flammini, and G. Melideo. Sharing the cost of muticast transmissions in wireless networks. In *Proceedings of SPAA*, pages 180–187, 2004.
2. B. Dutta and D. Ray. A concept of egalitarism under participation constraints. *Econometrica*, 57:615–635, 1989.
3. K. Jain and V.V. Vazirani. Applications of approximation algorithms to cooperative games. In *Annual ACM Symposium on Theory of Computing (STOC)*, pages 364–372, 2001.
4. K. Kent and D. Skorin-Kapov. Population monotonic cost allocation on MST's. In *Operational Research Proceedings KOI*, volume 43-48, 1996.
5. S. Leonardi and G. Schäfer. Cross-monotonic cost-sharing methods for connected facility location games. Technical Report MPI-I-2003-1-017, Max-Plankt-Institut für Informatik, September 2003.
6. N. Megiddo. Cost allocation for steiner trees. *Networks*, 8:1–6, 1978.
7. H. Moulin. Incremental cost sharing: characterization by coalition strategy-proofness. *Social Choice and Welfare*, 16:279–320, 1999.
8. H. Moulin and S. Shenker. Strategyproof sharing of submodular costs: budget balance versus efficiency. 1997. http://www.aciri.org/shenker/cost.ps.
9. M. Pál and É. Tardos. Group strategyproof mechanisms via primal-dual algorithms. In *Proceedings of FOCS*, 2003.
10. P. Penna and C. Ventre. More powerful and simpler cost-sharing methods. TR of the European Project CRESCCO, http://www.ceid.upatras.gr/crescco/, 2004.
11. L.S. Shapley. A value of *n*-person games. Princeton University Press, 1953. H. Kuhn and W. Tucker editors.

Approximation Schemes for Deal Splitting and Covering Integer Programs with Multiplicity Constraints

Hadas Shachnai[1,*], Oded Shmueli[1], and Robert Sayegh[2]

[1] Department of Computer Science, The Technion, Haifa 32000, Israel
{hadas, oshmu}@cs.technion.ac.il.
[2] 22 Anilevitch St. Haifa 35025, Israel
abufayez@hotmail.com.

Abstract. We consider the problem of splitting an order for R goods, $R \geq 1$, among a set of sellers, each having *bounded* amounts of the goods, so as to minimize the total cost of the deal. In *deal splitting with packages (DSP)*, the sellers offer packages containing combinations of the goods; in *deal splitting with price tables (DST)*, the buyer can generate such combinations using price tables. Our problems, which often occur in online reverse auctions, generalize *covering integer programs with multiplicity constraints (CIP)*, where we must fill up an R-dimensional bin by selecting (with bounded number of repetitions) from a set of R-dimensional items, such that the overall cost is minimized. Thus, both DSP and DST are NP-hard, already for a single good, and hard to approximate for arbitrary number of goods.

In this paper we focus on finding efficient approximations, and exact solutions, for DSP and DST instances where the number of goods is some fixed constant. In particular, we show that when R is fixed both DSP and DST can be optimally solved in pseudo-polynomial time, and develop *polynomial time approximation schemes (PTAS)* for several subclasses of instances of practical interest. Our results include a PTAS for CIP in fixed dimension, and a more efficient (combinatorial) scheme for CIP_∞, where the multiplicity constraints are omitted. Our approximation scheme for CIP_∞ is based on a non-trivial application of the fast scheme for the *fractional covering problem*, proposed recently by Fleischer [Fl-04].

1 Introduction

An increasing number of companies are using online *reverse auctions* for their sourcing activities. In reverse auctions, multiple sellers bid for a contract from a buyer for selling goods and/or services. We consider the *deal splitting* problems

---

* Part of this work was done while the author was on leave at Bell Laboratories, Lucent Technologies, 600 Mountain Ave., Murray Hill, NJ 07974.

G. Persiano and R. Solis-Oba (Eds.): WAOA 2004, LNCS 3351, pp. 111–125, 2005.

arising in these reverse auctions. Suppose that a buyer needs to order multiple units from a set of R goods. The number of units required from the j-th good, $1 \leq j \leq R$, is $n_j \geq 1$. The goods can be obtained from m sellers, $S_1, \ldots, S_m$. Each seller offers certain amount from each good (or some combination of the goods); the maximum number of units of the j-th good available from S_i is T_{ij}, $1 \leq j \leq R$, $1 \leq i \leq m$. In any deal, we may split the order for the goods among a subset of the sellers. We say that a deal is *feasible* if (*i*) the number of units obtained from the j-th good is at least n_j, $1 \leq j \leq R$, and (*ii*) the amount of the j-th good obtained from S_i does not exceed T_{ij}, its supply from that good, $1 \leq i \leq m$, $1 \leq j \leq R$. The goal is to find a feasible deal of minimum total cost. Deal splitting naturally models a procurement auction to obtain raw materials with flexible sized lots and many other services. We consider two variants of the problem.

In *deal splitting with packages (DSP)*, each of the sellers, S_i, offers a set of N_i packages. The ℓ-th package, p_ℓ^i, $1 \leq \ell \leq N_i$, has a non-negative cost $c(p_\ell^i)$ and is given by the R-tuple $(n_{\ell 1}^i, \ldots, n_{\ell R}^i)$; that is, S_i offers in this package $0 \leq n_{\ell j}^i \leq n_j$ units from the j-th good, $1 \leq j \leq R$. We need to find a feasible deal that minimizes the total cost.

In *deal splitting with price tables (DST)*, each seller S_i, has m_i price ranges. The minimal and maximal numbers of units of the j-th good available from S_i in the ℓ-th price range are $r_{\ell j}$ and $u_{\ell j}$, respectively. The unit cost for the j-th good in the ℓ-th range is $c_{\ell j}$, $1 \leq \ell \leq m_i$, $1 \leq j \leq R$.[1] Thus, the ℓ-th entry in the price table of S_i is given by the vector $\{(r_{\ell 1}, u_{\ell 1}, c_{\ell 1}), \ldots, (r_{\ell R}, u_{\ell R}, c_{\ell R})\}$. We need to find a feasible deal in which the sale of S_i, $1 \leq i \leq m$, corresponds to a valid entry in its price table, and the total cost is minimized.

We note that DSP is NP-hard already for $R = 1$, by reduction from Partition, and hard to approximate within factor $\ln R$ for arbitrary $R > 1$, as it includes as a special case the *multi-set multi-cover* problem.[2] For DST, we note that each price range of a seller "encodes" a possibly large number of packages (each formed by choosing the number of units from each good), as well as a simple rule for computing the price of a particular package (via the unit costs). Thus, in the special case where each price table consists of a *single* price range, which allows to form a *single* combination of the goods, we get an instance of the *constrained multi-set multi-cover*. It follows that DST is also hard to approximate within factor $\ln R$.

Note that DSP generalizes also *covering integer program with multiplicity constraints (CIP)*. In this core problem, we must fill up an R-dimensional bin by selecting (with bounded number of repetitions) from a set of R-dimensional items, such that the overall cost is minimized. Formally, let $A = \{a_{ji}\}$ denote the sizes of the items in the R dimensions, $1 \leq j \leq R$, $1 \leq i \leq n$; the cost of item i is $c_i \geq 0$. Let x_i denote the number of copies selected from item i, $1 \leq i \leq n$.

---

[1] See an example in the Appendix.

[2] We elaborate in [S$^+$04] on the relation of our problems to *set cover* and its generalizations.

We seek an n-vector $\mathbf{x}$ of non-negative integers, which minimizes $c^T\mathbf{x}$, subject to the R constraints given by $A\mathbf{x} \geq b$, where $b_j \geq 0$ is the size of the bin in dimension j. In addition, we have multiplicity constraints for the vector $\mathbf{x}$, given by $\mathbf{x} \leq d$, where $d \in \{1, 2, \ldots\}^n$. Recall that, in DSP, each seller S_i has T_{ij} units from the j-th good. Consider, for example, the case where $R = 2$, and suppose that S_i has $T_{i1} = 10$ units from the first good and $T_{i2} = 20$ units from the second good. S_i offers two possible packages: $p_1^i = (5, 7)$ and $p_2^i = (6, 2)$; then if we obtain two copies of p_1^i, no copies of p_2^i are available. This dependence among the packages makes DSP a generalization of CIP.[3] Indeed, an instance of CIP can be formulated as a special case of DSP, where each seller offers a *single* package, whose "multiplicity" reflects the precise amount that is available from each of the goods.

1.1   Our Results

Since our deal splitting problems are harder than set cover, the best approximation ratio that we can expect for arbitrary R is $O(\log R)$ (see, e.g., in [Va-01]); thus, we focus on finding efficient approximations, and exact solutions, for subclasses of instances in which R is a fixed constant. We summarize below our main results.

Deal Splitting with Packages: We show (in Section 2.1) that when R is fixed DSP can be solved in pseudo-polynomial time. In Section 2.2 we develop a PTAS for instances where the i-th seller offers a set of $N_i \geq 1$ packages, $p_1^i, \ldots, p_{N_i}^i$, and the buyer can obtain at most r_ℓ^i copies from p_ℓ^i, for some $r_\ell^i \geq 1$; the total amount of the j-th good available from S_i is $T_{ij} = \sum_{\ell=1}^{N_i} n_{\ell j}^i r_\ell^i$, $1 \leq j \leq R$, $1 \leq i \leq m$. Indeed, such instances can be formulated as CIP with $\sum_{i=1}^m N_i$ variables. Thus, we get a PTAS for CIP in fixed dimension. In Section 2.3 we consider DSP instances with unbounded supply. Such instances model deals in which the buyer's need is much smaller than the supply from each of the goods. For these instances we develop a faster (combinatorial) scheme. This gives a combinatorial approximation scheme for CIP_∞.

Deal Splitting with Price Tables: We show (in Section 3) that when R is fixed DST is solvable in pseudo-polynomial time. We then develop a PTAS for DST instances in which the price tables satisfy some natural properties such as *volume discount*, that is widely used in reverse auctions (see, e.g., in [KPS-03], [BK+02]).[4]

Techniques: Our PTAS for unbounded DSP (in Section 2.3) is based on a nontrivial application of a *fully polynomial time approximation scheme (FPTAS)* for the *fractional covering problem*, proposed recently by Fleischer [Fl-04]. We use this combinatorial scheme to obtain an *approximate* fractional solution for a lin-

---

[3] In the corresponding integer program, we get dependencies among the variables that give the number of copies obtained from each package.

[4] We elaborate on these properties in Section 3.

ear programming formulation of our problem, building on a technique of Chandra et al. [CHW-76]. We show that by rounding an approximate solution for the LP we increase the cost of the optimal (integral) solution for the DSP instance only by factor of ε. Thus, we get a fast combinatorial implementation for our LP-based scheme. The overall running time of the scheme is $O(N^{\lceil R/\varepsilon \rceil} \cdot \frac{1}{\varepsilon^2} \log C)$, where $N = \sum_{i=1}^{m} N_i$ is the total number of distinct packages offered by the sellers, and $C = \max_{1 \leq i \leq N} c_i$ is the maximal cost of any package. Since unbounded DSP is equivalent to CIP_∞, this yields a combinatorial approximation scheme for CIP_∞ in fixed dimension. With slight modification, we get the first combinatorial scheme for *multi-dimensional multiple choice knapsack*.

In our PTAS for DST (in Section 3), we combine the guessing technique of Chekuri and Khanna [CK-00] with a novel application of the technique of Frieze and Clarke [FC-84], to the minimum binary multiple choice knapsack problem in fixed dimension. Indeed, due to the constraints imposed on the solution for DST — the amounts chosen from the goods for each seller must correspond to a valid entry in its price table — we cannot apply the rounding technique of [FC-84] to the fractional solution obtained by our scheme; instead, we apply non-standard rounding, which relies heavily on the mathematical properties of the price tables.

1.2   Related Work

Procurement Auctions.: Our deal splitting problems belong to the class of *winner determination* problems in reverse auctions. Generally, in reverse auction we have a single buyer that needs to obtain multiple goods, and a set of sellers offers *bids* for selling the goods. Bidding may follow various mechanisms (a survey of common mechanisms is given in [W-96]). The DST problem with single good (i.e., $R = 1$) and price tables that satisfy the volume discount property[5] was studied in [KPS-03]. The paper shows that DST is NP-hard already in this case and presents an FPTAS for the problem. There has been some previous work on deal splitting with *multiple* goods, however, these papers present either experimental studies or software that implements a given mechanism (see, e.g., [BK+02]). Heuristic methods and preliminary analytic results related to deal splitting are given in [SG+02].

Multiple Choice Knapsack (MCK).: As shown in Section 2.2, DSP can be reduced to the *minimum R-dimensional binary MCK (R-MMCK)* problem. The *maximum* variant of this problem was studied since the mid-1970's (see, e.g., [Lu-75], [IH+78], [I-80]). For a single dimension, the best known result is a PTAS by Chandra et al. [CHW-76]. Most of the published work on the maximum multi-dimensional binary MCK presented heuristic solutions (see a survey in [AM+97]). Recently, Shachnai and Tamir developed in [ST-03] a PTAS for the problem in fixed dimension. Our scheme in Section 2.2 includes a PTAS for the minimum R-dimensional binary MCK in fixed dimension.

---

[5] See in Section 3.

In Section 2.3, we reduce unbounded DSP to the minimum (non-binary) R-dimensional MCK. Chandra et al. [CHW-76] gave a PTAS for the *maximum* version of this problem in fixed dimension; their scheme solves as a procedure a linear program. Our scheme yields the first combinatorial scheme for this problem.

Set Cover/Covering Integer Programs.: As mentioned above, our problems include as a special case the multi-set multi-cover problem. Set cover and its generalizations have been extensively studied. (A comprehensive survey is given in [Va-01].) Feige showed that in general set cover is hard to approximate within factor $\ln |E|$, where E is the set of elements to be covered. This hardness result carries over to multi-set multi-cover. The best approximation ratio for set cover is $(1 + \ln |E|)$ [C-79]. For multi-set multi-cover, the best ratio is $O(\log \max_S |S|)$, where $|S|$ is the size of the multi-set S when counting elements with multiplicity [RV-98]. This yields an $O(\log n)$-approximation algorithm for general instances of DSP with unbounded supply, where $n = \sum_{j=1}^{R} n_j$.

Covering integer programs form a large subclass of integer programs encompassing such NP-hard problems as minimum knapsack and set cover. This implies the hardness of CIP in fixed dimension (i.e., where R is a fixed constant). For general instances, the hardness of approximation results for set cover carry over to CIP. Dobson [D-82] gave an algorithm that outputs a solution of cost $O(\max_{1 \le i \le n} \{\log(\sum_{j=1}^{m} A_{ij})\})$ times the integral optimum. It was unknown until recently whether an $O(\log R)$-approximation existed. Kolliopoulos and Young [KY-01] settled this question. Their $O(\log R)$-approximation yields the first constant approximation for CIP in fixed dimension. A comprehensive survey of other results is given in [K-03] (see also in [KY-01]). The best known bounds for the CIP_∞ problem (that include existential improvements on the $O(\log R)$ factor) are due to Srinivasan ([S-99] and [S-96]). In this paper, we give the first pseudo-polynomial time algorithms and approximation schemes for CIP and CIP_∞ in fixed dimension.

Due to space limitations, we omit some of the proofs. Detailed proofs can be found in [S+04].

2 Deal Splitting with Packages

2.1 Exact Algorithms

When R is fixed DSP is solvable in pseudo-polynomial time. In particular,

Theorem 1. *DSP can be solved optimally in $O(m \cdot \max_{1 \le i \le m} N_i \cdot \prod_{j=1}^{R} n_j^3)$ steps, where n_j is the number of units required from the j-th good.*

This yields a pseudo-polynomial time algorithm for CIP in fixed dimension.

Corollary 1. *CIP in fixed dimension, R, and n variables can be solved optimally in $O(n \cdot \max_{i,j} (a_{ij} d_i)^{2R})$.*

Consider a restricted version of DSP, in which we require that the total number of packages used in the deal is bounded by some fixed constant, $k \ge 1$. It can be shown that the problem then becomes easy to solve.

Theorem 2. *Restricted DSP is solvable in polynomial time.*

2.2   DSP with Bounded Multiplicity

Approximation Scheme: Suppose that the packages offered by each of the sellers have bounded multiplicity. Specifically, there are r_ℓ^i copies available from p_ℓ^i, $1 \le \ell \le N_i$. In this case, if $p_\ell^i = (n_{\ell 1}^i, \ldots, n_{\ell R}^i)$, $1 \le \ell \le N_i$, then the number of units of the j-th good available from S_i is $T_{ij} = \sum_{\ell=1}^{N_i} n_{\ell j}^i r_\ell^i$, for $1 \le j \le R$, $1 \le i \le m$. We now develop a PTAS for these instances, assuming that R is fixed.

Reduction to the R-MMCK Problem: Assume that we know the optimal cost, C, for our instance, then we reduce our problem to the minimum R-dimensional binary multiple choice knapsack problem. Recall that for some $R \ge 1$, an instance of binary R-MMCK consists of a single R-dimensional knapsack, of size b_j in the j-th dimension, and m sets of items. Each item has an R-dimensional size and is associated with a cost. The goal is to pack a subset of items, by selecting at most one item from each set, such that the total size of the packed items in dimension j is at least b_j, $1 \le j \le R$, and the overall cost is minimized.

For given values of C and ε, we define an instance for R-MMCK, such that if there is an optimal solution for DSP with cost C, we can find a solution for the DSP instance, whose cost is at most $C(1 + \varepsilon)$. Note that C can be 'guessed' in polynomial time within factor $(1 + \varepsilon)$, using binary search over the range $(0, \sum_{i=1}^m \sum_{\ell=1}^{N_i} r_\ell^i c(p_\ell^i))$.

Formally, given the value of C, the parameter ε and a DSP instance with bounded multiplicity, we construct an R-MMCK instance in which the knapsack capacities in the R dimensions are $b_j = n_j$, $1 \le j \le R$. Also, we have $N = \sum_{i=1}^m N_i$ sets of items, denoted by A_ℓ^i, $1 \le i \le m$, $1 \le \ell \le N_i$. Let $\hat{K}_\ell^i$ be the value satisfying $r_\ell^i c(p_\ell^i) \in [\hat{K}_\ell^i \varepsilon C/N, (\hat{K}_\ell^i + 1)\varepsilon C/N)$, then the number of items in A_ℓ^i is $K_\ell^i = \min(\hat{K}_\ell^i, \lfloor N/\varepsilon \rfloor)$. The set A_ℓ^i represents a sale of the package p_ℓ^i which partially fulfills the order. In particular, the k-th item in A_ℓ^i, denoted (i, ℓ, k), represents a sale of at most r_ℓ^i copies of p_ℓ^i such that $c(i, \ell, k)$, the total cost incurred by these copies, is in $[k\varepsilon C/N, (k+1)\varepsilon C/N)$. This total cost is rounded down to the nearest integral multiple of $\varepsilon C/N$; thus, $c(i, \ell, k) = k\varepsilon C/N$. The size of the item (i, ℓ, k) in dimension j, $1 \le j \le R$, denoted by $s_j(i, \ell, k)$, is the total number of units of the j-th good that we can obtain, such that the total (rounded down) cost is $c(i, \ell, k)$.

Approximating the Optimal Solution for R-MMCK: Given an instance of R-MMCK, we 'guess' the set S of items of maximal costs in the optimal solution, where $|S| = h = \min(m, \lfloor \frac{2R(1-\varepsilon)}{\varepsilon} \rfloor)$. We choose the value of h such that the resulting solution is guaranteed to be within $1 + \varepsilon$ from the optimal, as computed below. Let $E(S)$ be the subset of items with costs that are larger than the minimal cost of any item in S, that is, $E(S) = \{(i, \ell, k) \notin S \mid c(i, \ell, k) > c_{min}(S)\}$, where $c_{min}(S) = \min_{(i,\ell,k) \in S} c(i, \ell, k)$. We select all the items $(i, \ell, k) \in$

S, and eliminate from the instance all the items $(i, \ell, k) \in E(S)$ and the sets A_ℓ^i from which an item has been selected. In the next step we find an optimal *basic solution* for the following linear program, in which $x_{i,\ell,k}$ is an indicator variable for the selection of the item $(i, \ell, k) \notin S \cup E(S)$.

$$(LP(S)) \quad \text{minimize} \quad \sum_{i=1}^{m} \sum_{\ell=1}^{N_i} \sum_{k=1}^{K_\ell^i} x_{i,\ell,k} \cdot c(i, \ell, k)$$

$$\text{subject to}: \quad \sum_{k=1}^{K_\ell^i} x_{i,\ell,k} \le 1 \text{ for } i = 1, \ldots, m, \quad \ell = 1, \ldots, N_i$$

$$\sum_{i=1}^{m} \sum_{\ell=1}^{N_i} \sum_{k=1}^{K_\ell^i} s_j(i, \ell, k) x_{i,\ell,k} \ge n_j \text{ for } j = 1, \ldots, R$$

$$0 \le x_{i,\ell,k} \le 1 \text{ for } (i, \ell, k) \notin S \cup E(S)$$

Rounding the Fractional Solution: Given an optimal fractional solution for R-MMCK, we get an integral solution as follows. For any i, $1 \le i \le m$ and ℓ, $1 \le \ell \le N_i$ let $k_{max} = k_{max}(\ell, i)$ be the maximal value of $1 \le k \le K_\ell^i$ such that $x_{i,\ell,k} > 0$; then we set $x_{i,\ell,k_{max}} = 1$ and, for any other item in A_ℓ^i, $x_{i,\ell,k} = 0$. Finally, we return to the DSP instance and take the maximum number of copies of the package p_ℓ^i whose total (rounded down) cost is $c(i, \ell, k_{max})$.

Analysis of the Scheme: We use the next three lemmas to show that the scheme yields a $(1+\varepsilon)$-approximation to the optimum cost, and that the resulting integral solution is feasible.

Lemma 1. *If there exists an optimal (integral) solution for DSP with cost C, then the integral solution obtained from the rounding for R-MMCK has the cost $\hat{z} \le (1 + \varepsilon)C$.*

Lemma 2. *The scheme yields a feasible solution for the DSP instance.*

Lemma 3. *The cost of the integral solution for the DSP instance is at most $\hat{z} + \varepsilon C$.*

Combining the above lemmas we get:

Theorem 3. *There is a polynomial time approximation scheme for DSP instances with fixed number of goods and bounded multiplicity.*

Consider an instance of CIP in fixed dimension, R. We want to minimize $\sum_{i=1}^{n} c_i x_i$ subject to the constraints $\sum_{i=1}^{n} a_{ij} x_i \ge b_j$ for $j = 1, \ldots, R$, and $x_i \in \{0, 1, \ldots d_i\}$ for $i = 1 \ldots, n$. We can represent such a program as an instance of DSP with $m = n$ sellers, each offering a single package i of multiplicity d_i. The number of units required from the j-th good is $n_j = b_j$.

Corollary 2. *The above is a PTAS for CIP in fixed dimension.*

2.3 Unbounded DSP

Consider now the special case where the sellers have unbounded supply from each of the goods. As before, we formulate our problem as a linear program, however, instead of applying standard techniques to solve this program, we use a fast combinatorial approximation scheme of [Fl-04] to get a fractional solution that is within factor of $(1 + \varepsilon)$ from the optimal; then, we round the solution to obtain an integral solution that is close to the optimal.

Overview of the Scheme. Our scheme, called *multi-dimensional cover with parameter ε (MDC_ε)*, proceeds in the following steps.

(i) For a given $\varepsilon \in (0, 1)$, let $\delta = \lceil R \cdot ((1/\varepsilon) - 1) \rceil$.

(ii) Let c_i denote the cost of package i. Recall that $N = \sum_{i=1}^{m} N_i$ is the total number of packages. We number the packages by $1, \ldots, N$, such that $c_1 \geq c_2 \geq \cdots \geq c_N$.

(iii) Denote by Ω the set of integer vectors $\mathbf{x} = (x_1, \ldots, x_N)$ satisfying $x_i \geq 0$ and $\sum_{i=1}^{N} x_i \leq \delta$. For any vector $\mathbf{x} \in \Omega$:

    – Let $d \geq 1$ be the maximal integer i for which $x_i \neq 0$. Find a $(1 + \varepsilon)$-approximation to the optimal (fractional) solution of the following linear program.

$$(LP') \quad \text{minimize} \quad \sum_{i=d+1}^{N} c_i z_i$$

$$\text{subject to} : \quad \sum_{i=d+1}^{N} a_{ij} z_i \geq n_j - \sum_{i=1}^{N} a_{ij} x_i \text{ for } j = 1, \ldots, R \quad (1)$$

$$z_i \geq 0, \text{ for } i = d+1, \ldots, N$$

The constraints (1) reflect the fact that we need to obtain from each of the goods at least $n_j - \sum_{i=1}^{N} a_{ij} x_i$, units, once we obtained the vector $\mathbf{x}$.

(iv) Let $\hat{z}_i$, $d + 1 \leq i \leq N$ be a $(1 + \varepsilon)$ -approximate solution for LP'. We take $\lceil \hat{z}_i \rceil$ as the integral solution. Denote by $C_{MDC}(\mathbf{x}) = \sum_{i=d+1}^{N} c_i \lceil \hat{z}_i \rceil$ the value obtained from the rounded solution, and let $c(\mathbf{x}) = \sum_{i=1}^{N} c_i x_i$.

(v) Select the vector $\mathbf{x}$ for which $C_{MDC_\varepsilon}(\mathbf{x}) = \min_{\mathbf{x}}(c(\mathbf{x}) + C_{MDC}(\mathbf{x}))$.

Analysis. We now show that MDC_ε is a PTAS for DSP with unbounded supply. Let C_o be the optimal cost for DSP (in which we take an *integral* number of units from each package).

Theorem 4. *(i) If $C_o \neq 0, \infty$ then $C_{MDC_\varepsilon}/C_o < 1 + \varepsilon$. (ii) The running time of algorithm MDC_ε is $O(N^{\lceil R/\varepsilon \rceil} \cdot \frac{1}{\varepsilon^2} \log C)$, where $C = \max_{1 \leq i \leq N} c_i$ is the maximal cost of any package, and its space complexity is $O(N)$.*

We use in the proof the next lemma.

Lemma 4. *For any $\varepsilon > 0$, a $(1 + \varepsilon)$-approximation to the optimal solution for LP' can be found in $O(1/\varepsilon^2 R \log(C \cdot R))$ steps.*

Proof. For a system of inequalities as given in LP', there is a solution in which at most R variables get non-zero values. This follows from the fact that the number of non-trivial constraints is R. Hence, it suffices to solve LP' for the $\binom{N-d}{R}$ possible subsets of R variables, out of $(z_{d+1}, \ldots, z_N)$. This can be done in polynomial time since R is fixed. Now, for each subset of R variables we have an instance of the *fractional covering problem*, for which we can find a $(1 + \varepsilon)$-approximate solution using, e.g., the fast scheme of Fleischer [Fl-04]. ∎

Proof of Theorem 4: For showing (i), assume that the optimal (integral) solution for the DSP instance is obtained by the vector $\mathbf{y} = (y_1, \ldots, y_N)$. If $\sum_{i=1}^N y_i \leq \delta$ then $C_{MDC_\varepsilon} = C_o$, since in this case $\mathbf{y}$ is a valid solution, and $\mathbf{y} \in \Omega$, therefore, in some iteration MDC_ε will examine $\mathbf{y}$. Suppose that $\sum_{i=1}^N y_i > \delta$, then we define the vector $\mathbf{x} = (y_1, \ldots, y_{d-1}, x_d, 0, \ldots, 0)$, such that $y_1 + \cdots + y_{d-1} + x_d = \delta$. (Note that $x_d \neq 0$.) Let $\tilde{C}_o(\mathbf{x}) = \sum_{i=d+1}^N c_i \hat{z}_i$ be the approximate fractional solution for LP'. We have that $\mathbf{x} \in \Omega$, therefore

$$C_{MDC}(\mathbf{x}) - \tilde{C}_o(\mathbf{x}) \leq R c_d, \tag{2}$$

Let $C_o(\mathbf{x})$ be the optimal fractional solution for LP' with the vector $\mathbf{x}$. Note that C_o, the optimal (integral) solution for DSP, satisfies $C_o > c(\mathbf{x}) + C_o(\mathbf{x})$, since $C_o(\mathbf{x})$ is a lower bound for the cost incurred by the integral values $y_{d+1}, \ldots, y_N$. In addition, $c(\mathbf{x}) + C_{MDC}(\mathbf{x}) \geq C_{MDC_\varepsilon}$. Hence, we get that

$$\frac{C_o}{C_{MDC_\varepsilon}} \geq \frac{c(\mathbf{x}) + C_o(\mathbf{x})}{c(\mathbf{x}) + C_{MDC}(\mathbf{x})} > \frac{c(\mathbf{x}) + \tilde{C}_o(\mathbf{x})(1 - \varepsilon)}{c(\mathbf{x}) + C_{MDC}(\mathbf{x})}$$

$$> (1 - \varepsilon)(1 - \frac{C_{MDC}(\mathbf{x}) - \tilde{C}_o(\mathbf{x})}{c(\mathbf{x}) + C_{MDC}(\mathbf{x}) - \tilde{C}_o(\mathbf{x})})$$

$$\geq (1 - \varepsilon)(1 - \frac{C_{MDC}(\mathbf{x}) - \tilde{C}_o(\mathbf{x})}{\delta c_d + C_{MDC}(\mathbf{x}) - \tilde{C}_o(\mathbf{x})})$$

The second inequality follows from the fact that $C_o(\mathbf{x}) \geq \tilde{C}_o(\mathbf{x})(1 - \varepsilon)$, and the last inequality follows from the fact that $c(\mathbf{x}) \geq \delta c_d$.

Let $f(w) = w/(a + w)$, then $f(w)$ is monotone increasing. Define $w = C_{MDC}(\mathbf{x}) - \tilde{C}_o(\mathbf{x})$, and $a = \delta c_d$; then, using (2), we get that $1 - w/(a + w) \geq 1 - R c_d/(\delta c_d + R c_d) \geq 1 - \varepsilon$. Thus, we get that $C_o/C_{MDC_\varepsilon} \geq (1 - \varepsilon)^2$. By taking in the scheme $\tilde{\varepsilon} = \varepsilon/2$ we get the statement of the theorem.

Next, we show (ii). Note that $|\Omega| = O(N^\delta)$, since the number of possible choices of N non-negative integers, whose sum is at most δ is bounded by $\binom{N+\delta}{\delta}$. Now, given a vector $\mathbf{x} \in \Omega$, we can compute $C_{MDC}(\mathbf{x})$ in $O(N^R)$ steps, since at most R variables out of $z_{d+1}, \ldots, z_N$ can have non-zero values. Multiplying by the complexity of the FPTAS for fractional covering, as given in Lemma 4, we get the statement of the theorem. ∎

Recall that DSP with unbounded supply is equivalent to CIP_∞.

Corollary 3. *There is a PTAS for CIP_∞ with n variables and fixed dimension, R, whose running time is $O(n^{R/\varepsilon} \cdot \frac{1}{\varepsilon^2} \log C)$.*

3   Deal Splitting with Price Tables

When R is fixed, DST can be solved in pseudo-polynomial time. In particular,

Theorem 5. *The DST problem can be optimally solved in $O(\sum_i m_i \cdot \prod_{j=1}^{R} n_j^2)$.*

3.1   A PTAS for DST

We now describe a PTAS for DS with price tables and fixed number of goods. Our scheme applies to any instance of DST satisfying the following properties. **(P1)** *Volume discount.* If we increase the quantity bought from each of the goods, the unit cost can only decrease; that is, let $(a_1^1, \ldots, a_R^1)$, $(a_1^2, \ldots, a_R^2)$ be two vectors representing feasible sales for S_i, for some $1 \le i \le m$. If $a_j^2 \ge a_j^1$ for all $1 \le j \le R$, then the unit costs corresponding to the two vectors satisfy $c_j^2 \le c_j^1$ for all j. **(P2)** *Dominance.* If the vectors $(d_1^1, \ldots, d_R^1)$, $(d_1^2, \ldots, d_R^2)$ represent valid sales (vis a vis the price table) for S_i, then the vector $\max((d_1^1, \ldots, d_R^1), (d_1^2, \ldots, d_R^2))$ also represents a valid sale for S_i, where the maximum is taken coordinate-wise. Table 1 (in the Appendix) satisfies the volume discount and the dominance properties.

We note that the properties (P1) and (P2) are quite reasonable in commercial scenarios, reflecting the desire of each seller to increase its part in the deal, by selling more units from each of the goods. (P1) implies that as the quantities increase, the unit prices decrease; (P2) allows for more combinations of the goods for the buyer.[6] It can be shown (by reduction from Partition) that DST is NP-hard even for instances that satisfy properties (P1) and (P2), already for $R = 1$.

Assume that we know the optimal cost, C, for our instance. Then, for a given value of $\varepsilon > 0$, we define an instance of R-MMCK, whose optimal solution induces a solution for DST with cost at most $(1 + \varepsilon)C$. We then find an optimal fractional solution for the R-MMCK instance. This gives an almost optimal fractional solution for the DST instance. Finally, we use non-standard rounding to obtain an integral solution whose cost is within factor $(1+\varepsilon)$ from the fractional solution. Note that C can be 'guessed' in polynomial time within factor $(1 + \varepsilon)$, using binary search over the range $(0, mR \cdot \max_{i,j} \max_{1 \le \ell \le m_i} u_{\ell j}c_{\ell j})$, i.e., we allow to take the maximum number of units from the j-th good in the ℓth range, for $1 \le \ell \le m_i$, $1 \le i \le m$, $1 \le j \le R$.

---

[6] It is easy to modify any price table to one that satisfies (P2). We elaborate on that in the full version of the paper.

Reduction to the R-MMCK Problem: Given the value of C, the parameter ε and a DST instance with m price tables, we construct an R-MMCK instance which consists of a single R-dimensional knapsack with capacities $b_j = n_j$, $1 \leq j \leq R$, and m sets of items; each set A_i has $m_i \cdot (m/\varepsilon)^R$ items, $1 \leq i \leq m$. Each of the items in A_i represents a sale of the i-th seller, which (partially) satisfies the order. Specifically, each item in A_i is an integer vector $(i, \ell, k_1, \ldots, k_R)$, where ℓ is the range in the i-th price table from which we choose the goods, and $0 \leq k_j \leq m/\varepsilon$ is the contribution of the j-th good, bought from the i-th seller, to the total cost. We take this contribution as an integral multiple of $\varepsilon C/m$; for each vector we find the maximal number of units of each good that can be bought with this vector. If for some integer $g \geq 1$, $k_j \varepsilon C/m < g c_{\ell j} \leq (k_j + 1)\varepsilon C/m$ then we buy g units from the good and round down the cost to $k_j \varepsilon C/m$. The cost of an item $(i, \ell, k_1 \ldots, k_R)$ in A_i is given by $c(i, \ell, k_1 \ldots, k_j) = \varepsilon C/m \sum_{j=1}^{R} k_j$. We denote by $s_j(i, \ell, k_1 \ldots, k_R)$ the maximum total number of units of the j-th good that can be bought from S_i at the cost $k_j \varepsilon C/m$, $1 \leq j \leq R$.

Approximating the Optimal Solution for R-MMCK: Given an instance of R-MMCK, we 'guess' the set S of items of maximal costs in the optimal solution, where $|S| = h = \min(m, \lfloor \frac{2R(1-\varepsilon)}{\varepsilon} \rfloor)$. Let $E(S)$ be the subset of items with costs that are larger than the minimal cost of any item in S, that is, $E(S) = \{(i, \ell, k_1, \ldots, k_R) \notin S \mid c(i, \ell, k_1, \ldots, k_R) > c_{min}(S)\}$, where $c_{min}(S) = \min_{(i, \ell, k_1, \ldots, k_R) \in S} c(i, \ell, k_1, \ldots, k_R)$.

We select all the items $(i, \ell, k_1, \ldots, k_R) \in S$ and eliminate from the instance all the items $(i, \ell, k_1, \ldots, k_R) \in E(S)$ and the sets A_i from which an item has been selected. In the next step we find an optimal *basic solution* for the following linear program, in which $x_{i, \ell, k_1, \ldots, k_R}$ is an indicator variable for the selection of an item $(i, \ell, k_1, \ldots, k_R) \notin S \cup E(S)$.

$$(LP(S)) \quad \min \sum_{i=1}^{m} \sum_{\ell=1}^{m_i} \sum_{k_1, \ldots, k_R} c(i, \ell, k_1, \ldots, k_R) x_{i, \ell, k_1, \ldots, k_R}$$

$$s.t. \quad \sum_{\ell=1}^{m_i} \sum_{k_1, \ldots, k_R} x_{i, \ell, k_1, \ldots, k_R} \leq 1 \text{ for } i = 1, \ldots, m$$

$$\sum_{i=1}^{m} \sum_{\ell=1}^{m_i} \sum_{k_1, \ldots, k_R} s_j(i, \ell, k_1, \ldots, k_R) x_{i, \ell, k_1, \ldots, k_R} \geq n_j, \quad 1 \leq j \leq R$$

$$0 \leq x_{i, \ell, k_1, \ldots, k_R} \leq 1 \text{ for } (i, \ell, k_1, \ldots, k_R) \notin S \cup E(S)$$

Rounding the Fractional Solution: Given an optimal fractional solution for R-MMCK, we now return to the DST instance and get an integral solution as follows. Suppose that we have $D = D(i)$ fractional variables for some set A_i, $x_{i, \ell_1, k_{11}, \ldots, k_{1R}}, \ldots, x_{i, \ell_D, k_{D1}, \ldots, k_{DR}}$, then we buy from the i-th seller $max_{1 \leq d \leq D}$ $s_j(i, \ell_d, k_{d1}, \ldots, k_{dR})$ units of the j-th good, $1 \leq j \leq R$.

3.2   Analysis

We now show that the above scheme yields a $(1 + \varepsilon)$-approximation for the optimum cost for DST, and that the resulting (integral) solution is feasible.

Lemma 5. *If there exists an optimal (fractional) solution with cost C for the R-MMCK instance, then there exists a (fractional) solution with cost at most $(1+\varepsilon)C$ for the DST instance.*

Proof. For any $\varepsilon' > 0$, in any fractional solution for R-MMCK with ε', the cost of each of the selected items $(i, \ell, k_1 \ldots, k_R)$ in the DST instance is at most $(c(i, \ell, k_1 \ldots, k_R) + R\varepsilon'C/m)x_{i,\ell,k_1\ldots,k_R}$. Since $\sum_{\ell=1}^{m_i} \sum_{k_1,\ldots,k_R} x_{i,\ell,k_1,\ldots,k_R} \leq 1$, for all $1 \leq i \leq m$ this yields an increase of at most $R\varepsilon'C/m$ for the seller S_i. By taking $\varepsilon' = \varepsilon/R$, we get that the overall increase in the cost is $R\varepsilon'C = \varepsilon C$.   ∎

Lemma 6. *The integral solution obtained from the fractional solution for LP(S) yields a ratio of at most $(1 + \varepsilon)$ to the optimal cost for the DST instance.*

Proof. Let $\mathbf{x}^*$ be an optimal (integral) solution for the linear program $LP(S)$, and let $S^* = \{(i, \ell, k_1, \ldots, k_R)\mid x^*_{i,\ell,k_1,\ldots,k_R} = 1\}$ be the corresponding subset of items. As in the proof of Lemma 1, we may assume that $|S^*| \geq h$, otherwise we are done. Let

$$S^* = \{(i_1, \ell_1, k_{11}, \ldots, k_{1R}), \ldots, (i_r, \ell_r, k_{r1}, \ldots, k_{rR})\},$$

such that $c(i_1, \ell_1, k_{11}, \ldots, k_{1R}) \geq \cdots \geq c(i_r, \ell_r, k_{r1}, \ldots, k_{rR})$, for some $r > h$, and let

$$S^*_h = \{(i_1, \ell_1, k_{11}, \ldots, k_{1R}), \ldots, (i_h, \ell_h, k_{h1}, \ldots, k_{hR})\}.$$

Let $\sigma = \sum_{t=1}^{h} c(i_t, \ell_t, k_{t1}, \ldots, k_{tR})$ be the total cost of the items in S^*_h. Then, for any item $(i, \ell, k_1, \ldots, k_R) \notin (S^*_h \cup E(S^*_h))$, $c(i, \ell, k_1, \ldots, k_R) \leq \sigma/h$.

We use below the notation $s_j(d)$ when referring to $s_j(i, \ell_d, k_{d1}, \ldots, k_{dR})$. Let $c(max_{1 \leq d \leq D} s_j(d))$ be the total cost of buying the j-th good in the entry of the price table where we obtain $max_{1 \leq d \leq D} s_j(d)$ units form good j, $1 \leq j \leq R$. The heart of the proof is the following claim.

Claim 1. For any $1 \leq i \leq m$, the cost of buying from the i-th seller satisfies

$$\sum_{j=1}^{R} c(max_{1 \leq d \leq D} s_j(d)) \leq \sum_{d=1}^{D} c(i, \ell_d, k_{d1}, \ldots, k_{dR}).$$

Proof. By our rounding technique, the vector giving the amounts bought from S_i from each of the goods satisfies $(max_{1 \leq d \leq D} s_1(d), \ldots, max_{1 \leq d \leq D} s_R(d)) \geq (s_1(i, \ell_d, k_{d1}, \ldots, k_{dR}), \ldots, s_R(i, \ell_d, k_{d1}, \ldots, k_{dR}))$, for all $1 \leq d \leq D$. By the volume discount property, the total cost of the rounded solution satisfies $c(max_{1 \leq d \leq D} s_1(d), \ldots, max_{1 \leq d \leq D} s_R(d)) \leq \sum_{d=1}^{D} c(i, \ell_d, k_{d1}, \ldots, k_{dR})$.   ∎

Let z^* denote the optimal (integral) solution for the R-MMCK instance. Denote by $\mathbf{x}^B(S^*_h)$ a basic solution for LP(S), and let $\mathbf{x}^I(S^*_h)$ be an integral solution obtained by setting $x_{i,\ell_d,k_{d1},...,k_{dR}} = 1$ for all $1 \le d \le D$. From Claim 1, we can bound the total cost of the solution output by the scheme, $\hat{z}$, by comparing z^* to the cost of $\mathbf{x}^I(S_h^*)$. In particular,

$$z^* \ge \sum_{i=1}^{m} \sum_{\ell=1}^{m_i} \sum_{k_1,...,k_R} c(i,\ell,k_1,...,k_R) x_{i,\ell,k_1,...,k_R}^B(S_h^*)$$

$$\ge \sum_{i=1}^{m} \sum_{\ell=1}^{m_i} \sum_{k_1,...,k_R} c(i,\ell,k_1,...,k_R) x_{i,\ell,k_1,...,k_R}^I(S_h^*) - \delta$$

where $\delta = \sum_{(i,\ell,k_1,...,k_R) \in F} c(i,\ell,k_1,...,k_R)$, and F is the set of items for which the basic variable was a fraction, i.e., $F = \{(i,\ell,k_1,...,k_R) | \ x_{i,\ell,k_1,...,k_R}^B(S_h^*) < 1\}$

Assume that in the optimal (fractional) solution of $LP(S_h^*)$ there are L tight constraints, where $0 \le L \le m + R$, then in the basic solution $\mathbf{x}^B(S_h^*)$, at most L variables can be strictly positive. Thus, at least $L - 2R$ variables get an integral value (i.e. '1'), and $|F| \le 2R$. Note that, for any $(i,\ell,k_1,...,k_R) \in F$, $c(i,\ell,k_1,...,k_R) \le \sigma/h$, since $F \cap (S_h^* \cup E(S_h^*)) = \emptyset$. Hence, we get that $z^* \ge \hat{z} + \frac{2R\sigma}{h} \ge \hat{z} + \frac{2R\hat{z}}{h} \ge \frac{\hat{z}}{1-\varepsilon}$.

Now, from Lemma 5, we have $(1+\varepsilon)^2$-approximation for DST, and since C is guessed within factor $1+\varepsilon$, we get a $(1+\varepsilon)^3$-approximation. By taking $\varepsilon' = \varepsilon/4$ we get the statement of the lemma. ∎

Lemma 7. *The integral solution obtained by the rounding is feasible for DST.*

Combining the above lemmas we get:

Theorem 6. *There is a polynomial time approximation scheme for any DST instance satisfying properties (P1) and (P2), with fixed number of goods.*

References

[AM+97]   M.M. Akbar, E.G., Manning, G.C. Shoja and S. Khan, "Heuristic Solutions for the Multiple-Choice Multi-Dimension Knapsack Problem". *Int. Conference on Computational Science*, (2), 2001: 659–668.

[BK+02]   M. Bichler, J. Kalagnanam, H.S. Lee, J. Lee, "Winner Determination Algorithms for Electronic Auctions: A Framework Design". *EC-Web* 2002: 37–46.

[CHW-76]   A.K. Chandra, D.S. Hirschberg and C.K. Wong, "Approximate Algorithms for Some Generalized Knapsack Problems". *Theoretical Computer Science* 3, pp. 293–304, 1976.

[CK-00]   C. Chekuri and S. Khanna, "A PTAS for the Multiple Knapsack Problem". In *Proc. of SODA*, pp. 213–222, 2000.

[C-79]   V. Chvátal, V. "A Greedy Heuristic for the Set Covering Problem", *Math. Oper. Res.* 4, 1979, 233–235.

[D-82]   G. Dobson, "Worst-case Analysis of Greedy for Integer Programming with Nonnegative Data". *Math. of Operations Research*, 7, pp. 515–531, 1982.

[FC-84]   A. M. Frieze and M.R.B. Clarke, Approximation Algorithms for the m-dimensional 0-1 knapsack problem: worst-case and probabilistic analyses. In *European J. of Operational Research*, 15(1):100–109, 1984.

[Fe-96]   U. Feige. "A threshold of ln n for approximating set cover". In *Proc. of 28th Symposium on Theory of Computing*, pp. 314–318, 1996.

[Fl-04]   L. Fleischer, "A Fast Approximation Scheme for Fractional Covering Problems with Variable Upper Bounds". In *Proc. of SODA*, pp. 994–1003, 2004.

[GJ-79]   M.R. Garey and D.S. Johnson, Computers and intractability: *A Guide to the Theory of NP-Completeness*. W.H. Freeman, 1979.

[I-80]   T. Ibaraki, "Approximate algorithms for the multiple-choice continuous knapsack problems", *J. of Operations Research Society of Japan*, 23, 28–62, 1980.

[IH$^+$78]   T. Ibaraki, T. Hasegawa, K. Teranaka and J. Iwase. "The Multiple Choice Knapsack Problem". *J. Oper. Res. Soc. Japan* 21, 59–94, 1978.

[K-03]   S. G. Kolliopoulos, "Approximating covering integer programs with multiplicity constraints", *Discrete Applied Math.*, 129:2–3, 461 –473, 2003.

[KY-01]   S. G. Kolliopoulos and N. E. Young, "Tight Approximation Results for General Covering Integer Programs". In *Proc. of FOCS*, 522–528, 2001.

[KPS-03]   A. Kothari, D. Parkes and S. Suri, "Approximately-Strategyproof and Tractable Multi-Unit Auctions". *Proc of ACM-EC*, 2003.

[La-76]   E. L. Lawler, *Combinatorial Optimization: Networks and Metroids*. Holt, Reinhart and Winston, 1976.

[Lu-75]   G. S. Lueker, "Two NP-complete problems in nonnegative integer programming". Report # 178, Computer science Lab., Princeton Univ., 1975.

[RV-98]   S. Rajagopalan and V. V. Vazirani, "Primal-Dual RNC Approximation Algorithms for Set Cover and Covering Integer Programs". *SIAM J. Comput.*, 28(2) pp. 525–540, 1998.

[S$^+$04]   H. Shachnai, O. Shmueli and R. Sayegh, "Approximation Schemes for Deal Splitting and Covering Integer Programs with Multiplicity Constraints", full version. http://www.cs.technion.ac.il/ ∼hadas/PUB/ds.ps.

[ST-03]   H. Shachnai and T. Tamir, "Approximation Schemes for Generalized 2-dimensional Vector Packing with Application to Data Placement". In *proc of Random-APPROX*, 2003.

[S-96]   A. Srinivasan, "An Extension of the Lovász Local Lemma, and its Applications to Integer Programming". In *Proc. of SODA*, 6–15, 1996.

[S-99]   A. Srinivasan, "Improved Approximation Guarantees for Packing and Covering Integer Programs". *SIAM J. Comput.*, 29(2): 648–670, 1999.

[SG$^+$02]   O. Shmueli, B. Golany, R. Sayegh, H. Shachnai, M. Perry, N. Gradovitch and B. Yehezkel, "Negotiation Platform". *International Patent Application* WO 02077759, 2001-2.

[Va-01]   V. V. Vazirani, Approximation Algorithms. Springer-Verlag, 2001.

[W-96]   E. Wolfstetter, "Auctions: An Introduction". *J. of Economic Surveys*, 10, 1996: 367–420.

A   Deal Splitting with Price Tables - An Example

Suppose that $R = 3$ and the goods are printers, cartridges and paper boxes. Table 1 gives the possible combinations of goods for the seller S_1, specified by amounts and unit costs, in 3 price ranges (i.e., $m_1 = 3$).

Table 1. A price table for multiple (3) goods

Price range	Printers	Cartridges	Paper
1	$(0, 2, 300)$	$(0, 5, 30)$	$(0, 9, 15)$
2	$(3, 5, 280)$	$(7, 9, 25)$	$(10, 100, 10)$
3	$(6, 20, 250)$	$(10, 50, 23)$	$(10, 100, 10)$

Thus, if we buy 2 printers or less, the unit cost is 300, whereas the unit cost for buying $3 \leq p \leq 5$ printers is 280. A valid sale for S_1 is the combination $(1, 0, 7)$, in which we obtain a printer and 7 paper boxes. The cost of this sale, which corresponds to the first price range, is 405.

Priority Algorithms for Graph Optimization Problems

Allan Borodin[1], Joan Boyar[2],[*],[**], and Kim S. Larsen[2],[*]

[1] Department of Computer Science, University of Toronto
`bor@cs.toronto.edu`, fax: +1 416 978 1931.
[2] Department of Mathematics and Computer Science,
University of Southern Denmark, Odense
{`joan, kslarsen`}`@imada.sdu.dk`, fax: +45 6550 2325.

Abstract. We continue the study of priority or "greedy-like" algorithms as initiated in [6] and as extended to graph theoretic problems in [9]. Graph theoretic
problems pose some modelling problems that did not exist in the original applications of [6] and [2]. Following [9], we further clarify these concepts. In the graph
theoretic setting there are several natural input formulations for a given problem
and we show that priority algorithm bounds in general depend on the input formulation. We study a variety of graph problems in the context of arbitrary and
restricted priority models corresponding to known "greedy algorithms".

1   Introduction

The concept of a greedy algorithm was explicitly articulated in a paper by Edmonds [11],
following a symposium on mathematical programming in 1967 although one suspects
that there are earlier references to this concept. Since that time, the greedy algorithm
concept has taken on a broad intuitive meaning and a broader set of applications beyond
combinatorial approximation. The importance of greedy algorithms is well motivated by
Davis and Impagliazzo [9] and constitutes an important part of many texts concerning
algorithm design and analysis. New greedy algorithms keep emerging, as, for instance,
in [18], which considers mechanisms for combinatorial auctions, requiring solutions
to difficult optimization problems. Given the importance of greediness as an algorithm
design "paradigm", it is somewhat surprising that a rigorous framework, as general
as priority algorithms, for studying greedy algorithms is just emerging. Of course, the
very diversity of algorithms purported to be greedy makes it perhaps impossible to find
one definition that will satisfy everyone. The goal of the priority algorithm model is to
provide a framework which is sufficiently general so as to capture "most" (or at least
a large fraction) of the algorithms we consider to be greedy or greedy-like while still
allowing good intuition and rigorous analysis, e.g., being able to produce results on the
limitations of the model and suggesting new algorithms.

The priority model has two forms, fixed priority and the more general adaptive priority
model. The general form of fixed and adaptive priority algorithms is presented in Figures

---

[*] Partially supported by the Danish Natural Science Research Council (SNF) and the IST Pro-
gramme of the EU under contract number IST-1999-14186 (ALCOM-FT).
[**] Contact author.

G. Persiano and R. Solis-Oba (Eds.): WAOA 2004, LNCS 3351, pp. 126–139, 2005.

Determine an allowable ordering of the set of possible input items
(without knowing the actual input set S of items)
while not empty(S)
  $next$: = index of input item I in S that comes first in the ordering
  Make an irrevocable decision concerning I_{next} and remove I_{next} from S

Fig. 1. The form of a fixed priority algorithm

while input set S not empty
  Determine a total ordering of all possible input items
  (without knowing the input items in S not yet considered)
  $next$: = index of item I in S that comes first in the ordering
  Make an irrevocable decision concerning I_{next} and remove I_{next} from S

Fig. 2. The form of an adaptive priority algorithm

1 and 2. To make this precise, for each specific problem we need to define the nature and representation (the type) of the input items and the nature of the allowable (irrevocable) decisions. Surprisingly, the issue as to what orderings are allowed has a rather simple and yet very inclusive formalization. Namely, the algorithm can use any total ordering on some sufficiently large set of items from which the actual set of input items will come. (For adaptive algorithms, the ordering can depend on the items already considered.) The priority framework was first formulated in Borodin, Nielsen and Rackoff [6] and applied to (worst case approximation algorithms for) some classical scheduling problems such as Graham's makespan problem and various interval scheduling problems. In a subsequent paper, Angelopoulos and Borodin [2] applied the framework to the set cover and uncapacitated facility location problems. These problems were formulated so that the data items were "isolated" in the sense that one data item did not refer to another data item and hence any set of valid data items constituted a valid input instance. For example, in the makespan problem on identical machines with no precedence constraints, a data item is represented by a processing time and the items are unrelated. The version of facility location studied in [2] was for the "disjoint model" where the set of facilities and the set of clients/cities are disjoint sets and a facility is represented by its opening cost and a vector of distances to each of the cities. In contrast, in the "complete model" for facility location, there is just a set of cities and every city can be a facility. Here a city is represented by its opening cost and a vector of distances to every other city. In the complete model for facility location, an input item (a city) directly refers to other input items. This is similar to the standard situation for graph theoretic problems when vertices are, say, represented by adjacency lists.

The work of Davis and Impagliazzo [9] extends the priority formulation to graph theoretic problems. Davis and Impagliazzo consider a number of basic graph theory problems (single source shortest path, vertex cover, minimum spanning tree, Steiner trees, maximum independent set) with respect to one of two different input models depending on the problem and known "greedy algorithms". For the shortest path, minimum spanning tree and Steiner tree problems, the model used is the "edge model", where input items are edges represented by their weights, the names of the endpoints, and in the case of the Steiner tree problem by the types (required or Steiner) of the edge endpoints. Note that in this edge representation, input items are isolated and all of the definitions in [6] can be applied. In particular, the definition of a greedy decision is well defined. In contrast, for the vertex cover and maximum independent set problems, Davis and Impagliazzo use a vertex adjacency list representation, where input items are vertices, represented by the names of the vertices to which they are adjacent, and in some problems also the weight of the vertex. This representation presents some challenges for defining priority algorithms and greedy decisions. These definitional issues have helped to clarify the nature and usefulness of "memoryless priority algorithms".

Noting that lower bounds for graph theoretic priority algorithms appear to be hard to obtain in (say) the vertex adjacency model, Angelopoulos has recently [1] proposed a reasonable change to the model by restricting what priority algorithms can do, thus increasing the power of the adversary. The basic effect of his change is to force items which are indistinguishable (except for their different identification labels) to receive the same priority. Angelopoulos proves lower bounds for the complete facility location problem (for both fixed and adaptive priority algorithms) and the dominating set problem (for the more general adaptive priority algorithms). It is not clear if Angelopoulos' results can be obtained in the model which we use, but even if they can, this simple restriction on priority algorithms should make it easier to derive lower bound proofs.

In this paper, we continue the study of priority algorithms for graph problems using two models (again motivated by current algorithms), namely the vertex adjacency model as in Davis and Impagliazzo and an "edge adjacency model", where input items are vertices now represented by a list of adjacent edge names (rather than a list of adjacent vertex names) and possible vertex weights where appropriate. It should be clear that the vertex adjacency model is more general in the sense that any priority algorithm in the edge adjacency model can be simulated in the vertex adjacency model (making exactly the same set of decisions). Most existing priority algorithms can function in the edge adjacency model; the authors were unable to recall one which does not. However, we show (using an example in Davis and Impagliazzo showing that memorylessness is restrictive) that the edge adjacency model can be restrictive. We also introduce an "acceptances-first" model and clarify the relation of memoryless algorithms to this "acceptances-first" model rather than to greediness. We prove a number of new results within these models. Due to space limitations, some proofs have been omitted, but they can be found in [7].

2 Priority Algorithms for Graph Problems

As mentioned in the introduction, we consider two input formulations. In the common vertex adjacency formulation, an input item is a vertex, represented by the tuple

$(v, w, v_1, v_2, \ldots, v_d)$, where v is the name of the vertex, w is the weight (if any) of vertex v and $v_1, \ldots, v_d$ is a list of adjacent vertices. In the more restrictive edge adjacency formulation (but still a model sufficient to capture most known greedy graph algorithms), an input item is a vertex $(v, w, e_1, e_2, \ldots, e_d)$ where again v is the vertex name, w is the weight (if any) of v and $e_1, e_2, \ldots, e_d$ is a list of adjacent edges.

In either of the above models, we have the situation that not every set of valid input items constitutes a valid input instance. Clearly, a valid input instance cannot have the same vertex appear as two different items. And in the vertex adjacency model, if a vertex v is an input item and v' is in its adjacency list then v' must also be an input item with v in its adjacency list. Similarly, if an edge e appears in some input item then e must appear in exactly one other input item. Although the priority algorithm framework is designed to model greedy algorithms, it is possible to define priority algorithms where the irrevocable decisions do not seem greedy. As noted by Davis and Impagliazzo, the definition of "greedy decision" (as formulated in [6]) is no longer well defined when the algorithm "knows" that the current item is not the last. More specifically, in [6], a greedy priority algorithm is one in which all of the irrevocable decisions are "greedy" in the sense that the algorithm acts as if the current item being considered is the last item in the input. In more colloquial terms, greediness is defined by the motto "live for today". We would like to formulate a general concept of a greedy decision that also makes sense when the input items are not isolated. (We would like such a definition to also make sense for non-graph problems such as scheduling problems with precedence relations amongst the jobs where one can have non-isolated input items.) We offer one such definition in [7]. We note, however, that in the context of priority algorithms the greedy versus non greedy distinction is not that important and to the extent that it is important it is only because greedy is such a widely used (albeit mostly undefined) concept. We do argue that the priority algorithm formulation is important as it captures such a wide variety of existing algorithms which might be called "greedy-like" extending the concept of greedy and including (for example) all online algorithms.

One can always make an ad hoc definition of a greedy decision in the context of any given problem. For example, for the vertex coloring problem, one might define a greedy decision to be one that never assigns a new color to a vertex if an existing color could be used *now*. But for a given input and history of what has been seen, it may be known to the algorithm that any valid completion of the input sequence will force an additional color and it might be that in such a case one would also allow a new color to be used before it was needed. This can, of course, all be considered as a relatively minor definitional issue and one is free to choose whatever definition seems to be more natural and captures known "greedy algorithms".

Perhaps a more meaningful distinction is the concept of "memoryless" priority algorithms. Although motivated by the concept of memoryless online algorithms, especially in the context of the k-server problem, the concept of memorylessness takes on a somewhat different meaning as applied in [6] and [2]. Namely these papers apply the concept to problems where the irrevocable decision is an accept/reject decision (or at least that acceptance/rejection is part of the irrevocable decision). In this context memoryless priority algorithms are defined as priority algorithms in which the irrevocable decision for the current item (and the choice of next item in the case of adaptive algorithms) depends only on the set of previously accepted items. That is, in the words of [9], a rejected item is treated as a NO-OP. In the accept/reject context, memoryless algorithms are equivalent to *acceptances-first* algorithms which do not accept any items after the first rejected item. As observed[1] in [6] and [2], we have the following:

Theorem 1. *Let $\mathbb{A}$ be a memoryless priority algorithm for a problem with accept/reject decisions. Then there exists an "acceptances-first" adaptive priority algorithm $\mathbb{A}'$ that "simulates" $\mathbb{A}$ in the sense that it accepts the same set of items and makes the same irrevocable decisions.*

We observe that many graph theoretic algorithms called greedy may or may not satisfy some generic general definition of greedy. But many of these algorithms are indeed memoryless (or equivalently, acceptances-first) according to the above definition. (By the definition of memoryless, the converse of the above theorem holds trivially.)

To prove negative results, showing that no priority algorithm in some model can achieve an approximation ratio better than ρ for a given problem P, we use an adversary. The adversary initially chooses a set S of valid input items. It interacts with an algorithms $\mathbb{A}$, maintaining the invariant that the items remaining in S, together with the items already selected by $\mathbb{A}$, contain at least one valid input instance. At each step, the adversary removes the item i remaining in S, to which $\mathbb{A}$ has given the highest priority. It may also remove more items from S at this point, as long as the invariant is maintained.

In most cases the initial set S contains multiple copies of each vertex and possibly additional vertices than in the final input graph. After the algorithm chooses a vertex, the adversary removes the other copies of that vertex from S, since its adjacency list is now determined. An adaptive priority algorithm in the edge-adjacency model knows the names of the edges adjacent to the vertices already chosen, so it can give vertices with the same edges in their lists either high or low priority. The adversary may still have more than one copy of the neighbors at this time, though. In the vertex adjacency model, an adaptive priority algorithm has even more power; it can give the neighbors high or low priority and it can also give the neighbors of the neighbors high or low priority, since it knows the names of the neighbors. Although the adversary may still retain multiple copies of the neighbors, it cannot make arbitrary decisions as to whether or not a vertex chosen by the algorithm is or is not at distance at most two from any chosen vertices.

For some scheduling results in [6], the adversaries assume that the algorithm does not know (or use information concerning) the final number of jobs to be processed. The same holds here for graph problems; in some cases the adversary creates final input graphs that have different sizes for different algorithms. In practice, most priority algorithms do not seem to use the total number of vertices or edges in the graph in assigning priorities or in making the irrevocable decisions, so the results based on adversaries of this type are widely applicable. Unless otherwise stated, the results below assume the algorithm does not know the total number of vertices or edges in the graph.

---

[1] In [6], this fact is stated in terms of memoryless algorithms being simulated by greedy algorithms, but the essence of that observations really concerns the acceptance-first restriction and not greediness.

3   Independent Set

Maximum Independent Set is the problem of finding a largest subset, I, of vertices in a graph such that no two vertices in I are adjacent to each other.

The independent set problem and the clique problem, which finds the same set in the complement of the graph, are well studied NP-hard problems, where approximation also appears to be hard. The bounded degree maximal independent set problem is one of the original MAX SNP-Complete problems [19]. Håstad [13] has shown a general lower bound on the approximation ratio for the independent set problem of $n^{1-\epsilon}$, for all ϵ, provided that NP $\neq$ ZPP, where ZPP is the class of languages decidable by a random expected polynomial-time algorithm that makes no errors. A general upper bound of $O(n/\log^2 n)$ was presented by Boppana and Halldórsson [5], and an upper bound for graphs of degree 3 of $6/5$ was shown by Berman and Fujito [4]. These algorithms are not priority algorithms.

Davis and Impagliazzo [9] have shown that no adaptive priority algorithm (in the vertex adjacency model) can achieve an approximation ratio better than $\frac{3}{2}$ for the maximum independent set problem[2], and their proof used graphs with maximum degree 3. We consider algorithms in more restrictive models. We again note that many known greedy-like graph algorithms are acceptances-first priority algorithms.

In the proofs of Theorems 2, 3, and 8, the adversary uses a modification of a construction due to Hochbaum [15].

Construction $\mathbb{G}$: There are two sets of vertices, U and V. The set U consists of k independent $(k+1)$-cliques, and the set V is an independent set consisting of k^2 vertices, each of which is adjacent to every vertex in every $(k+1)$-clique.

Note that all vertices in $\mathbb{G}$ have degree $k^2 + k$. Thus, initially, $\mathbb{A}$ cannot distinguish between the vertices when assigning priorities. The optimum independent set includes every vertex in V and has size k^2. If n is the total number of vertices in $\mathbb{G}$, $k \in \Theta(\sqrt{(n)})$.

Theorem 2. *No acceptances-first adaptive algorithm $\mathbb{A}$ in the vertex adjacency model for independent set can achieve an approximation ratio better than $\Omega(\sqrt{n})$, where n is the number of vertices (even if the number of vertices and edges in the graph is known to the algorithm).*

The proof of this result depends on the first vertex being accepted. One can obtain a similar result, removing the acceptances-first assumption, if the algorithm $\mathbb{A}$ is a fixed priority algorithm in the edge adjacency model.

Theorem 3. *No fixed priority algorithm $\mathbb{A}$ in the edge adjacency model for independent set (or clique) can achieve an approximation ratio better than $\Omega(n^{1/3})$, where n is the number of vertices.*

Proof. The adversary uses possibly several copies of the construction, $\mathbb{G}$. Since $\mathbb{A}$ is a fixed priority algorithm and all vertices have the same degree, $\mathbb{A}$ cannot distinguish the vertices when assigning priorities.

---

[2] We have defined the approximation ratio so that all ratios are at least one.

The adversary arranges that the selected vertices are independent during the first phase. We let n' denote the number of vertices processed so far. The first phase continues until either $\mathbb{A}$ has accepted at least $c = \lceil \frac{n'}{k} \rceil$ vertices or $n' = k^2$; whichever happens first.

If the first phase stopped because at least c vertices were accepted, then the adversary creates c copies of the construction $\mathbb{G}$. There are enough cliques so that each of the n' vertices can be placed in distinct cliques in U. The accepted vertices are placed such that at most one is in each construction. This means that in each construction, $\mathbb{G}$, all vertices in V must be rejected. In addition, the algorithm can accept at most 1 vertex in every clique in U. This gives a ratio of at least $\frac{c \cdot k^2}{c \cdot k} = k = \Omega(n^{1/3})$.

If the first phase stopped because $n' = k^2$, the adversary uses a single copy of the construction, $\mathbb{G}$. The n' vertices are in V. Note that the number of accepted vertices is strictly smaller than $\lceil \frac{k^2}{k} \rceil = k$, since otherwise the algorithm would have terminated for that reason. If any of the n' vertices are accepted, then no vertices from U can be; otherwise at most one vertex from each clique can be accepted. Thus, the best ratio is when all n' vertices from phase one are rejected: $\frac{k^2}{k} = k = \Omega(\sqrt{n})$.   □

Combining the acceptances-first requirement with the fixed priority requirement, gives a model which is so weak that it appears to be uninteresting. Consider, for example, a complete bipartite graph with n vertices in each part. All vertices look the same to the algorithm as it assigns priorities, so the adversary can decide that the two vertices with highest priority are adjacent. If the algorithm is acceptances-first, since it must reject the second vertex, it cannot accept more than one vertex in all.

Our next result is based on the example used in Davis and Impagliazzo to show that memoryless priority algorithms are less powerful than those which use memory. Namely, we consider WIS(k), the weighted maximum independent set problem when restricted to cycles whose vertex weights are either 1 or k. In their proof separating the power of memoryless algorithms from those which use memory, Davis and Impagliazzo show that in the vertex adjacency model there is an adaptive priority algorithm whose approximation ratio approaches one as k goes to infinity. We now show a lower bound of $\frac{3}{2}$ for the approximation ratio for this same problem in the edge adjacency model, thus showing that the edge adjacency model can be restrictive when compared to the vertex adjacency model.

Theorem 4. *For the WIS(k) problem with $k \geq 4$, no adaptive priority algorithm in the edge adjacency model can obtain an approximation ratio better than $\frac{3}{2}$.*

Proof. We will represent the cycles by lists of weights. Two neighbors in the list are also neighbors in the cycle. In addition, the first and last element in the list are also neighbors in the cycle.

We use w^+ to denote a vertex accepted by the priority algorithm and w^- to denote a vertex rejected by the priority algorithm. To demonstrate a best possible result which the priority algorithm can obtain given the accept/reject actions it has already made, we use w^c to mark vertices which could be included in addition to the already accepted vertices. Finally, we indicate an optimal vertex cover by marking vertices in one such cover by $\underline{w}$. Neither the vertices marked w^c nor $\underline{w}$ can in general be chosen uniquely, but their total weight will be unique.

The argument is structured according to the choices made by the priority algorithm, beginning with whether the first vertex has weight 1 or k and whether the priority algorithm accepts or rejects that vertex. In all but one case, the adversary can immediately guarantee a specific approximation ratio, but in one case, the next vertex chosen by the algorithm must also be used by the adversary:

First accept weight k vertex: $(k^+, \underline{k}, 1^c, \underline{k})$ gives $\frac{2k}{k+1}$.

First reject weight k vertex: $(\underline{k}^-, 1^c, 1)$ gives $\frac{k}{1}$.

First accept weight 1 vertex: $(1^+, \underline{k}, 1^c, \underline{k})$ gives $\frac{2k}{2}$.

First reject weight 1 vertex: We now ensure that no vertex of weight k will appear as a neighbor of the rejected vertex. All the remaining cases are subcases of the current case.

Next accept non-neighbor weight k vertex: $(\underline{1}^-, 1^c, \underline{k}, k^+, \underline{k}, 1^c)$ gives $\frac{2k+1}{k+2}$.

Next accept non-neighbor weight 1 vertex: $(1^-, 1^c, \underline{k}, 1^+, \underline{1})$ gives $\frac{k+1}{2}$.

Next accept neighbor weight 1 vertex: $(\underline{1}^-, 1^+, \underline{k}, 1^c)$ gives $\frac{k+1}{2}$.

Next reject non-neighbor weight k vertex: $(\underline{1}^-, 1^c, \underline{k}^-, 1^c)$ gives $\frac{k+1}{2}$.

Next reject non-neighbor weight 1 vertex: $(\underline{1}^-, 1^c, \underline{1}, 1^-, \underline{1}, 1^c)$ gives $\frac{3}{2}$.

Next reject neighbor weight 1 vertex: $(\underline{1}^-, 1^-, \underline{1}, 1^c)$ gives $\frac{2}{1}$.

Choosing $k \geq 4$ ensures the stated approximation ratio lower bound of $\frac{3}{2}$.   □

The following result shows that a $\frac{3}{2}$ approximation ratio for WIS(k) can be achieved in the edge adjacency model[3].

Theorem 5. *For the WIS(k) problem, there is an adaptive priority algorithm in the edge adjacency model with approximation ratio $\frac{3}{2}$ for $k \geq 2$.*

Proof. The algorithm proceeds as follows:

I. Give highest priority to vertices with weight 1 which are not adjacent to anything processed yet, as long as this is possible. Reject them all.

II. If there were no vertices of weight 1, accept one of weight k. Then follow it around the cycle, accepting every other vertex until finding a vertex adjacent to two already processed vertices. That last vertex must be rejected.

III. Repeat the next two steps as long as possible:

1. If there is a vertex with both neighbors already processed, accept it. (The neighbors have been rejected.)
2. If there is vertex with weight k adjacent to exactly one vertex which was already processed, accept it. Then, reject its other neighbor.

IV. If there are any vertices remaining, there must be a vertex of weight 1 adjacent to only one already processed vertex. Reject this vertex of weight 1 and accept its unprocessed neighbor. Follow this around the cycle, accepting every other vertex until reaching a vertex which has already been processed. Repeat this step until all processed chains have been joined.

---

[3] In contrast, for the WIS problem in the vertex adjacency model, Davis and Impagliazzo show a 2-approximation lower bound for memoryless algorithms.

Note that this algorithm maintains the invariant that for any maximal chain of vertices already processed, the endpoints have been rejected. The remainder of the proof is a case analysis and is given in [7].   □

4   Vertex Cover

Minimum Vertex Cover is the problem of finding a smallest subset, C, of vertices in a graph such that all edges are incident to some vertex in C.

The unweighted vertex cover problem is one of the most celebrated open problems in the area of worst case approximation algorithms. The naive algorithm (taking both adjacent vertices in any maximal matching) provides a 2-approximation. This is essentially the best known polynomial time approximation bound in the sense that there are no known polynomial time $2 - \epsilon$ approximation algorithms (for a fixed $\epsilon > 0$), although various algorithms are known that guarantee an approximation better than 2 but converging to 2 as some parameter grows. This maximal matching algorithm is easily seen to be an acceptances-first adaptive priority algorithm in the edge adjacency model. Surprisingly, Johnson [16] showed that the greedy algorithm which chooses the vertex with highest degree in the remaining graph is only a H_n-approximation, and that this bound is tight in that there are arbitrarily large graphs on which the algorithm produces a vertex cover whose size is H_n times the size of the optimal cover. Although the weighted vertex cover problem can be essentially reduced (in polynomial time) to the unweighted case (by making multiple copies of vertices), this reduction does not preserve the property of being a priority algorithm and hence the study of the unweighted and weighted vertex cover problems may be substantially different problems in the context of priority algorithms. It turns out that there are several priority algorithms for the weighted case that also achieve a 2-approximation algorithm (or slightly better). One such algorithm is Johnson's "greedy algorithm" (the layered algorithm as given in Vazirani's excellent text on approximation algorithms [21]). Essentially for the vertex cover problem this algorithm chooses all maximum degree vertices and removes them simultaneously. Another simple to state (and also called greedy) algorithm is given by Clarkson [8]. This algorithm achieves the approximation bound $\frac{\Delta}{\Delta-2}(2 - \frac{2n}{\Delta \cdot OPT(I)})$ where Δ is the maximum degree in the graph and n is the number of vertices [4]. Both the layered algorithm and Clarksons's algorithm can be expressed as acceptances-first adaptive algorithms in the edge adjacency model. In terms of complexity lower bounds, Dinur and Safra [10] provide a sophisticated proof that it is NP-hard to have a c-approximation algorithm for the (unweighted) vertex cover problem for $c < 1.36$.

Davis and Impagliazzo show that for the weighted case, priority algorithms (in the vertex adjacency model) cannot essentially do better than a 2-approximation. Priority lower bounds for the unweighted case seem more difficult.

The following $\frac{4}{3}$ lower bound matches the upper bound by Clarkson for the case $n = 7, \Delta = 3$, and $OPT(I) = 3$. In this case, Clarkson's algorithm on our graph 2 in the

---

[4] The stated bound is not defined for $\Delta \leq 2$. The more general bound that applies to all Δ is that $w(C_{MG}) \leq w(C_{OPT}) - \frac{2(n - w(C_{MG}))}{\Delta}$.

Fig. 3. Graph 1 to the left and graph 2 to the right

proof of the theorem below would give a vertex cover with four vertices. The results hold for arbitrarily large graphs, since disjoint copies of the constructions can be used.

Theorem 6. *No adaptive priority algorithm in the vertex adjacency model can achieve an approximation ratio better than 4/3 for the vertex cover problem.*

Proof. First note that both graphs in Fig. 3 have vertex covers of size 3. We will now force any adaptive priority algorithm to choose at least 4 vertices.

In the first step, $\mathbb{A}$ must choose either a degree 2 or a degree 3 vertex, and it can choose to accept or reject. We treat these four cases.

If $\mathbb{A}$ rejects a degree 2 vertex first, we let it be vertex A in graph 1. If $\mathbb{A}$ accepts a degree 2 vertex first, we let it be vertex B in graph 1. If $\mathbb{A}$ rejects a degree 3 vertex first, we let it be vertex C in graph 1. If $\mathbb{A}$ accepts a degree 3 vertex first, we let it be vertex A in graph 2.   □

Note that the numbers of vertices in the two graphs used in the proof of the above theorem are the same, so the theorem holds true in a model where the algorithms know the number of vertices.[5] Notice that with graph 2 in the proof, as long as the algorithm accepts the first vertex it processes, it will accept at least four vertices. Thus, only the one graph is necessary, when the algorithm is acceptances-first, so the algorithm can be given the number of vertices and the number of edges.

In more restrictive models, we obtain stronger lower bounds.

Theorem 7. *In the vertex adjacency model, no acceptances-first adaptive priority algorithm can achieve an approximation ratio better than 3/2 for the vertex cover problem (even if the number of edges and vertices in the graph is known to the algorithm).*

The proof of the following theorem is very similar to that of Theorem 3.

Theorem 8. *No fixed priority algorithm $\mathbb{A}$ in the edge adjacency model for vertex cover can achieve an approximation ratio better than 2.*

The 2-approximation algorithms for vertex cover are adaptive rather than fixed priority, so the above result may not be tight. (We do not know of a fixed priority algorithm which is an $O(1)$-approximation algorithm for vertex cover.)

---

[5] If the number of edges should also be the same, we can add a cycle of 4 vertices to graph 2 and a cycle of 4 vertices with one diagonal to graph 1 and obtain a bound of 6/5.

5   Vertex Coloring

Minimum Vertex Coloring is the problem of coloring the vertices in a graph using the minimum number of different colors in such a way that no two adjacent vertices have the same color. The problem is also known as Graph k-Colorability and as Chromatic Number.

Hardness results are known for minimum vertex coloring under various complexity theoretical assumptions: minimum vertex coloring is NP-hard to approximate within $\Omega(n^{1-\epsilon})$, for all ϵ, provided that NP $\neq$ ZPP [12]. It is NP-hard to approximate within $n^{\frac{1}{5}}$ provided that NP $\neq$ coRP and within $n^{\frac{1}{7}}$ provided that P $\neq$ NP [3].

From [17], it is known that it is NP-hard to 4-color a 3-chromatic graph, NP-hard to color a k-chromatic graph with at most $k + 2\lceil k/3 \rceil - 1$ colors, and NP-hard to approximate within n^ϵ for some fixed ϵ as the chromatic number of graphs tend towards infinity.

On the positive side, a general upper bound of $O(n \log \log^2 n / log^3 n)$ is shown by Halldórsson [14]. In [20], an upper bound of $\lambda(G) + 1$ is established, where λ is any function of graphs $G = (V, E)$ such that

$$(G' \subset G \;\Rightarrow\; \lambda(G') \leq \lambda(G)) \;\wedge\; \lambda(G) \geq \min_{v \in V} deg(v).$$

Let $d(G)$ be the maximum over all vertex-induced subgraphs of the minimum degree in that subgraph. The result in [20] constructively establishes that any graph is $d(G) + 1$ colorable, so a corollary of the theorem below is that the algorithm from [20] is not a priority algorithm. This theorem is proven using an adversary which is defined using a lengthy case analysis.

Theorem 9. *No priority algorithm in the edge adjacency model can 3-color all graphs G with $d(G) = 2$.*

In more restrictive models, we obtain stronger lower bounds. The following two results apply to models which include the simplest and most natural greedy algorithm; namely, order the vertices in any way and then color vertices using the lowest possible numbered color.

Theorem 10. *Any fixed priority algorithm in the edge adjacency model must use at least $d + 1$ colors on a bipartite graph of maximum degree d.*

Proof. The adversary will create many independent portions of a bipartite graph, each with the same number of vertices and the same colors in each part. These portions will grow in size and it may be necessary to join two portions, making the correct decision as to which partition of the one portion is placed with which partition of the other. At the end all vertices will have degree d, so in assigning priorities, the fixed priority algorithm will continually choose vertices of degree d. Its only choice is which color to give after it is told which already colored vertices the chosen vertex is adjacent to.

Initially, the adversary will arrange that all vertices chosen are independent. The number chosen at this stage will be large enough so that there are either $d + 1$ colors given or enough vertices given the same color to make the remainder of the proof possible.

It will be clear that some large number will be sufficient. This stage 1 ends when there are enough vertices given the same color, which we call color 1.

In stage i, we have a large number of independent bipartite graphs, where both sides contain vertices with colors $1, 2, ..., i - 1$, but no other colors. The vertices chosen are made adjacent to one vertex of each color $1, 2, ..., i - 1$, all from one partition of one of the graphs. If there are a large enough number of graphs which get the same additional color on both sides, this color is called color i and the adversary proceeds to stage $i + 1$. Otherwise, there will eventually be enough graphs given the same two additional colors, which will be called i and $i + 1$. Graphs of this type can be joined in pairs. For each pair, the adversary joins them so that both partitions in the resulting bipartite graph have both colors i and $i + 1$. Then, the adversary proceeds to stage $i + 2$.

The adversary stops this process as soon as $d + 1$ colors have been used, and more vertices are included to create a bipartite graph where all vertices have degree d. Note that if fewer than $d + 1$ colors are used before stage $d + 1$, a $d + 1$st color will be used then, since the vertices in that stage will be adjacent to each of the colors $1, 2, ..., d$. If there is no stage $d + 1$, because the adversary went from stage d to $d + 2$, the $d + 1$st color was used in stage d.   □

In the next result, we consider adaptive priority algorithms which use different information in its two main phases. When assigning priorities to vertices, it only considers the number of uncolored neighbors a vertex has and the vector $(n_1, \ldots, n_k)$ of the k colors used so far where n_i is the number of nodes that have already been colored with color i. In this phase, the algorithm may not use information about how many of a vertex's neighbors have already been colored or what colors these neighbors have been given. For the irrevocable decision of coloring a vertex, the color given will simply be a function of the set of colors already given to the neighbors. This could, for example, be the lowest possible numbered color.

Theorem 11. *Any adaptive priority algorithm in the edge adjacency model, which gives priorities based only on the current degree of the vertex and the already processed subgraph, must use $d + 1$ colors on a d-colorable graph of maximum degree d, when the color given is a function of the set of currently adjacent colors (no state information).*

Proof. The adversary uses the following graph. It creates two K_d cliques A and B. Two selected vertices, a in A and b in B are then connected by an additional edge. The remaining vertices in A and B may or may not be connected via a single edge to additional copies of K_d cliques. This will depend on the degree of vertices chosen by $\mathbb{A}$. At any point in time during the execution of $\mathbb{A}$, $\mathbb{A}$ will have a choice of two consecutive degrees within A, and two (possibly different) consecutive degrees within B. Whenever it chooses the higher degree, the chosen vertex will be connected to one of the additional K_d cliques. At most $2d - 2$ of the additional K_d cliques may be necessary. The adversary must present this many originally. When a vertex with the lower of the two possible degrees from A or B is chosen, one of the additional K_d cliques is removed from the possibilities the adversary gives $\mathbb{A}$, so that vertices not present in the graph are never chosen.

The adversary will ensure that a is the last of the vertices in A which is colored, and b is the last in B. Thus, the last one of them colored will be adjacent to d different colors and get the $d+1$st color. When there is a choice between choosing vertices in A or B or in the additional cliques, vertices from A or B are chosen. For the additional K_d cliques, if the connecting vertex is chosen after the adjacent vertex in A or B, then there is no problem; the additional K_d, G, cannot have any influence on A or B. If the connecting vertex is chosen before the adjacent vertex in A or B, there will be fewer colors used in G than in A or B, whichever it is adjacent to. So the connecting vertex will be assigned a color which is already among the neighbors of the connecting vertex in A or B; again there will be no influence on how A or B are colored. As soon as the connecting vertex has been chosen, it is connected to some vertex in A or B which has not been colored yet, further restricting the number of possible vertices of the lower degree.

Note that any graph constructed in this manner is easily d colorable, since the cliques can be connected via vertices of different colors.   □

6   Conclusions and Open Problems

We have considered priority algorithms in the vertex adjacency and edge adjacency models, and it was shown that the edge adjacency model can be more restrictive than the vertex adjacency model. Most known priority algorithms, however, can be implemented in the edge adjacency model, so it would be interesting to find natural problems (especially well studied problems) for which the priority input models are provably different with respect to the best approximation ratio attainable.

Maximum Independent Set and Vertex Cover were studied using both models, and Vertex Coloring was studied using the edge adjacency model. For problems where a priority algorithm makes only accept/reject decisions for each vertex, acceptances-first algorithms are equivalent to memoryless algorithms. The acceptances-first model was introduced and applied to the Maximum Independent Set and Vertex Cover problems.

Most of the lower bound results do not meet the upper bounds provided by known algorithms. It would be interesting to close some of these gaps. For example, in the result for the unweighted vertex cover, our adaptive priority $4/3$ lower bound meets Clarkson's result in the case when the maximum degree is three. But what if the maximum degree is larger than three? Can one prove a better lower bound? It has long been an open problem whether or not the optimal (polynomial time) approximation ratio for vertex cover is $2 - o(1)$. More generally, establishing tight priority approximation bounds for unweighted graph optimization problems remains a challenging area for future research.

References

1. S. Angelopoulos. Ordering-preserving transformations and greedy-like algorithms. In *2nd Workshop on Approximation and Online Algorithms*, 2004. This volume.
2. S. Angelopoulos and A. Borodin. On the power of priority algorithms for facility location and set cover. In *Proceedings of the 5th International Workshop on Approximation Algorithms for Combinatorial Optimization*, volume 2462 of *Lecture Notes in Computer Science*, pages 26–39. Springer-Verlag, 2002.
3. Mihir Bellare, Oded Goldreich, and Madhu Sudan. Free bits, PCPs and non-approximability—towards tight results. *SIAM Journal on Computing*, 27:804–915, 1998.

4. Piotr Berman and Toshihiro Fujito. On approximation properties of the independent set problem for degree 3 graphs. In *Fourth International Workshop on Algorithms and Data Structures*, volume 955 of *Lecture Notes in Computer Science*, pages 449–460. Springer-Verlag, 1995.

5. Ravi Boppana and Magnús M. Halldórsson. Approximating maximum independent sets by excluding subgraphs. *Bit*, 32:180–196, 1992.

6. A. Borodin, M. Nielsen, and C. Rackoff. (Incremental) priority algorithms. *Algorithmica*, 37:295–326, 2003.

7. Allan Borodin, Joan Boyar, and Kim S. Larsen. Priority algorithms for graph optimization problems. Preprint PP-2004-10, Department of Mathematics and Computer Science, University of Southern Denmark, 2004.

8. Kenneth L. Clarkson. A modification of the greedy algorithm for vertex cover. *Information Processing Letters*, 16:23–25, 1983.

9. S. Davis and R. Impagliazzo. Models of greedy algorithms for graph problems. In *Proceedings of the 15th Annual ACM-SIAM Symposium on Discrete Algorithms*, 2004.

10. Irit Dinur and Shmuel Safra. The importance of being biased. In *Proceedings of the 34th Symposium on Theory of Computing*, pages 33–42. ACM Press, 2002.

11. Jack Edmonds. Matroids and the greedy algorithm. *Mathematical Programming*, 1:127–136, 1971.

12. Urid Feige and Joe Kilian. Zero knowledge and the chromatic number. *Journal of Computer and System Sciences*, 57:187–199, 1998.

13. Johan Håstad. Clique is hard to approximate within $n^{1-\varepsilon}$. *Acta Mathematica*, 182:105–142, 1999.

14. Magnús M. Halldórsson. A still better performance guarantee for approximate graph coloring. *Information Processing Letters*, 45:19–23, 1993.

15. D. Hochbaum. Efficient bounds for the stable set, vertex cover and set packing problems. *Discrete Applied Mathematics*, 6:243–254, 1983.

16. David S. Johnson. Approximation algorithms for combinatorial problems. *Journal of Computer and System Sciences*, 9(3):256–278, 1974.

17. Sanjeev Khanna, Nathan Linial, and Shmuel Safra. On the hardness of approximating the chromatic number. *Combinatorica*, 20(3):393–415, 2000.

18. D. Lehmann, L. O'Callaghan, and Y. Shoham. Truth revelation in approximately efficient combinatorial auctions. *Journal of the ACM*, 49(5):1–26, 2002.

19. C. Papadimitriou and M. Yannakakis. Optimization, approximation and complexity classes. *Journal of Computer and System Sciences*, 43:425–440, 1991.

20. G. Szekeres and Herbert S. Wilf. An inequality for the chromatic number of graphs. *Journal of Combinatorial Theory*, 4:1–3, 1968.

21. Vijay V. Vazirani. *Approximation Algorithms*. Springer-Verlag, 2001.

Pricing Network Edges to Cross a River[*]

Alexander Grigoriev[1], Stan van Hoesel[1], Anton F. van der Kraaij[1],
Marc Uetz[1], and Mustapha Bouhtou[2]

[1] Maastricht University, Quantitative Economics, P.O.Box 616,
NL–6200 MD Maastricht, The Netherlands
{a.grigoriev, s.vanhoesel, a.vanderkraaij, m.uetz}@ke.unimaas.nl
[2] France Télécom R&D, 39-40 rue du Général Leclerc,
F–92131 Issy-Les-Moulineaux, France
mustapha.bouhtou@francetelecom.com

Abstract. We consider a Stackelberg pricing problem in directed networks. Tariffs have to be defined by an operator, the leader, for a subset of the arcs, the tariff arcs. Clients, the followers, choose paths to route their demand through the network selfishly and independently of each other, on the basis of minimal cost. Assuming there exist bounds on the costs clients are willing to bear, the problem is to find tariffs such as to maximize the operator's revenue. Except for the case of a single client, no approximation algorithm is known to date for that problem. We derive the first approximation algorithms for the case of multiple clients. Our results hold for a restricted version of the problem where each client takes at most one tariff arc to route the demand. We prove that this problem is still strongly $\mathcal{NP}$-hard. Moreover, we show that uniform pricing yields both an m–approximation, and a $(1 + \ln D)$–approximation. Here, m is the number of tariff arcs, and D is upper bounded by the total demand. We furthermore derive lower and upper bounds for the approximability of the pricing problem where the operator must serve all clients, and we discuss some polynomial special cases. A computational study with instances from France Télécom suggests that uniform pricing performs better than theory would suggest.

1 Introduction

The general setup for the tarification problem that we study involves two non-cooperative groups, an operator that sets tariffs, the *leader* of the Stackelberg game, and n clients that have to pay these tariffs, the *followers* of the Stackelberg game. More precisely, we assume that a network is given, and a subset of m arcs, the tariff arcs, are owned by an operator. The operator can set the tariffs on these arcs for renting capacity to one or several clients. Each client wishes to route a certain amount of a commodity on a path connecting two vertices. Such a path can involve one or several arcs belonging to the operator, and we assume

[*] This research was partially supported by France Télécom Research & Development.

that each client selfishly selects a path with minimum cost to route his demand. Before the clients select their paths, the operator has to set the tariffs, which he does in order to maximize total revenue. In order to avoid non-boundedness, we assume that clients always have the alternative of routing on a path without using any of the operators arcs.

The problem we consider here is different in two aspects from the network congestion problems studied recently, e.g., by Roughgarden and Tardos [11], and Cole et al. [2, 3]. First, we assume that there is no congestion, hence the clients do not influence each other. They choose minimum cost paths to route their commodities, independent of each other. The Game Theoretic setting is only introduced by the fact that there exist an operator trying to maximize revenue using high tariffs, and the clients try to avoid high tariffs by choosing minimal cost paths. Second, the pricing takes place before the users choose their paths, so we are faced with a Stackelberg game, where the operator (leader) first sets the tariffs, and then, subject to these tariffs, the clients (followers) react selfishly.

A natural formulation of the problem, referred to as the (general) tarification problem, is the bilevel linear formulation of Labbé et al. [9]. They show that already the problem with a single client is (strongly) NP-hard, given that also negative tariffs are allowed. Roch et al. [10] show that the single client problem remains (strongly) $\mathcal{NP}$-hard, even when restricted to nonnegative tariffs. In the same paper, a polynomial time $\mathcal{O}(\ln m)$–approximation algorithm for the problem with a single client is proposed, where m is the number of tariff arcs.

Our Results. We derive the first approximation results for the problem with multiple clients. However, we consider a restricted variant of the problem, since we assume that the path taken by any client utilizes at most one tariff arc. Several applications of this particular tarification problem, to which we refer as the *river tarification problem* (RTP) are briefly discussed in Section 2. Section 3 describes the model in detail. In Section 4, we show that the river tarification problem is (strongly) $\mathcal{NP}$-hard.

The quality of uniform tarification policies, where all arcs are priced with the same tariff, is analyzed in Section 5. The problem to find an optimal uniform tariff is well-known to be solvable in polynomial time, even for the general tarification problem [12]. We show that uniform tarification is an m–approximation, and this is tight. Using a simple geometric argument, we also show that uniform tarification is a $(1 + \ln D)$–approximation, which is tight up to a constant factor. Here, D is the total demand that is served by the operator in an optimal solution, which is upper bounded by the total demand. Hence, whenever the clients have unit demand, this yields a $(1 + \ln n)$–approximation.

We also consider another variant of the problem where the operator is forced to serve all clients. We show in Section 6, by a reduction from INDEPENDENT SET, that this problem is not approximable to within a factor $\mathcal{O}(m^{1-\varepsilon})$ or $\mathcal{O}(n^{1/2-\varepsilon})$, unless $\mathcal{ZPP} = \mathcal{NP}$. (Recall that m is the number of tariff arcs and n is the number of clients.) On the positive side, we can show that the problem admits an n-approximation.

We briefly discuss some polynomially solvable special cases of the river tarification problem in Section 7. Finally, we empirically analyze the quality of uniform tarification policies in Section 8, using instances from France Télécom.

2   Applications

As an illustration, consider transportation networks that resemble the situation of a town that is divided by a river. Different traversal possibilities exist, and some of these are to be priced by an operator. These traversal possibilities are the tariff arcs in the network. Customers want to route certain commodities from one side of the river to the other.

Such a network topology may be assumed (after a simple transformation described below) in telecommunication networks where we know a priori that the path of each client takes at most one tariff arc. This occurs, e.g., in the international interconnections market, where several operators offer connections to a particular country. If we focus on the market for this particular country, we can assume that it is not profitable for any client to enter the country twice.

For another motivation, consider the internet. Whenever an autonomous system (represented by some subnetwork) has to transit data, the data may enter and exit the autonomous system at different points. Clients have to pay a price for transmitting data through the autonomous system, yielding revenue for its owner. The data flow can be modelled such that once it is routed through the autonomous system, it does not pass a second time, thereby creating an instance of the river tarification problem.

Finally, in point-to-point markets, a telecommunications operator is offering bandwidth capacity between two points in different qualities of service (QoS). In that setting, it is often the case that information is available concerning the prices customers are willing to pay for different levels of QoS. That pricing problem can be modelled easily as a river tarification problem, too.

3   Model

An instance of the general tarification problem is a directed graph $G = (N, A)$, where the arc set A is partitioned into a set of m tariff arcs $T \subseteq A$ and a set of fixed cost arcs $F = A \setminus T$. There are n clients (or commodities) $k \in \{1, \ldots, n\}$, where each client k has a demand d_k that has to be routed from source node s_k to target node t_k. Because there is no congestion involved, we may assume without loss of generality that all demand values d_k are scaled to be integral. We define for a commodity k the set of all possible paths from s_k to t_k by P_k. The tariff on a tariff arc $a \in T$ is denoted by τ_a, and the vector of all tariffs is given by $\tau = (\tau_a)_{a \in T}$. The cost of a fixed cost arc $a \in F$ is denoted by c_a.

The clients route their demands from source to destination according to a path with minimal total cost, where the total cost of a path is defined as the sum of the tariffs and fixed costs on the arcs of the path. Whenever the client has a choice among multiple paths with the same total cost but with different revenues for the operator, we assume that the client takes the path that is most profitable to the operator. This can always be achieved with arbitrary precision by reducing all tariffs by some small value ε. We assume that an $\{s_k, t_k\}$-path exists that consists only of fixed cost arcs for every client $k \in \{1, \ldots, n\}$, since the problem is otherwise unbounded.

Without going into further details, we mention that this tarification problem is a classical Stackelberg Game that can be modelled as (linear-linear) bilevel program [9,1]. It follows from Jeroslow [7] that (linear-linear) bilevel programs are $\mathcal{NP}$-hard in general. For annotated bibliographies on bilevel programming, see Vicente and Calamai [13] or Dempe [4].

We next describe a simple transformation of the given graph G that allows us to restrict to very specific graphs (although probably losing certain graph properties, such as planarity). When replacing shortest paths using only fixed cost arcs by direct arcs, and possibly introducing some dummy arcs with zero or infinite cost, one obtains a shortest path graph model (SPGM) as defined by Bouhtou et al. [1]. In that model, all tariff arcs are disjoint, and there exists an arc from any source node s_k to the tail node of any tariff arc, and from the head node of any tariff arc to any target node t_k. Moreover, there exists a fixed cost arc (s_k, t_k) for all $k = 1, \ldots, n$, and the cost c_k is the highest acceptable price for client k.

The additional assumption in the problem considered in this paper (to which we refer as the river tarification problem) is the following: Independent of the tariffs, any client routes his demand only on a path that includes at most one tariff arc. In the shortest graph path model, that is equivalent to the deletion of any backward-arc that might exist between the head nodes of tariff arcs back to tail nodes of other tariff arcs. Figure 1 shows the shortest path graph model of an instance of the river tarification problem with three tariff arcs and two clients. The tariff arcs $a_i, i \in \{1, 2, 3\}$ are given by the dashed arcs in the network. We may also assume without loss of generality that all fixed cost arcs incident with the target nodes t_k have zero cost (by adding their costs to the fixed cost

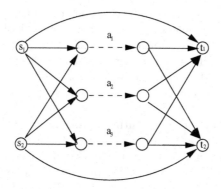

Fig. 1. River tarification problem with $n = 2$ and $m = 3$

arcs incident with s_k). Notice that the only difference to the general tarification problem described above is the non-existence of backward arcs.

The essential parameters that define an instance of a (river) tarification problem are therefore the number of tariff arcs m, the number of clients n, the demand values d_k, $k \in \{1, \ldots, n\}$, and the costs c_a of the fixed cost arcs $a \in F$. Due to the fact that any path taken by a client involves exactly one fixed cost arc with non-zero cost, we may assume without loss of generality that the costs c_a of the fixed cost arcs $a \in F$ are integral. Moreover, due to the integrality of the costs of the fixed cost arcs, it is immediate that any reasonable solution will adopt only tariffs which are integral, too. Notice that this might not hold for the general tarification problem, where a path chosen by a client can consist of more than one tariff arc.

4   Complexity

Roch et al. [10] show that the general tarification problem is $\mathcal{NP}$-hard in the strong sense, even when restricted to a single client, using a reduction from the $\mathcal{NP}$-complete problem 3-SAT [5]. Their reduction works for tarification problems where paths are allowed to use (and indeed, must use) several tariff arcs. We show that the tarification problem with multiple clients, but restricted to at most one tariff arc per path, is $\mathcal{NP}$-hard in the strong sense, too.

We also use a reduction from 3-SAT. Therefore, consider a boolean function $f : \{0,1\}^n \to \{0,1\}$ on n variables $x_1, \ldots, x_n$, in conjunctive normal form. Such a function f is the conjunction of m clauses C_k,

$$f = \bigwedge_{k=1}^{m} C_k, \tag{1}$$

each clause C_k being the disjunction of three literals, $C_k = (\ell_{k1} \vee \ell_{k2} \vee \ell_{k3})$. Any literal ℓ_{kj} represents either a variable x_i, or its negation $\bar{x}_i$, $i \in \{1, \ldots, n\}$. Then f is satisfiable if there exists a truth assignment $x_1, \ldots, x_n$ such that at least one literal per clause is true.

Any function of the form (1) can be polynomially transformed to an instance of the river tarification problem as follows. For each variable x_i, $i \in \{1, \ldots, n\}$, we construct a constant-size subnetwork as shown in Figure 2. Each of these subnetworks has three clients with unit demand, with origin-destination pairs $\{s_{ij}, t_{ij}\}$, $j \in \{1, 2, 3\}$. Moreover, each subnetwork has two tariff arcs, a_i representing the truth assignment $x_i = 1$, and $\bar{a}_i$ representing $x_i = 0$, as depicted in Figure 2.

An upper bound on the cost of routing commodities 1 and 3 is given by fixed cost arcs (s_{i1}, t_{i1}) and (s_{i3}, t_{i3}), both with cost 3. For commodity 2, the upper bound on the cost is given by a fixed cost arc (s_{i2}, t_{i2}), with cost 2. The maximal revenue for each subnetwork is thus given by setting one of the tariffs to 2, and the other to 3, yielding a revenue of $2 \cdot 2 + 3 = 7$. In all other cases, the revenue is not more than 6.

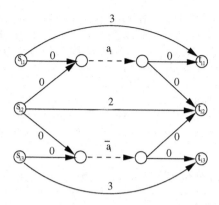

Fig. 2. Subnetwork for variable x_i, $i \in \{1, \ldots, n\}$

Next, for each clause C_k, $k \in \{1, \ldots, m\}$, we create a *clause-commodity* k with origin destination pairs $\{s_k, t_k\}$, with unit demand. Whenever a variable x_i ($\bar{x}_i$, respectively) appears as one of clause C_k's literals, we connect s_k to s_{i1} (s_{i3}, respectively), and t_{i1} (t_{i3}, respectively) to t_k, using arcs of zero cost. In addition, we introduce a fixed cost arc (s_k, t_k) with cost 2, defining an upper bound of 2 for the cost of routing clause-commodity k. The so-defined instance of the river tarification problem has $2n$ tariff arcs, $3n + m$ commodities (or clients), and $7m + 11n$ fixed cost arcs, hence the transformation is indeed polynomial.

Theorem 1. *The river tarification problem is strongly* $\mathcal{NP}$*-hard.*

Proof. Consider the polynomial transformation defined previously. It is now straightforward to show that a satisfying truth assignment for f exists if and only if the revenue for the river tarification problem is equal to $2m + 7n$. $\quad\square$

The reduction used for the proof of Theorem 1 shows that the river tarification problem remains $\mathcal{NP}$-hard even for unit demands, a fixed number of tariff values and when the operator is forced to use tariffs such that he serves (a given subset of) all clients.

5   The Quality of Uniform Tarification Policies

The uniform tarification problem (UTP) is the same as the general tarification problem, with the additional restriction that all tariffs are required to be identical. As shown by van Hoesel et al. [12], the uniform tarification problem can be solved in polynomial time, even in the general setting where clients may use paths with several tariff arcs. The algorithm described in van Hoesel et al. [12] uses the parametric shortest path algorithm of Young et al. [14] and Karp and Orlin [8] to determine the tariff values (i.e. breakpoints) for which the shortest path tree changes for any client. Calculating the revenue for the operator at each breakpoint and maintaining the best solution yields the optimal uniform tarification policy in polynomial time.

We next analyze the loss that can be experienced by adopting such a uniform tarification policy for the river tarification problem. Apart from theoretical interest, the question is motivated by the interest in efficient tarification strategies in a more general setting with more than one operator. In addition, although it is quite easy to think of smarter tarification policies, so far all these policies, except uniform tarification, resisted our attempts of a worst-case analysis.

Therefore, denote by Π^{UTP} the revenue for an optimal uniform tarification, and by Π^{OPT} the revenue for an optimal non-uniform tarification. By definition, $\Pi^{UTP} \leq \Pi^{OPT}$.

Lemma 1. *If an optimal tarification for the river tarification problem with revenue Π^{OPT} utilizes at most r different tariffs, then for the optimal uniform tarification, $\Pi^{UTP} \geq \Pi^{OPT}/r$.*

The proof of this lemma is indeed trivial. To this end, consider an optimal non-uniform tarification with tariffs $\tau_1 \leq \cdots \leq \tau_m$, and let D_i be the total demand on an arc a_i with tariff τ_i, $i \in \{1, \ldots, m\}$. By $D = \sum_{k=1}^{n} D_k$ we denote the total demand served by the operator. Then the revenue created by this solution is the area under the following 'staircase' function $f : [0, D] \to [0, \infty[$.

$$f(x) = \tau_i \quad \text{for all } x \text{ with } \sum_{j<i} D_j \leq x < \sum_{j\leq i} D_j, \quad i \in \{1, \ldots, m\}. \tag{2}$$

Proof (of Lemma 1). Consider any of the rectangles inscribed under the graph of function $f(x)$, with area $T_i := \tau_i \cdot \sum_{j\geq i} D_j$. Then it holds that $\Pi^{UTP} \geq T_i$ for all $i \in \{1, \ldots, m\}$, since the area of any such rectangle is a lower bound for the revenue yielded by the optimal uniform tariff Π^{UTP}. (Notice that this does not hold for the general tarification problem.) Hence, if only r different tariffs are utilized, we consider the r (inclusion-)maximal rectangles under function f, say $T_{i_1}, \ldots, T_{i_r}$, and get $r \cdot \Pi^{UTP} \geq \sum_{j=1}^{r} T_{i_j} \geq \Pi^{OPT}$. $\qquad\square$

Since $r \leq m$, Lemma 1 yields the following theorem. Tightness of the result will be shown below, using Example 1.

Theorem 2. *Uniform tarification is an m–approximation for the river tarification problem.*

We next derive an another bound on the quality of uniform tarification policies, using a geometric argument.

Theorem 3. *Uniform tarification is a $(1 + \ln D)$–approximation for the river tarification problem, where $D \leq \sum_{k=1}^{n} d_k$ is the total demand that is served by the operator in an optimal solution.*

Proof. Indeed, we will even prove a slightly stronger result than claimed in Theorem 3. Consider an optimal non-uniform tarification, and recall the definition of the corresponding staircase function f in (2), as well as the inscribed rectangles, with areas $T_i = \tau_i \cdot \sum_{j\geq i} D_j$. Let ℓ be the index of the maximal area rectangle among all T_i, with area T_ℓ. Let $x_\ell := \sum_{j \geq \ell} D_j = T_\ell / \tau_\ell$. Moreover, denote by $\tau_{\max}$ the maximal tariff utilized in that optimal solution. We show

$$\Pi^{\mathrm{UTP}} \geq \frac{\Pi^{\mathrm{OPT}}}{1 + \ln(D\tau_{\max}/T_\ell)} \,. \tag{3}$$

Theorem 3 then follows, because $T_\ell \geq \tau_{\max}$ by definition of T_ℓ. To prove (3), we define the function

$$g(x) := \frac{T_\ell}{D - x} \text{ for } x \in [0, D)\,. \tag{4}$$

We claim that $g(x) \geq f(x)$ for $x \in [0, D)$. To see this, take any x with $\sum_{j<i} D_j \leq x < \sum_{j \leq i} D_j$, then $f(x) = \tau_i$ by definition. Now

$$g(x) = \frac{T_\ell}{D - x} \geq \frac{T_\ell}{D - \sum_{j<i} D_j} = \frac{T_\ell}{\sum_{j \geq i} D_j} = \frac{T_\ell}{T_i/\tau_i} \geq \tau_i = f(x)\,,$$

where the first inequality follows by choice of x, and the last follows by choice of ℓ as the index of the largest rectangle.

Hence, the area under the staircase function, which equals Π^{OPT}, can be upper bounded in terms of the area defined by the function $g(x)$, as depicted in Figure 3. To compute this area, we partition it into three parts, namely the rectangle T_ℓ itself, the area under $g(x)$ on the domain $x \in [0, D - x_\ell]$, as well as the area to the right of $g(x)$ on the domain $\tau \in [\tau_\ell, \tau_{\max}]$. The latter is the integral of the function $D - g^{-1}(\tau) = T_\ell/\tau$ on the domain $[\tau_\ell, \tau_{\max}]$. We thus obtain the following.

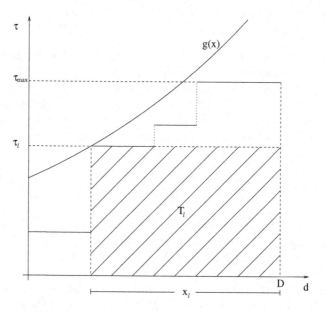

Fig. 3. Illustration for the proof of Theorem 3

$$\Pi^{\mathrm{OPT}} \le T_\ell + \int_0^{D-x_\ell} \frac{T_\ell}{D-x} \, dx + \int_{T_\ell}^{\tau_{\max}} \frac{T_\ell}{\tau} \, d\tau$$
$$= T_\ell[1 + \ln D + \ln \tau_{\max} - \ln T_\ell - \ln x_\ell]$$
$$= T_\ell [1 + \ln(D\tau_{\max}/T_\ell)] ,$$

and since $T_\ell \le \Pi^{\mathrm{UTP}}$, claim (3) follows.   □

Notice that claim (3) confirms the following geometric intuition: The closer the staircase function $f(x)$ is to the straight line $x \mapsto (\tau_{\max}/D) \cdot x$, the closer is T_ℓ to $D\tau_{\max}/4$, which yields an approximation ratio of $(1 + \ln 4) \approx 2.4$ for uniform tarification. Geometric intuition indeed suggests a ratio of roughly 2, the additional 0.4 being caused be the difference between the functions $g(x)$ and $f(x)$. In Section 8, we compare the quality of uniform versus non-uniform tarification, based on instances obtained from France Télécom.

In the case of unit demands of the clients, that is, if $d_k = 1$ for all clients $k = 1, \ldots, n$, we obtain the following.

Corollary 1. *Whenever clients have unit demands, uniform tarification is a $(1 + \ln n)$-approximation for the river tarification problem.*

Finally, let us show tightness of the bounds in Theorems 2 and 3.

Example 1. Given $n=m$ commodities and m tariff arcs. Every commodity is operating its own subnetwork with one tariff arc, thus the entire network consists of m disjoint subnetworks and each of them contains one commodity and one tariff arc. Fix $b > 1$ and let the demand in subnetwork k be given by $d_k = b^k - b^{k-1}$, $k \in \{1, \ldots, m\}$. This way, the total demand D equals $b^m - 1$. Moreover, the maximal revenue for subnetwork k is limited by a fixed cost arc (s_k, t_k), with cost $c_k = b^{2m-k}$. Hence, the maximal tariff $\tau_{\max}$ equals b^{2m-1}.   □

In the optimal solution, the tariff for each subnetwork k is set to its maximal value, b^{2m-k}. Each subnetwork therefore contributes a revenue of $b^{2m} - b^{2m-1}$, and $\Pi^{\mathrm{OPT}} = m(b^{2m} - b^{2m-1})$. The optimal uniform tarification consists in setting the tariff on all tariff arcs to b^m. This way, every unit of demand creates a profit of b^m, yielding a total revenue of $b^{2m} - b^m$. Other (reasonable) uniform tariffs would be values b^{2m-k}, $k \in \{1, \ldots, m-1\}$. This yields a total revenue of $b^{2m} - b^{2m-k}$, which is less. Therefore, we obtain

$$\Pi^{\mathrm{UTP}}/\Pi^{\mathrm{OPT}} = \frac{b^{2m} - b^m}{m(b^{2m} - b^{2m-1})} \le \frac{b^{2m}}{m(b^{2m} - b^{2m-1})} = \frac{1}{m} \cdot \frac{b}{b-1} .$$

Now, observe that in the optimal solution m different tariffs are utilized. Lemma 1 (Theorem 2, respectively) suggests that uniform tarification provides an m-approximation. Example 1 proves that this is best possible, since b can be chosen arbitrarily large.

Moreover, Theorem 3 suggests that uniform tarification is a $(1 + \ln D)$-approximation. In Example 1, we have $D = (b^m - 1)$ and thus $(1 + \ln D) = 1 + \ln(b^m - 1) \le 1 + m \ln b$. Hence, Theorem 3 yields that uniform tarification is a $\mathcal{O}(m)$–approximation on this example. The same Example 1 shows that $\mathcal{O}(m)$ is indeed best possible. Summarized, we thus get the following.

Theorem 4. *For uniform tarification, the performance bound of Theorem 2 is best possible, and the performance bound of Theorem 3 is best possible up to a constant factor.*

6   All-Service River Tarification Problem

In this section, we consider the following variation of the river tarification problem. The operator must set tariffs in order to capture the demand of all clients, that is, tariffs must be such that no client k is forced to use the arc (s_k, t_k). We refer to this problem as the *all-service* river tarification problem. NP-hardness of this problem follows by our previous reduction presented in Section 4.

It follows from trivial examples that the maximal revenue for the all-service problem can be an arbitrary factor away from the maximal revenue without the all-service constraint. Hence, we have an arbitrarily high 'cost of regulation'. In addition, we can show that the maximal revenue for the all-service problem cannot be approximated well.

Theorem 5. *For any $\varepsilon > 0$, the existence of a polynomial time approximation algorithm for the all-service river tarification problem with with n clients and m tariff arcs with worst case ratio $\mathcal{O}(m^{1-\varepsilon})$ or $\mathcal{O}(n^{1/2-\varepsilon})$ implies $\mathcal{ZPP} = \mathcal{NP}$.*

Proof. The proof uses an approximation preserving reduction from INDEPENDENT SET [5] to the all-service RTP. So assume we are given a graph $G = (V, E)$, and the problem is to find a maximum cardinality subset $V' \subseteq V$ of vertices such that no two vertices in V' are connected by an edge. The transformation works as follows. For every vertex $v \in V$ we introduce a client with origin-destination pair $\{s_v, t_v\}$ and demand $d_v = |E|$, and a corresponding tariff arc a_v. We connect the source s_v to the tail of the tariff arc a_v, and the head of a_v to the destination t_v, using zero cost fixed cost arcs. Moreover, there is a fixed cost arc (s_v, t_v) with cost $(|V| + 1)$ for all vertices $v \in V$. For every edge $e \in E$ we introduce a client with origin-destination pair $\{s_e, t_e\}$ and unit demand. The upper bound on the cost of routing this demand is given by the fixed cost arc (s_e, t_e) with cost 1. For all edges $e \in E$ and all vertices $v \in V$ with $v \in e$, we furthermore introduce fixed cost arcs $(s_e, \text{tail}(a_v))$ and $(\text{head}(a_v), t_e)$, with zero cost. This transformation results in an instance of the all-service RTP with $|V|$ tariff arcs, and $|V| + |E|$ clients. Figure 4 gives an example of such a transformation for a graph $G = (V, E)$ with 3 nodes and 2 edges.

We claim that G has an independent set of cardinality at least k if and only if there exists a tariff policy for the all-service RTP with a total revenue of $|V||E|(k + 1) + |E|$.

First, assume that G has an independent set V' of cardinality k. For all $v \in V'$, set the tariff on the corresponding tariff arc a_v to $|V| + 1$, and all other tariffs to 1. By the definition of an independent set, for any edge $e = (v, u) \in E$

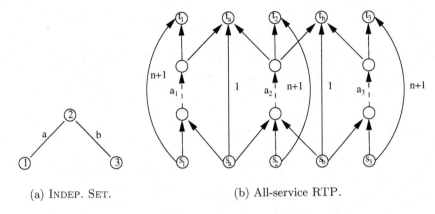

(a) INDEP. SET.   (b) All-service RTP.

Fig. 4. Reduction of INDEPENDENT SET to all-service RTP

at least one of the vertices, v or u, is not in V'. Therefore, the tariff of at least one of the tariff arcs, a_v or a_u is 1. All clients corresponding to an edge e can thus be served, using one of the tariff arcs a_v or a_u. The clients (s_v, t_v) corresponding to the vertices $v \in V$ are also served, since the upper bound of $|V|+1$ is not exceeded with the so-defined tariffs. Hence, all demands are served. The revenue consists of $|E|$ from all clients corresponding to the edges E of G, $|E|(|V|+1)k$ from the clients corresponding to the independent set V', and $|E|(|V|-k)$ from the clients corresponding to $V \setminus V'$. That yields a total revenue of $|E||V|(k+1) + |E|$.

Conversely, assume that there exists a set of tariffs that captures all demands, such that the revenue is $|E||V|(k+1) + |E|$. We will show that this implies that the graph G has an independent set of cardinality at least k. Since all demands are captured at this tarification strategy, for any edge $e = (v, u) \in E$, the tariff on at least one of the arcs, a_v or a_u, is 1. Consider the set of vertices $V' := \{v \in V : t_{a_v} > 1\}$. By definition, no pair of nodes $v, u \in V'$ is connected by an edge. Hence, V' is an independent set in G. Let $k' := |V'|$. The revenue is equal to $|E| + |E|(|V| - k') + |E|(|V|+1)k' = |E||V|(k'+1) + |E|$, which by assumption is at least as large as $|E||V|(k+1) + |E|$. This implies that $k' \geq k$ and thus that V' is an independent set in G of cardinality $k' \geq k$.

Now, let us assume that we have an α-approximation algorithm $\mathcal{A}$ for the all-service RTP, with $\alpha \geq 1$. Consider any instance $G = (V, E)$ of INDEPENDENT SET, and the all-service RTP resulting from the above reduction. We can assume that both the optimal solution and the solution produced by $\mathcal{A}$ only utilize tariff values 1 or $|V|+1$, because any tariff greater than 1 and not equal to $|V|+1$ can be turned into $|V|+1$ with a revenue gain. So $\Pi^{\text{OPT}} = |E||V|(k+1) + |E|$ for some k, and $\Pi^{\mathcal{A}} = |E||V|(k'+1) + |E|$ for some k'. The first part of the proof yields that the maximal independent set of G has size k, and algorithm $\mathcal{A}$ can be used to find an independent set of size at least k'. Moreover,

$$\frac{1}{\alpha} \leq \frac{|E||V|(k'+1) + |E|}{|E||V|(k+1) + |E|} = \frac{1 + \frac{1}{|V|} + k'}{1 + \frac{1}{|V|} + k} \leq \frac{2 + k'}{1 + k},$$

hence $k' \geq (k+1)/\alpha - 2$. In other words, we have an $\mathcal{O}(\alpha)$–approximation algorithm for the INDEPENDENT SET problem.

It is now well known from work of Håstad [6] that the INDEPENDENT SET problem cannot have a polynomial time approximation algorithm with worst case guarantee $\mathcal{O}(|V|^{1/2-\varepsilon})$ unless $\mathcal{P} = \mathcal{NP}$, and that it cannot have a polynomial time approximation algorithm with worst case guarantee $\mathcal{O}(|V|^{1-\varepsilon})$ unless $\mathcal{ZPP} = \mathcal{NP}$. Since the number of tariff arcs m in our transformation equals $|V|$, the first claim of the theorem follows. Since the number of clients n in our transformation equals $|V| + |E| \in \mathcal{O}(|V|^2)$, the second claim follows. $\square$

On the positive side, we can show the following.

Theorem 6. *There exists an n-approximation algorithm for the all-service river tarification problem.*

The proof works by enumeration over all $m \cdot n$ possibilities for a maximum revenue client using a specific arc. Given that arc-client pair, we can find a corresponding optimal tariff for that arc in polynomial time using binary search, in each step solving a system of linear inequalities. We skip the details due to space limitations.

7   Polynomially Solvable Special Cases

Several polynomially solvable special cases of the (general) tarification problem are discussed by Labbé et al. [9] and van Hoesel et al. [12]. Clearly, these results hold for the problem considered in this paper, too.

In addition, the river tarification problem is also polynomially solvable if the number of clients n is bounded from above by a constant. In that case, the number of assignments of clients to tariff arcs is bounded by m^n which is a polynomial for fixed n. Consider therefore the following formulation, where we use notation as given next. The path taken by each client in the network is denoted by $p_k^* \in P_k$, and P_k represents the set of all possible paths taken by a client $k \in K$. The revenue associated with a path $p \in P_k$ induced by a client k with demand d_k is defined by $\pi_p(\tau, d_k) = d_k \cdot \tau_a$, where a is the (unique) tariff arc on path p. The fixed cost of a path p is given by $c_p(d_k) = d_k \sum_{a \in F \cap p} c_a$. Then $l_p(\tau, d_k) := c_p(d_k) + \pi_p(\tau, d_k)$ is the total cost of the path $p \in P_k$ for a client k.

$$\begin{aligned}
\max_{\tau} \ & \sum_{k \in K} \pi_{p_k^*}(\tau, d_k) \\
\text{s.t.} \ & l_p(\tau, d_k) \geq l_{p_k^*}(\tau, d_k) \ \forall k \in K, \forall p \in P_k \\
& \tau_a \geq 0 \qquad\qquad\qquad \forall a \in T
\end{aligned} \tag{5}$$

Since for each client, there are at most $m + 1$ paths in the network, $|P_k|$ is bounded by $m + 1$. Hence, the number of constraints is polynomial in the input data. Therefore, if we solve m^n instances of (5), we can retrieve the optimal solution in polynomial time.

8   Numerical Results

As stated previously, whenever the function that describes the total revenue in an optimal non-uniform solution, i.e. the staircase function defined in (2), is close to a straight line, geometric intuition suggests a worst-case ratio for uniform tarification of approximately 2. The worst case Example 1 crucially hinges on a (staircase) function that approximates a hyperbola. Thus, it can be conjectured that the empirical performance of uniform tarification policies outperforms the theoretical bounds we have found. This is indeed confirmed in the following numerical experiments, displayed in Table 1. The study is based on instances obtained from France Télécom.

Table 1. Quality of Uniform Tarification on France Télécom instances

Instance	$|N|$	$|A|$	m	n	Π^{OPT}	Π^{UTP}	%
RTN1	29	94	7	15	841	624	74%
RTN2	29	98	6	21	4099	3496	85%
RTN3	59	206	10	13	1118	880	79%
RTN4	59	204	10	20	2217	1512	68%
RTN5	49	120	9	21	74948	55968	74%
RTN6	33	116	15	12	28166	20328	72%

These instances represent telecommunication networks for the international interconnections market, as described in Section 2. We compare the optimal solutions for uniform tariffs (Π^{UTP}) and non-uniform tariffs (Π^{OPT}). The optimal non-uniform solution is calculated using the model and mixed integer programming formulation described in Bouhtou et al. [1]. The value of Π^{UTP} is calculated using the same formulation, requiring that all tariffs be equal. As such, we do not compare the actual computation times, but are just interested in effectiveness of the optimal uniform tarification policies. Table 1 gives a brief description of each network, stating the number of nodes, arcs, tariff arcs and clients. The optimal non-uniform and uniform solution values are displayed in the columns Π^{OPT} and Π^{UTP}. The final column is the approximation ratio.

Acknowledgements

We thank Maxim Sviridenko for an insightful discussion and the suggestion for the proof of Theorem 6, and the anonymous referees for some helpful remarks.

References

1. M. Bouhtou, S. van Hoesel, A. F. van der Kraaij, and J.L. Lutton. Tariff optimization in networks. *Meteor Research Memorandum* RM03011, Maastricht University, 2003.

2. R. Cole, Y. Dodis, and T. Roughgarden. How much can taxes help selfish routing? In *Proceedings of the 4th Annual ACM Conference on Electronic commerce*, pages 98–107, 2003.
3. R. Cole, Y. Dodis, and T. Roughgarden. Pricing network edges for heterogeneous selfish users. In *Proceedings of the 35th Annual ACM Symposium on Theory of Computing*, pages 521–530, 2003.
4. S. Dempe. Annotated bibliography on bilevel programming and mathematical programs with equilibrium constraints. *Optimization*, 52:333–359, 2003.
5. M.R. Garey and D.S. Johnson. *Computers and Intractability: A Guide to the Theory of NP-Completeness*. W.H. Freeman, San Francisco, 1979.
6. J. Håstad. Clique is hard to approximate within $n^{1-\varepsilon}$. *Acta Mathematica*, 182:105–142, 1999.
7. R.G. Jeroslow. The polynomial hierarchy and a simple model for competitive analysis. *Mathematical Programming*, 32:146–164, 1985.
8. R.M. Karp and J. B. Orlin. Parametric shortest path algorithms with an application to cyclic staffing. *Discrete Applied Mathematics*, 3:37–45, 1981.
9. M. Labbé, P. Marcotte, and G. Savard. A bilevel model of taxation and its application to optimal highway pricing. *Management Science*, 44:1608–1622, 1998.
10. S. Roch, G. Savard, and P. Marcotte. Design and Analysis of an approximation algorithm for Stackelberg network pricing. *Optimization Online*, 2003.
11. T. Roughgarden and É. Tardos. How bad is selfish routing? *Journal of the Association for Computing Machinery*, 49(2):236–259, 2002.
12. S. van Hoesel, A. F. van der Kraaij, C. Mannino, G. Oriolo, and M. Bouhtou. Polynomial cases of the tarification problem. *Meteor Resarch Memorandum* RM03053, Maastricht University, 2003.
13. L.N. Vicente and P. H. Calamai. Bilevel and multilevel programming: A bibliography review. *Journal of Global Optimization*, 5:291–306, 1994.
14. N. E. Young, R. E. Tarjan, and J. B. Orlin. Faster parametric shortest path and minimum-balance algorithms. *Networks*, 21:205–221, 1991.

Submodular Integer Cover and Its Application to Production Planning

Toshihiro Fujito* and Takatoshi Yabuta**

Graduate School of Information Science, Nagoya University,
Furo, Chikusa, Nagoya 464-8603 Japan
fujito@nuee.nagoya-u.ac.jp

Abstract. Suppose there are a set of suppliers i and a set of consumers j with demands b_j, and the amount of products that can be shipped from i to j is at most c_{ij}. The amount of products that a supplier i can produce is an integer multiple of its capacity κ_i, and every production of κ_i products incurs the cost of w_i. The *capacitated supply-demand (CSD)* problem is to minimize the production cost of $\sum_i w_i x_i$ such that all the demands (or the total demand requirement specified separately) at consumers are satisfied by shipping products from the suppliers to them.

To capture the core structural properties of CSD in a general framework, we introduce the *submodular integer cover (SIC)* problem, which extends the *submodular set cover (SSC)* problem by generalizing submodular constraints on subsets to those on integer vectors. Whereas it can be shown that CSD is approximable within a factor of $O(\log(\max_i \kappa_i))$ by extending the greedy approach for SSC to CSD, we first generalize the primal-dual approach for SSC to SIC and evaluate its performance. One of the approximation ratios obtained for CSD from such an approach is the maximum number of suppliers that can ship to a single consumer; therefore, the approximability of CSD can be ensured to depend only on the network (incidence) structure and not on any numerical values of input capacities κ_i, b_j, c_{ij}.

The CSD problem also serves as a unifying framework for various types of covering problems, and any approximation bound for CSD holds for set cover generalized *simultaneously* into various directions. It will be seen, nevertheless, that our bound matches (or nearly matches) the best result for each generalization *individually*. Meanwhile, this bound being nearly tight for standard set cover, any further improvement, even if possible, is doomed to be a marginal one.

1   Introduction

1.1   Capacitated Supply-Demand Problem

Suppose there are a set A of factories, that all produce the same product, and a set B of customers that use the product. Each factory $i \in A$ has capability

---

* Supported in part by a Grant in Aid for Scientific Research of the Ministry of Education, Science, Sports and Culture of Japan.
** currently at Toyota Motor Corporation

of producing products in the units of κ_i tons, incurring the cost of w_i dollars per unit. Customer $j \in B$ requests b_j tons of the product every month. ¿From factory i to customer j at most c_{ij} tons of the product can be transported every month. Moreover, the total demand b_{total}, with $0 \leq b_{\text{total}} \leq \sum_{j \in B} b_j$, specifies, out of the total requested amount $\sum_{j \in B} b_j$, what amount of products need to be supplied in total. A monthly production plan specifies the number of units to be produced a month at each factory. The problem is: what is the most economical monthly production plan to fulfill at least b_{total} tons of all the needs requested by the customers ? We call this problem the *capacitated supply-demand (CSD)* problem.

Using a supply-demand model network, this problem can be defined equivalently as follows. Let $\mathcal{N} = (V, E)$ be a network, where $V = A \cup B \cup \{s, t, t'\}$, with source s, subsink t, sink t', a set A of supply nodes, and a set B of demand nodes. Each supply node $i \in A$ has an incoming arc from source s, each demand node has an outgoing arc to subsink t, one arc goes from t to sink t', and all the other arcs go from A to B (thus, $E = (\{s\} \times A) \cup E' \cup (B \times \{t\}) \cup \{(t, t')\}$ where $E' \subseteq (A \times B)$). All the arcs other than those in $\{s\} \times A$ are a priori associated with integral capacities; each capacity b_j on arc (j, t) specifies the "demand" requested by node $j \in B$, c_{ij} on arc $(i, j) \in A \times B$ limits the amount of supply that can be shipped from $i \in A$ to $j \in B$, and b_{total} on (t, t') is the total amount of demands to be supplied. The capacities on the remaining arcs $(s, i) \in \{s\} \times A$ are not given initially; rather, they need to be "purchased" as follows. For each $i \in A$ the unit capacity of κ_i is available for the unit cost of w_i, and it costs $w_i x_i$ to install the capacity of $a_i = \kappa_i x_i$ on (s, i), where a nonnegative integer x_i specifies the number of unit capacities to be used on (s, i). With all the capacities fully specified, the network is denoted by $\mathcal{N} = (A \cup B \cup \{s, t, t'\}, E, a, b, c, b_{\text{total}})$, where $a_i = \kappa_i x_i$ for each $i \in A$. The problem is then to install capacities $a \in \mathbb{Z}_+^A$ of minimum total cost on arcs in $\{s\} \times A$ by purchasing the available unit capacities κ, so that the total demand of b_{total} can be shipped in $\mathcal{N}$, or in other words, the max flow value reaches b_{total} in $\mathcal{N}$.

It should be clear by now that, denoting by f_{ij} the amount of a flow going from $i \in A$ to $j \in B$ in $\mathcal{N}$, CSD can be succinctly formulated by the following integer program:

$$\min \sum_{i \in A} w_i x_i$$

subject to:
$$\sum_{j \in B} f_{ij} \leq \kappa_i x_i \qquad i \in A$$
$$\sum_{i \in A} f_{ij} \leq b_j \qquad j \in B$$
$$\sum_{j \in B} \sum_{i \in A} f_{ij} \geq b_{\text{total}}$$
$$x_i \in \mathbb{Z}_+ \qquad i \in A$$
$$0 \leq f_{ij} \leq c_{ij} \quad i \in A, \; j \in B$$

where $w \in \mathbb{Q}_+^A, C = [c_{ij}] \in \mathbb{Z}_+^{A \times B}, \kappa \in \mathbb{Z}_+^A, b \in \mathbb{Z}_+^B$. To capture the core structural properties of CSD in a general framework, however, we next introduce the *submodular integer cover* problem, and will later derive an IP formulation for it.

1.2   Submodular Integer Cover

Definition 1. *A function* $g : \mathbb{R}^n \to \mathbb{R}$ *is said to be*

1. *nondecreasing if* $x \leq y$ *implies* $g(x) \leq g(y)$, *and*
2. *submodular if* $g(x) + g(y) \geq g(x \vee y) + g(x \wedge y)$ *for all* $x, y \in \mathbb{R}^n$,

where $x \vee y$ *and* $x \wedge y$ *are, respectively, the vectors of componentwise maxima and minima of* x *and* y: $(x \vee y)_i = \max\{x_i, y_i\}, (x \wedge y)_i = \min\{x_i, y_i\}$.

Let $G = (V, E)$ be a network with a nonnegative capacity $c(a)$ for each arc $a \in E$. Two arcs are said to be *parallel* if every simple cycle containing both of them orients them in the opposite direction, and a set of arcs is *parallel* if it consists of pairwise parallel arcs. Let P be a parallel arc set and denote the vector of capacities on arcs in P by c_P.

Proposition 2 (Gale-Politof [12]) *The maximum flow value* F *of* G *as a function in* c_P *is submodular.*

Recall the network $\mathcal{N}$ constructed in Sect. 1.1. The arc set $\{s\} \times A$ in $\mathcal{N}$ is then parallel, and hence, the max flow value F of $\mathcal{N}$ is submodular in $c_{\{s\} \times A}$. Define a function $\rho : \mathbb{Z}_+^A \to \mathbb{R}_+$ s.t.

$$\rho(x) = (\text{max flow value of } \mathcal{N} \text{ with } a_i = \kappa_i x_i, \forall i)$$

for a CSD instance of $(\mathcal{N}, \kappa, w)$. Then, $x \in \mathbb{Z}_+^A$ is a feasible CSD solution for $(\mathcal{N}, \kappa, w)$ iff $\rho(x) = b_{\text{total}} = \rho(\vec{\infty})$. Using ρ, CSD can be thus formulated by $\min\{\sum_{i \in A} w_i x_i \mid \rho(x) = \rho(\vec{\infty})\}$.

One can easily observe that ρ is nondecreasing as well as submodular by Proposition 2 since $\rho(x) = F((\cdots, \kappa_i x_i, \cdots))$, and CSD thus fits into the following general framework:

Definition 3. *Given a finite set* A, *a weight function* $w : A \to \mathbb{R}$, *and a nondecreasing submodular function* $g : \mathbb{Z}_+^A \to \mathbb{R}$, *the problem of computing* $\min\{\sum_{i \in A} w_i x_i \mid g(x) = g(\vec{\infty})\}$ *is called the* Submodular Integer Cover (SIC) *problem. Any* $x \in \mathbb{Z}_+^A$ *satisfying* $g(x) = g(\vec{\infty})$ *is a solution for the instance* (A, w, g) *of SIC.*

NOTE: In case when g is a *set* function (i.e., a function defined on 0-1 vectors), the problem is known as *submodular set cover* [29].

1.3   Related Problems and Previous Work

When all the unit capacities κ_i equal to one, CSD can be seen reducible to the minimum-cost flow problem. Another basic problem arising as a special case is the *set cover (SC)* or *vertex cover (VC)* problems, classic NP-hard problems of which polynomial time approximability has been intensively studied in the literature. In fact CSD serves as a unifying framework for various types of covering problems as will be seen below. In SC, given a family $\mathcal{F}$ of subsets of some base

Table 1. Covering Problems in the framework of CSD

	κ_i	c_{ij}	b_j	b_{total}
Set Cover	$|\delta^+(i)|$	1	1	$|B| = \sum_{j \in B} b_j$
Multicover	$|\delta^+(i)|$	1		$\sum_{j \in B} b_j$
Capacitated Set Cover		1	1	$|B| = \sum_{j \in B} b_j$
(with Demands)		1		$\sum_{j \in B} b_j$
Partial Set Cover	$|\delta^+(i)|$	1	1	

set U with associated nonnegative costs, it is required to compute a minimum cost subfamily $\mathcal{C}$ such that every element of U is "covered by" (i.e., contained in) some subset in $\mathcal{C}$. Defined analogously on graphs G, VC is to compute a minimum cost vertex subset C in G such that every edge of G is incident to some vertex in C. In the *multicover (MC)* problem, each element j of U in an SC instance is associated with "demand" b_j, and each j now needs to be covered b_j times (by different subsets). One of the most general problems previously considered along this line is the *multiset multicover (MMC)* problem, that can be defined by the following integer program: $\min\{w^T x \mid x^T C \geq b^T, x \leq u, x \in \mathbb{Z}_+^n\}$. It also has a version without explicit upper bounds u on x. In *capacitated SC*, each subset $S \in \mathcal{F}$ in an SC instance is associated with capacity κ_S and cost w_S. A single copy of S can cover only κ_S elements among those contained in S, and by paying w_S per copy, more copies of S can be used to cover more elements of S. In case when each element e is associated with demand b_e in capacitated SC, e has to be covered b_e times. In yet another direction of generalizations of SC, it is required to cover only b_{total} elements or more (instead of all) in *partial SC* when an additional integer b_{total} is given to SC.

For a node $v \in V$ in network $\mathcal{N}$, let $\delta^+(v)$ ($\delta^-(v)$, resp.) denote the set of arcs leaving v (entering v, resp.). For a finite set J, $J' \subseteq J$, and a vector $z \in \mathbb{Z}_+^J$ in general, $z(J')$ will be used as an abbreviation for $\sum_{j \in J'} z_j$. Those covering problems listed above can be realized in CSD by fixing some of problem parameters appropriately (see Table 1). In MMC $\min\{w^T x \mid x^T C \geq b^T, x \leq u, x \in \mathbb{Z}_+^n\}$, explicit upper bounds $x \leq u$ are called *multiplicity constraints*, and if it is non-existent, a trivial upper bound on x_i is $\max_j \lceil b_j/c_{ij} \rceil$. When cast in CSD, each $i \in A$ is replaced (not explicitly) by u_i copies, $i_1, \dots, i_{u_i}$, each $c_{i_l j}$ is set to $\min\{c_{ij}, \max\{0, b_j - \sum_{k=1}^{l-1} c_{i_k j}\}\}$, and the unit capacity κ_{i_l} to $c(\delta^+(i_l))$. (We remark that possibly non-polynomial expansion of problem instances in this reduction causes no trouble in our algorithm.)

It was (or can be) shown in all the cases that a greedy heuristic yields a factor $H(\max_{i \in A} \kappa_i) = O(\log(\max_{i \in A} \kappa_i))$ approximation, where $H(k) = \sum_{i=1}^k (1/i)$ is the kth harmonic number; see [22, 24, 6] for SC, [8] for MMC, [26] for partial SC, and [5] for capacitated MMC. Other approximation bounds known for these problems include:

- $\max_{j \in B} |\delta^-(j)|$ [20, 2] and $\max_{j \in B} |\delta^-(j)| - (1 - o(1)) \frac{(\max_{j \in B} |\delta^-(j)| - 1) \ln \ln n}{\ln n}$ [18] for SC.

- $\max_{j \in B} |\delta^-(j)|$ for MC [17], and $\max_{j \in B} |\delta^-(j)| - b_{\min} + 1$ for unweighted MC [25], where $b_{\min} = \min_{j \in B} b_j$.
- $O(\log |B|)$ [23] and $\max_{j \in B} |\delta^-(j)| =$ "max # of nonzero entries in a row of C" [4] for MMC.
- $\max_{j \in B} |\delta^-(j)|$ for capacitated SC, and 3 for capacitated VC with "unsplittable" demands [16].
- $\max_{j \in B} |\delta^-(j)|$ for partial SC [1, 11, 13], 2 for partial VC [10, 3, 21], and $2 - \Theta(\frac{\ln \ln |\delta^+(i)|}{\ln |\delta^+(i)|})$ for unweighted partial VC on graphs of maximum degree $|\delta^+(i)|$ [19].

On the other hand, the following lower bounds are known for approximability of SC: $(1 - \epsilon) \ln |B|$ for any $\epsilon > 0$ [9] (assuming $NP \not\subset \text{DTIME}\,(n^{O(\log \log n)})$), and $|\delta^-(j)| - 1 - \epsilon$ [7].

CSD can be seen also related to the *capacitated facility location* problem, the *network loading* problem, and the *single-sink buy-at-bulk* problem among others. In fact, if shipping a unit product from $i \in A$ to $j \in B$ incurs a certain cost and the objective is to minimize total cost of production and shipping, CSD corresponds to the capacitated facility location problem having "flow constraints" in it. To the best of our knowledge, however, no covering-type problem with covering capacities *and* flow constraints explicitly given as in CSD, has been previously considered.

1.4   Summary of Results

In designing approximation algorithms for CSD or SIC, it is natural to consider extending known approximation algorithms for SSC. There are two such algorithms, one greedy [29] and the other primal-dual [11]. It is rather straightforward to extend the former to CSD resulting in the approximation ratio of $H(\max_{i \in A} \kappa_i) = O(\log(\max_{i \in A} \kappa_i))$. It will be seen, on the other hand, that the primal-dual approach for SIC yields an approximation algorithm for CSD running in time $O(nM(n, m))$, where $M(n, m)$ denotes time complexity of computing an $s - t$ max flow in a network with n nodes and m arcs. It requires much more intricate analysis based on reasoning on the relationship between flow values and capacity settings, however, to estimate its performance ratio, and to describe it, let $\delta^{D-}(j) = \delta^-(j) - D, b_j^D = b_j - c(D)$, and $c_{ij}^D = \min\{c_{ij}, b_j^D\}$ for any $D \subset \delta^-(j)$. It will be shown that the approximation factor guaranteed is

$$\max\left\{2, \max_{j \in B,\, D \subset \delta^-(j)} \left\{\frac{c^D(\delta^{D-}(j))}{b_j^D}\right\}\right\} \tag{1}$$

in its general form. Various consequences can be drawn immediately from (1), e.g.,

1. By assuming w.l.o.g. that $c_{ij} \le b_j, \forall i, j$, (1) reduces to $\max_{j \in B} |\delta^-(j)|$; hence, the approximability of CSD can be ensured to dependent only on the network (incidence) structure of a given instance, and *not* on any numerical values of input capacities κ_i, b_j, c_{ij}.

The primal-dual algorithm is thus more effective than greedy when the number of suppliers that can ship products to each consumer is relatively small (such as in "vertex cover type" problems).

To measure the effectiveness of these bounds, it is instructive to compare them with existing results for the covering problems previously considered. The bound in 1. matches the best ones for capacitated SC [16], partial SC [1, 11, 13], weighted MC [17], and MMC [4] (see 3. below), respectively. Meanwhile, this bound nearly matching the best one for SC [18] as well, any further improvement would be necessarily a marginal one, even if possible, due to the strong lower bound of $\max_{j \in B} |\delta^-(j)| - 1 - \epsilon$ for standard SC [7]. Further consequences implied by (1) (or the bound in 1.) include:

2. Capacitated VC with splittable demands is approximable within a factor of 2, in contrast with the 3-approximation of [16] for unsplittable demands.
3. In MMC, with or without multiplicity constraints, c_{ij}^D's $(= \min\{c_{i_l j}, b_j^D\}$'s$)$ in (1) are evaluated s.t., when summed up over all i_l's for some i, they never exceed b_j^D (more details will be given in the full version). And then, the obtained bound (assuming $\max_j |\delta^-(j)| \geq 2$) is at most

$$\max\left\{2, \max_{j,D}\left\{\frac{|\delta^{D-}(j)|b_j^D}{b_j^D}\right\}\right\} = \max\{2, \max_{j,D} |\delta^-(j) - D|\} = \max_{j \in B} |\delta^-(j)|,$$

which coincides with the approximation factor, "max # of nonzero entries in a row of C", given in [4].
4. Depending on actual values of c_{ij} and b_j, the estimation could be further lowered. If $c_{ij} = c_j$, $\forall i, j$, let $b_j = s_j c_j + t_j$ s.t. $0 \leq s_j$, $0 < t_j \leq c_j$. Then, the value of (1) can be seen reducible to $\max_{j \in B} \{\max\{|\delta^-(j)| - s_j,$ $(|\delta^-(j)| - s_j + 1)\frac{c_j}{c_j + t_j}\}\}$. Thus, when $c_{ij} = 1, \forall i, j$, for instance, CSD is approximable within a factor of $\max_{j \in B} (|\delta^-(j)| - b_j) + 1$. This CSD bound with such restrictions on c_{ij}'s alone already improves e.g. $\max_{j \in B} |\delta^-(j)| - b_{\min} + 1$ for unweighted MC observed in [25].

In the *multi-capacitated* version of CSD, multiple types of unit capacities are available at different prices for each $i \in A$. Such a generalization enables CSD to reflect e.g. an "economy of scale" (or "volume discount"). Our algorithm still works with this version providing the same approximation guarantee of (1) (details given in the full version).

2   Approximating Submodular Integer Cover

It is rather straightforward to obtain the greedy bound of $H(\max_{i \in A} \kappa_i)$ for CSD by extending the greedy algorithm for SSC and its performance analysis given by Wolsey [29]. Therefore, we focus on the primal-dual approach for SIC and its application to CSD for the rest of the paper.

2.1   Preliminaries and IP Formulation

Definition 4. *Let* $\mathbf{e}^k$ *denote the kth unit vector in* $\mathbb{Z}_+^A$ *s.t.* $\mathbf{e}_i^k = \begin{cases} 1 & \text{if } i = k \\ 0 & \text{otherwise} \end{cases}$ *for* $i \in A$. *For nondecreasing* g *we also assume it is bounded, i.e., there exists* $u \in \mathbb{R}$ *s.t.* $g(x) \leq u, \forall x \in \mathbb{Z}_+^A$. *Then, there must exist an integer* u_i *for each* $i \in A$ *s.t., for any* x *with* $x_i = u_i$ *and* $x_i' \geq u_i$, $g(x') = g(x)$ *if* $x_j' = x_j$ *for* $j \in A - \{i\}$. *Let* u_i *be minimal such an integer for each* $i \in A$, *and let* χ^S *denote the vector in* $\mathbb{Z}_+^A$ *s.t.* $\chi_i^S = \begin{cases} u_i & \text{if } i \in S \\ 0 & \text{otherwise} \end{cases}$ *for* $S \subseteq A$. *Thus,* $g(\chi^A) = \sup_{x \in \mathbb{Z}_+^A} g(x)$.

Let L_i be a lattice for $i = 1, \cdots, n$, and L be a sublattice of $\prod_{i=1}^n L_i$. It was shown by Topkis that a submodular function on L has antitone (i.e., nonincreasing) differences on L:

Proposition 5 (Topkis [27]) *Let* $(x_1, \cdots, x_n)$ *be an element of* L. *If* g *is a real-valued submodular function on* L, *for all* $j \neq k$ *with each* x_i *fixed for* $i \neq j$ *and* $i \neq k$, $g(x_j, z) - g(x_j, x_k)$ *is nonincreasing in* x_j *on* $L^z \cap L^{x_k}$ *for each* $x_k < z$ *in* L_k, *where* $L^t = \{x \mid (x_1, \cdots, x_j = x, \cdots, x_k = t, \cdots, x_n) \in L\}$.

Lemma 6. *If* g *is submodular and nondecreasing,*

$$g(x) \leq g(\chi^S) + \sum_{j \in A - S} (g(\chi^S + \mathbf{e}^j) - g(\chi^S)) x_j$$

for $x \in \mathbb{Z}_+^A$ *and* $S \subseteq A$.

Proof. Let $A - S = \{j_1, \ldots, j_r\}$. Then,

$$g(x + \chi^S) - g(\chi^S) = \sum_{t=1}^{r} \sum_{l=1}^{x_{j_t}} \{g(\chi^S + \sum_{k=1}^{t-1} x_{j_k} \mathbf{e}^{j_k} + l \mathbf{e}^{j_t})$$
$$-g(\chi^S + \sum_{k=1}^{t-1} x_{j_k} \mathbf{e}^{j_k} + (l-1) \mathbf{e}^{j_t})\}$$
$$\leq \sum_{t=1}^{r} \sum_{l=1}^{x_{j_t}} \{g(\chi^S + \mathbf{e}^{j_t}) - g(\chi^S)\}$$
$$= \sum_{t=1}^{r} \{g(\chi^S + \mathbf{e}^{j_t}) - g(\chi^S)\} x_{j_t}$$

where the inequality holds due to Proposition 5. Since g is also nondecreasing,

$$g(x) \leq g(x + \chi^S) = g(\chi^S) + (g(x + \chi^S) - g(\chi^S))$$
$$\leq g(\chi^S) + \sum_{t=1}^{r} \{g(\chi^S + \mathbf{e}^{j_t}) - g(\chi^S)\} x_{j_t}$$

□

Define function $\Delta_S : A - S \to \mathbb{Z}_+$ for $S \subseteq A$ s.t. $\Delta_S(i) = g(\chi^S + \mathbf{e}^i) - g(\chi^S)$.

Theorem 7. *If x is a solution for an SIC instance (A, w, g), it is feasible in the following integer program:*

$$\text{(IP)} \qquad \text{subject to:} \qquad \begin{aligned} &\min \sum_{i \in A} w_i x_i \\ &x \in \mathbb{Z}_+^A \\ &\sum_{i \in A - S} \Delta_S(i) x_i \geq g(\chi^A) - g(\chi^S), S \subseteq A \end{aligned}$$

Consequently, the optimum for (A, w, g) is lower bounded by that of the corresponding (IP).

Proof. If x is a solution for (A, w, g), $g(x) = g(\vec{\infty}) = g(\chi^A)$, and by Lemma 6,

$$g(\chi^A) = g(x) \leq g(\chi^S) + \sum_{i \in A - S} (g(\chi^S + \mathbf{e}^i) - g(\chi^S)) x_i = g(\chi^S) + \sum_{i \in A - S} \Delta_S(i) x_i$$

for any $S \subseteq A$; hence, x satisfies all the constraints of (IP). $\qquad \square$

2.2   Primal-Dual Schema

To design a primal-dual based approximation algorithm for SIC, we begin with relaxing the integral constraints $x \in \mathbb{Z}_+^A$ of (IP) to the linear constraints $x \geq 0$. By taking the dual of the resulting LP relaxation (LP) of (IP), we next obtain

$$\text{(D)} \qquad \text{subject to:} \qquad \begin{aligned} &\max \sum_{S \subseteq A} \left(g(\chi^A) - g(\chi^S) \right) y_S \\ &\sum_{S : i \in A - S} \Delta_S(i) y_S \leq w_i, \qquad i \in A \\ &y_S \geq 0, \qquad\qquad\qquad S \subseteq A \end{aligned}$$

The primal-dual schema for "set cover type" problems (i.e., "covering by a 0-1 vector" type) is by now a well established algorithmic technique (see surveys given in e.g. [14, 28]); we here extend it to the primal-dual pair of (IP) and (D) in designing an algorithm called PD. It consists of the following two phases; the phase in which a maximal dual solution y is constructed in a greedy fashion, and an integral primal solution x is correspondingly chosen s.t. it satisfies the primal complementary slackness conditions with y, followed by the phase called *reverse delete* in which x is ensured to satisfy minimality conditions in a certain order.

More specifically, starting with $F = \emptyset$ and the dual solution $y = 0$, a variable y_F in (D) is iteratively increased maximally without violating dual feasibility in the first phase, so that the dual constraint for i becomes newly binding for some i in $A - F$; that is, $\sum_{S : i \in A - S} \Delta_S(i) y_S = w_i$. This amounts to finding i in $A - F$ (say, i') at the lth iteration minimizing

$$\frac{w_i - \sum_{S : i \notin S, S \neq F} \Delta_S(i) y_S}{\Delta_F(i)} = \frac{w_i - \sum_{1 \leq t \leq l-1} \Delta_{F_t}(i) y_{F_t}}{\Delta_F(i)}$$

and setting y_F to

Initialize: $x = 0, y = 0, F = \emptyset, \mathsf{STACK} = \emptyset, \bar{w}_i \leftarrow w_i \ (\forall i \in A)$.
While x is not an SIC solution for (A, w, g) (i.e., $g(x) < g(\chi^A)$) do
    Let $y_F \leftarrow \min_{i \in A-F}\{\bar{w}_i/\Delta_F(i)\}$ and $i' \leftarrow \operatorname{argmin}_{i \in A-F}\{\bar{w}_i/\Delta_F(i)\}$.
    Add i' into F, push it onto STACK, and set $x_{i'} \leftarrow u_{i'}$.
    For each $i \in A - F$ do
        Reset $\bar{w}_i \leftarrow \bar{w}_i - \Delta_F(i)y_F$ $\left(= \bar{w}_i - \frac{\Delta_F(i)}{\Delta_F(i')}\bar{w}_{i'}\right)$.
While $\mathsf{STACK} \neq \emptyset$ do
    Let $i' \leftarrow \operatorname{pop}(\mathsf{STACK})$.
    Set $x_{i'} \leftarrow \min\{x_{i'} \mid g(x) = g(\chi^A)\}$.
Output x.

Fig. 1. Algorithm PD for SIC

$$\min_{i \in A-F} \left\{ \frac{w_i - \sum_{S:i \notin S, S \neq F} \Delta_S(i)y_S}{\Delta_F(i)} \right\} = \frac{w_{i'} - \sum_{S:i' \notin S, S \neq F} \Delta_S(i')y_S}{\Delta_F(i')}$$
$$= \frac{w_{i'} - \sum_{1 \leq t \leq l-1} \Delta_{F_t}(i')y_{F_t}}{\Delta_F(i')}$$

where $F_0 = \emptyset \subseteq F_1 \subseteq \ldots \subseteq F_T$ denote (intermediate) F's constructed, in this order, by each iteration of the first phase (So, $y_S > 0$ iff S is one of these F's). This i' (or any other binding i's) is then added to F so that F remains as the set of all i's whose corresponding constraints are binding. At the same time it is kept track of in what order dual constraints become binding during this process (or, in what order i's enter F) in the first phase. Let x represent χ^F. Then, x eventually becomes an SIC solution as F may grow up to A if necessary.

In the second phase an actual SIC solution x is constructed based on F and the ordering of i's in F computed as above. Starting with $x = \chi^F$, each $i \in F$ is processed, one by one, in reversal of the order in which they were added to F during the first phase, and x_i is set to the minimal value needed for x to remain as a solution for (A, w, g). The SIC solution x constructed in this manner satisfies a certain minimality property: For all $t = 1, \ldots, T$, the values of x_i's with $i \in F - F_t$ are the ones "minimally" required, in addition to χ^{F_t}, to increase its g-value from $g(\chi^{F_t})$ up to the required $g(\chi^A)$; in other words, the value of $g(x + \chi^{F_t})$ drops below that of $g(\chi^A)$ whenever any x_i with $i \in F - F_t$ is decremented. The pseudo-code description of this algorithm is given in Fig. 1.

2.3   Analysis

In the algorithm PD any element $i \in A$ enters F (in the first phase) iff the corresponding dual constraint becomes binding; that is, $w_i = \sum_{S:i \in A-S} \Delta_S(i)y_S$. Therefore, the weight of computed x is

$$\sum_{i \in F} w_i x_i = \sum_{i \in F} \left(\sum_{S:i \in A-S} \Delta_S(i)y_S \right) x_i = \sum_{S \subseteq A} \left(\sum_{i \in F-S} \Delta_S(i)x_i \right) y_S.$$

When the RHS in this equation is compared with the objective function of (D), of which value gives a lower bound for the optimal weight, it can be seen that the ratio of the weight of x to the optimal weight is bounded above by

$$\max_{S \subseteq A, y_S > 0} \frac{\sum_{i \in F - S} \Delta_S(i) x_i}{g(\chi^A) - g(\chi^S)} = \max_{1 \leq t \leq T} \frac{\sum_{i \in F - F_t} \Delta_{F_t}(i) x_i}{g(\chi^A) - g(\chi^{F_t})}. \tag{2}$$

Recall now how an actual SIC solution x is constructed in the second phase.

Definition 8. *We call a solution x for (A, w, g) minimal w.r.t. $S \subseteq A$ if $x + \chi^{A-S}$ becomes a non-solution when any $x_i > 0$ with $i \in S$ is decremented.*

It is ensured that x be minimal w.r.t. $A - F_t$ for each t $(1 \leq t \leq T)$. Therefore, x can be restricted to such solutions for (A, w, g) in estimating the bound in (2), and we have the following general approximation bound for SIC:

Theorem 9. *When applied to an SIC instance (A, w, g), the algorithm PD computes a solution such that the ratio of its weight to the optimal weight is bounded by*

$$\max \left\{ \frac{\sum_{i \in A - S} \Delta_S(i) x_i}{g(\chi^A) - g(\chi^S)} \right\}, \tag{3}$$

where max is taken over any $S \subseteq A$ and such a solution x for (A, w, g) that is minimal w.r.t. $A - S$.

3   Application to CSD Problem

To model CSD in the framework of SIC, let $u_i = \lceil c(\delta^+(i))/\kappa_i \rceil$, $\forall i \in A$, and define a function $\rho : \mathbb{Z}_+^A \to \mathbb{Z}_+$ s.t. $\rho(x) = $ (max flow value of $\mathcal{N}$ with $a_i = x_i \kappa_i, \forall i$) for a CSD instance of $(\mathcal{N}, \kappa, w)$. As we will need to consider $\mathcal{N}$ specified with flow capacities of our own choice, such a network with new capacities a', b', c' will be denoted by $\mathcal{N}(a = a', b = b', c = c')$. Let $\mathcal{N}_S$ denote the network $\mathcal{N}$ in which capacity a_i on $(s, i) = \infty$ if $i \in S$ and $a_i = 0$ otherwise, for $S \subseteq A$ (i.e., $\mathcal{N}_S = \mathcal{N}(a_i = \infty$ for $i \in S, a_i = 0$ for $i \in A - S))$. In estimating the bound of (3) in the context of CSD, the next lemma is useful:

Lemma 10. *Let f be any max flow in the network $\mathcal{N}_S$ for $S \subseteq A$. When f is augmented up to a max flow in $\mathcal{N}_A$, no augmenting path passes through any $i \in S$.*

Proof. Consider the residual network $\mathcal{N}_A^r$ with respect to $\mathcal{N}_A$ and f. Assuming that f is not yet a max flow in $\mathcal{N}_A$, observe that f saturates either arc (i, j) or (j, t) in $\mathcal{N}_S$ for each $i \in S$ and all j with $(i, j) \in \delta^+(i)$. Or, in other words, for each j reachable from some $i \in S$, either arc (j, t) is saturated, or all (i, j)'s going from S to j are saturated. Therefore, $\mathcal{N}_A^r$ has no s-t dipath in it passing through any $i \in S$, and it remains so even if f is augmented to a larger flow in $\mathcal{N}_A$; even after augmentations, the original f is nowhere decremented, and saturated arcs continue to block any augmentation through nodes in S, due to the network structure $(A \cup B \cup \{s, t, t'\}, E)$ of $\mathcal{N}$.   □

The next is a key lemma of this paper, yet due to the space limitation, its proof is omitted here (will be given in the full version):

Lemma 11. *Let* $x \in \mathbb{Z}_+^{\bar{A}}$ *be a minimal CSD solution for* $(\bar{\mathcal{N}} = (\bar{A} \cup \bar{B} \cup \{s, t, t'\}, \bar{E}, \bar{b}, c, \bar{b}_{\text{total}}), \kappa, w)$. *Then,*

$$\frac{\sum_{i \in \bar{A}} \kappa_i x_i}{\bar{b}_{\text{total}}} \leq \max \left\{ 2, \max_{j \in \bar{B}} \frac{c(\delta^-(j))}{\bar{b}_j} \right\}. \tag{4}$$

Consider the network $\mathcal{N}_S'$ obtained from $\mathcal{N}_S$ and a max flow f in it by removing all the nodes in S along with incident arcs, and adjusting the capacities on (j, t) to $b_j - f(j, t)$. It follows from Lemma 10 that the values of $\Delta_S(i)$ and $\rho(\chi^A) - \rho(\chi^S)$ in (3) coincide with the max flow value in $\mathcal{N}_S'(a_i = \kappa_i)$ and $b_{\text{total}} - |f|$, respectively, where the former is always bounded by κ_i. Recall that, x being minimal w.r.t. $A - S$, the vector $(x_i)_{i \in A - S}$ specifies minimal capacities on arcs $(s, i), i \in A - S$, required to increase the max flow value $|f|$ of $\mathcal{N}_S$ to b_{total}; again from Lemma 10, it means that $(x_i)_{i \in A - S}$ is by itself a minimal CSD solution for $\mathcal{N}_S'$ with b_{total} adjusted to the necessary increments of $b_{\text{total}} - |f|$. Therefore, the value of (3) can be evaluated by taking the maximal value of (4) over all possible $\mathcal{N}_S'$ (subject to $y_S > 0$) used in place of $\bar{\mathcal{N}}$ of Lemma 11. It follows from these observations and Theorem 9 that

Theorem 12. *For any* $D \subset \delta^-(j)$, *let* $\delta^{D-}(j) = \delta^-(j) - D, b_j^D = b_j - c(D)$, *and* $c_{ij}^D = \min\{c_{ij}, b_j^D\}$. *The algorithm* PD *computes a CSD solution to an instance* $(\mathcal{N} = (A \cup B \cup \{s, t, t'\}, E, b, c, b_{\text{total}}), \kappa, w)$, *s.t. the ratio of its cost to the optimal cost is bounded above by*

$$\max \left\{ 2, \max_{j \in B, D \subset \delta^-(j)} \left\{ \frac{c^D(\delta^{D-}(j))}{b_j^D} \right\} \right\}. \tag{5}$$

By assuming w.l.o.g. that $c_{ij} \leq b_j, \forall i, j$, it can be shown that (5) reduces to $\max\{2, \max_{j \in B} |\delta^-(j)|\}$. If $\max_{j \in B} |\delta^-(j)| < 2$, however, $|\delta^-(j)| = 1, \forall j \in B$, and this occurs only in a trivial case.

Corollary 13. *The algorithm* PD *computes a CSD solution to an instance* $(\mathcal{N} = (A \cup B \cup \{s, t, t'\}, E, b, c, b_{\text{total}}), \kappa, w)$, *s.t. the ratio of its cost to the optimal cost is bounded above by* $\max_{j \in B} \{|\delta^-(j)|\}$.

Running Time. Each iteration in the first phase (first while-loop) can be carried out in time $O(m)$, whereas time complexity of the second phase (second while-loop) is dominated by that of computing a max flow in each iteration. The running time of PD, when applied to CSD, is thus $O(nM(n, m))$, and $M(n, m) = O(nm \log(n^2/m))$, for instance, when the Goldberg-Tarjan's algorithm is used [15].

References

1. Bar-Yehuda, R.: Using homogeneous weights for approximating the partial cover problem. In: Proc. 10th SODA. ACM-SIAM (1999) 71–75

2. Bar-Yehuda, R., Even, S.: A linear time approximation algorithm for the weighted vertex cover problem. J. Algorithms **2** (1981) 198–203
3. Bshouty, N.H., Burroughs, L.: Massaging a linear programming solution to give a 2-approximation for a generalization of the vertex cover problem. In: Proc. 15th STACS. (1998) 298–308
4. Carr, R.D., Fleischer, L.K., Leung, V.J., Phillips, C.A.: Strengthening integrality gaps for capacitated network design and covering problems. In: Proc. 11th SODA. ACM-SIAM (2000) 106–115
5. Chuzhoy, J., Naor, J.: Covering problems with hard capacities. In: Proc. 43rd FOCS. IEEE (2002) 481–489
6. Chvátal, V.: A greedy heuristic for the set-covering problem. Math. Oper. Res. **4**(3) (1979) 233–235
7. Dinur, I., Guruswami, V., Khot, S., Regev, O.: A new multilayered PCP and the hardness of hypergraph vertex cover. In: Proc. 35th STOC. ACM (2003) 595–601
8. Dobson, G.: Worst-case analysis of greedy heuristics for integer programming with nonnegative data. Math. Oper. Res. **7**(4) (1982) 515–531
9. Feige, U.: A threshold of $\ln n$ for approximating set cover. In: Proc. 28th STOC. ACM (1996) 314–318
10. Fujito, T.: A unified local ratio approximation of node-deletion problems. In: Proc. ESA'96. (1996) 167–178
11. Fujito, T.: On approximation of the submodular set cover problem. Oper. Res. Lett. **25** (1999) 169–174
12. Gale, D., Politof, T.: Substitutes and complements in network flow problems. Discrete Appl. Math. **3** (1981) 175–186
13. Gandhi, R., Khuller, S., Srinivasan, A.: Approximation algorithms for partial covering problems. In: Proc. 28th ICALP. (2001) 225-236
14. Goemans, M.X., Williamson, D.P.: The primal-dual method for approximation algorithms and its application to network design problems. In: Hochbaum, D. (ed.): Approximation Algorithm for NP-Hard Problems. PWS, Boston (1996) 144–191
15. Goldberg, A.V., Tarjan, R.E.: A new approach to the maximum-flow problem. J. ACM **35** (1988) 921–940
16. Guha, S., Hassin, R., Khuller, S., Or, E.: Capacitated vertex covering with applications. In: Proc. 13th SODA. ACM-SIAM (2002) 858–865
17. Hall, N.G., Hochbaum, D.S.: A fast approximation algorithm for the multicovering problem. Discrete Appl. Math. **15** (1986) 35–40
18. Halperin, E.: Improved approximation algorithms for the vertex cover problem in graphs and hypergraphs. In: Proc. 11th SODA. ACM-SIAM (2000) 329–337
19. Halperin, E., Srinivasan, A.: Improved approximation algorithms for the partial vertex cover problem. In: Proc. APPROX 2002. (2002) 161–174
20. Hochbaum, D.S.: Approximation algorithms for the set covering and vertex cover problems. SIAM J. Comput. **11**(3) (1982) 555–556
21. Hochbaum, D.S.: The t-vertex cover problem: Extending the half integrality framework with budget constraints. In: Proc. APPROX'98. (1998) 111–122
22. Johnson, D.S.: Approximation algorithms for combinatorial problems. J. Comput. System Sci. **9** (1974) 256–278
23. Kolliopoulos, S. G., Young, N. E.: Tight approximation results for general covering integer programs. In: Proc. 42nd FOCS. IEEE (2001) 522–528
24. Lovász, L.: On the ratio of optimal integral and fractional covers. Discrete Math. **13** (1975) 383–390
25. Peleg, D., Schechtman, G., Wool, A.: Randomized approximation of bounded multicovering problems. Algorithmica **18** (1997) 44–66

26. Slavík, P.: Improved performance of the greedy algorithm for partial cover. Inform. Process. Lett. **64**(5) (1997) 251–254
27. Topkis, D.M.: Minimizing a submodular function on a lattice. Operations Res. **26**(2) (1978) 305–321
28. Vazirani, V.: Approximation Algorithms. Springer, Berlin (2001)
29. Wolsey, L.A.: An analysis of the greedy algorithm for the submodular set covering problem. Combinatorica **2**(4) (1982) 385–393

Stochastic Online Scheduling on Parallel Machines*

Nicole Megow[1], Marc Uetz[2], and Tjark Vredeveld[3]

[1] Technische Universität Berlin, Institut für Mathematik, Strasse des 17. Juni 136,
10623 Berlin, Germany
nmegow@math.tu-berlin.de
[2] Maastricht University, Department of Quantitative Economics, P.O.Box 616,
6200 MD Maastricht, The Netherlands
m.uetz@ke.unimaas.nl
[3] Konrad-Zuse-Zentrum für Informationstechnik Berlin, Department Optimization,
Takustr. 7, 14195 Berlin, Germany
vredeveld@zib.de

Abstract. We consider a non-preemptive, stochastic parallel machine scheduling model with the goal to minimize the weighted completion times of jobs. In contrast to the classical stochastic model where jobs with their processing time distributions are known beforehand, we assume that jobs appear one by one, and every job must be assigned to a machine online. We propose a simple online scheduling policy for that model, and prove a performance guarantee that matches the currently best known performance guarantee for stochastic parallel machine scheduling. For the more general model with job release dates we derive an analogous result, and for NBUE distributed processing times we even improve upon the previously best known performance guarantee for stochastic parallel machine scheduling. Moreover, we derive some lower bounds on approximation.

1 Introduction

Non-preemptive parallel machine scheduling to minimize the weighted completion times of jobs, $P \mid \mid \sum w_j C_j$ in the three-field notation of Graham et al. [6], is one of the classical problems in scheduling theory. This problem plays a role whenever many jobs must be processed on a limited number of machines, with typical applications, e.g., in parallel computing [2] or compiler optimization [3]. The main characteristic of the model of *stochastic scheduling* is the fact that the processing times of jobs are subject to fluctuations, and become known only upon completion of the jobs. Their respective distributions are assumed to be given beforehand. This usually requires the notion of *scheduling policies* instead of simple schedules.

---

* Research partially supported by the DFG Research Center "Mathematics for key technologies" (FZT 86) in Berlin.

G. Persiano and R. Solis-Oba (Eds.): WAOA 2004, LNCS 3351, pp. 167–180, 2005.

Stochastic scheduling. Stochastic machine scheduling models have been addressed mainly since the 1980s [4]. Let us briefly recall the concept of a scheduling policy as introduced by Möhring et al. [10]. Roughly spoken, at any time t, such a policy specifies which action to perform, in particular which jobs to start at t. In order to decide, it may utilize the complete information contained in the partial schedule up to time t. However, it must not utilize any information about the future. An *optimal* scheduling policy is one that minimizes the objective function value in expectation. Notice that, in general, a scheduling policy need not yield a fixed assignment of jobs to machines.

With the exception of the papers by Weiss [18, 19], the first approximation algorithms for stochastic machine scheduling have been derived only recently [11, 13, 14, 16]. In the papers [11, 14], the expected performance of a linear programming based list scheduling policy is compared against the expected performance of an optimal scheduling policy. The results are constant-factor approximations for problems with or without release dates [11], and also with precedence constraints [14]. The approach is based upon the solution of linear programming relaxations, and for the models with release dates or precedence constraints, their solutions are used in order to define corresponding list scheduling policies. Recently, another type of analysis has been pursued by Steger et al. in the papers [13, 16], where the expected *ratio* of the performance of the (W)SEPT rule[1] over the optimum solution is analyzed. This approach may indeed have advantages over the previous approach, namely in terms of averaging over different realizations of processing times, and we refer to [13, 16] for a discussion. One of the main differences, however, is the fact that it uses a comparison against the off-line optimum, whereas the approach in [11, 14] compares against the on-line optimum. Nevertheless, restricted to models without release dates or precedence constraints, constant-factor approximation results for the expected ratio have been obtained for the (W)SEPT rule on parallel machines [13, 16].

Stochastic online scheduling. In this paper, we follow the approach taken by Möhring et al. [11]. In other words, we compare the expected outcome of a certain scheduling policy against the expected outcome of an optimal scheduling policy. In contrast to the previously mentioned work on stochastic scheduling, however, we consider a model where jobs have to be assigned to machines online. More precisely, jobs j are presented to the scheduler one by one, with their weights w_j and expected processing times $\mathbb{E}[P_j]$, and without knowledge about jobs that might appear in the future, or their number, they must be assigned to a machine. This assignment cannot be revised later. Once all jobs have been assigned this way, there is freedom concerning the scheduling of jobs on every single machine; of course, still within the restrictions that jobs must not be preempted, and

---

[1] In the WSEPT rule, jobs are scheduled greedily in the order of non-increasing ratios $w_j/\mathbb{E}[P_j]$, where w_j is the weight of job j, P_j its processing time distribution, and $\mathbb{E}[P_j]$ is its expectation. For unit weights, this equals SEPT; shortest expected processing time first.

that their actual processing times become known only upon completion. For convenience, let us denote this model SOS, for stochastic online scheduling.

Discussion of the model. As a matter of fact, the solution of LP relaxations is crucial for the work of Möhring et al. [11] or Skutella and Uetz [14]. For models with release dates or precedence constraints, optimal LP solutions are not only required for the purpose of analysis, but also to define the corresponding list scheduling policies. In order to set up these LP relaxations, it is required to know beforehand the set of jobs, their expected processing times $\mathbb{E}[P_j]$, as well as a uniform upper bound Δ on the squared coefficient of variation of all processing times distributions

$$\mathrm{CV}[P_j]^2 = \mathrm{Var}[P_j]/\mathbb{E}[P_j]^2 \leq \Delta \quad \text{for all jobs } j.$$

One critique of this approach is the fact that in practical applications, parts of this data might not be available. Even worse, in an online setting there is no knowledge about jobs that might appear in the future. In that case, algorithms that first require the solution of sophisticated LP relaxations might be useless.

The SOS model as proposed in this paper can be seen as a first step in the direction of simpler, combinatorial algorithms for models with stochastic processing times. It is a two-phase model, where the first phase consists of an online assignment of jobs to machines. In this phase, whenever a job j is presented to the scheduler, the only information that is disclosed is its weight w_j and its expected processing time $\mathbb{E}[P_j]$. In the model with release dates, it is also the release time r_j. The second phase consists of the actual process of scheduling the jobs over time, processing times being realized according to the respective distributions. Yet, we compare the expected outcome of the online stochastic scheduling policy to the expected outcome of an optimal scheduling policy, according to the definition of general scheduling policies by Möhring et al. [10].

In comparison to classical online models, we make two remarks. First, like in classical online optimization, the adversary in the SOS model may choose an arbitrary sequence of jobs in the first phase. These jobs are assumed to be stochastic, with corresponding processing time distributions (deterministic jobs being a special case). However, in the second phase, the actual processing times are realized according to the exogenous probability distributions, thus they are not under control of the adversary. Moreover, given the exogenously controlled processing times, the best the adversary can do is in fact to use an optimal stochastic scheduling policy. In this view, our model indeed incorporates some of the ideas by Koutsoupias and Papadimitriou in [8].

Results and methodology. We derive worst case performance guarantees for the expected performance of very simple, combinatorial online scheduling policies for models with and without release dates. For the model without release dates, $\mathrm{P}|\,|\mathbb{E}[\sum w_j C_j]$, this is a performance guarantee of

$$1 + \frac{(\Delta+1)(m-1)}{2m},$$

matching the previously best known performance guarantee of [11] for the performance of the WSEPT rule. Note, however, that this bound holds even though we use a restricted scheduling policy that first has to assign jobs to machines online, without knowledge of the jobs to come. For the model with release dates, $P|r_j|\mathbb{E}\left[\sum w_j C_j\right]$ we prove a more complicated performance guarantee for a class of processing time distributions that we call δ-NBUE. They generalize NBUE distributions[2], which are contained as a special case. For NBUE distributions, we obtain a performance bound strictly less than

$$\frac{5+\sqrt{5}}{2} - \frac{1}{2m},$$

where $(5+\sqrt{5})/2 \approx 3.62$. Thereby, we improve upon the previously best known performance guarantee of $4 - 1/m$ for NBUE distributions, which was derived for an LP based list scheduling policy [11]. Again, notice that this improved bound holds even though we use a restricted policy that first has to assign jobs to machines online, without knowledge of the jobs to come.

Our results are achieved by the following, quite simple SOS policy. Once the jobs have been assigned to the machines, we assume that on every machine the jobs are processed in the WSEPT order[3]. To make the online decisions on machine assignments in the first phase, at any time when a job is presented, we assign it to that machine where it causes the minimal increase in total expected objective value; given the jobs that have been assigned so far, and given that the jobs on each machine will be scheduled in WSEPT order. Intuitively, the reason why we can recover (or improve, respectively) the previous best known results in stochastic machine scheduling is the following: On the one hand, we restrict the full power of scheduling policies by fixing machine assignments beforehand. On the other hand, it is precisely this fixed machine assignment, together with an averaging argument over the number of machines, that allows an improvement in the analysis in comparison to general scheduling policies. We mention that, to obtain our results, we in fact utilize one of the LP based lower bounds of [11].

2   Model Definition, Notation, and Preliminaries

Let n be the number of jobs, index $j \in \{1,\ldots,n\}$ denote a job, with associated weight w_j and processing time distribution P_j. By $\mathbb{E}\left[P_j\right]$ we denote its expected processing time, and p_j denotes a particular realization of P_j. In the model with release dates, r_j denotes the earliest point in time when job j can be started. Given a schedule of start times $S_1,\ldots,S_n$ for a particular realization $p = (p_1,\ldots,p_n)$ of processing times, $C_j = S_j + p_j$ is the completion time of

---

[2] A distribution X is called NBUE, new better than used in expectation, if $\mathbb{E}\left[X - t \mid X > t\right] \leq \mathbb{E}\left[X\right]$ for all $t > 0$.

[3] In the case with release dates, this is in fact a modified version of the WSEPT order, that will be explained later.

job j, $j = 1, \ldots, n$. Each job must be processed non-preemptively, on any of the m machines, and each machine can process at most one job at a time. The goal is to find a scheduling policy that minimizes the expected value of the weighted completion times of jobs, $\mathbb{E}\left[\sum w_j C_j\right]$.

A scheduling policy eventually yields a feasible m-machine schedule for each realization p of the processing times. For a given policy, denoted by Π, let $S_j^\Pi(p)$ and $C_j^\Pi(p)$ denote the start and completion times of job j for a given realization p, and let $S_j^\Pi(P)$ and $C_j^\Pi(P)$ denote the associated random variables. Unless there is danger of ambiguity, we also write S_j and C_j, for short. We let

$$\mathbb{E}\left[Z^\Pi\right] = \mathbb{E}\left[\sum_j w_j C_j^\Pi(P)\right]$$

denote the expected performance of a scheduling policy Π. Then, if OPT is an optimal scheduling policy according to the most general definition of stochastic scheduling policies in [10], we say that a policy Π is a ρ–approximation if, for some $\rho \geq 1$,

$$\mathbb{E}\left[Z^\Pi\right] \leq \rho\, \mathbb{E}\left[Z^{\mathrm{OPT}}\right].$$

We assume that the jobs are presented to the scheduler one by one, in the order $1, \ldots, n$. However, the number of jobs n is not known before. When a job is presented, the scheduler is informed about its weight w_j and its expected processing time $\mathbb{E}[p_j]$ (in the case with release dates, also its release date r_j). When job j appears, it must be assigned to a machine $i \in \{1, \ldots, m\}$ immediately, and this decision must not be revised later. For a given job $j \in W$, and a given subset of jobs W, let us define by $H(j)$ the jobs in W that have a higher priority in the WSEPT ordering, that is

$$H(j) = \left\{k \in W \mid \frac{w_k}{\mathbb{E}[P_k]} \geq \frac{w_j}{\mathbb{E}[P_j]}\right\}.$$

Notice that, by convention, $H(j)$ contains job j, too. Accordingly, define

$$L(j) = \left\{k \in W \mid \frac{w_k}{\mathbb{E}[P_k]} < \frac{w_j}{\mathbb{E}[P_j]}\right\}$$

as those jobs that have lower priority in the WSEPT order.

It is clear that the online scheduling policies for the Sos model can in fact be interpreted as a subclass of stochastic scheduling policies in general. This because, assuming a classical input for a stochastic scheduling problem where all (stochastic) input data is disclosed at the outset, the only thing we require in the Sos model is a fixed assignment of jobs to machines beforehand. Therefore, the expected performance of an optimal Sos policy is no less than the expected performance of an optimal policy for a corresponding classical stochastic problem. (The latter being defined by the input after the online phase.) Hence, lower bounds on the expected value of an optimal policy known from stochastic scheduling carry over to the online setting. We crucially exploit that fact, and will utilize the following lower bound on the expected performance $\mathbb{E}\left[Z^{\mathrm{OPT}}\right]$ of an optimal stochastic scheduling policy. It is a generalization of a lower bound by Eastman et al. [5] for the deterministic setting.

Lemma 1 (Möhring et al. [11]). *For any instance of* $\mathrm{P}|\,|\mathbb{E}\left[\sum w_j\, C_j\right]$, *we have that*

$$
\mathbb{E}\left[Z^{\mathrm{OPT}}\right] \geq \sum_j w_j \sum_{k \in H(j)} \frac{\mathbb{E}\left[P_k\right]}{m} - \frac{(m-1)(\Delta-1)}{2m} \sum_j w_j \mathbb{E}\left[P_j\right],
$$

where Δ bounds the squared coefficient of variation of the processing times, that is, $\mathrm{Var}[P_j]/\mathbb{E}\left[P_j\right]^2 \leq \Delta$ *for all jobs $j = 1, \ldots, n$ and some $\Delta \geq 0$.*

This lemma indeed plays a crucial role in achieving performance guarantees for the SOS policies presented in the following sections. Clearly, the claim of Lemma 1 also applies to the setting with release dates $\mathrm{P}|r_j|\mathbb{E}\left[\sum w_j\, C_j\right]$.

3   Lower Bounds on Approximation

The requirement of a fixed assignment of jobs to machines beforehand may be interpreted as ignoring the additional information on the outcome of the stochastic process (that is, the actual realization of processing times), at least with respect to assignments of jobs to machines. In the following, we therefore give a lower bound on the expected performance $\mathbb{E}\left[Z^{\mathrm{FIX}}\right]$ of an optimal stochastic scheduling policy FIX that assigns jobs to machines beforehand. A fortiori, this lower bound holds for the best possible SOS policy, too.

Theorem 1. *For stochastic parallel machine scheduling with unit weights and i.i.d. exponential processing times,* $\mathrm{P}|p_j \sim \exp(1)|\mathbb{E}\left[\sum C_j\right]$, *there exist instances such that*

$$
\mathbb{E}\left[Z^{\mathrm{FIX}}\right] \geq 3(\sqrt{2}-1) \cdot \mathbb{E}\left[Z^{\mathrm{OPT}}\right] - \varepsilon,
$$

for any $\varepsilon > 0$. Here, $3(\sqrt{2}-1) \approx 1.24$. Hence, no policy that uses fixed assignments of jobs to machines can perform better in general.

Notice that the Theorem is formulated for the special case of exponentially distributed processing times. Stronger bounds could probably be obtained for arbitrary distributions. However, since our performance guarantees, as in [11], will depend on the coefficient of variation of the processing times, we are particularly interested in lower bounds for classes of distributions where this coefficient of variation is small. The coefficient of variation of exponentially distributed random variables equals 1. For example, for the case of $m = 2$ machines, we get a lower bound of $8/7 \approx 1.14$, and for that case our performance bound equals $2 - 1/m = 1.5$.

Proof (of Theorem 1). For simplicity, we will prove a slightly worse lower bound. Let us consider an instance with m machines and $n = m + \lceil m/2 \rceil$ exponentially distributed jobs, $p_j \sim \exp(1)$. The optimal stochastic scheduling policy is SEPT, shortest expected processing time first [1, 20], and the expected performance is (see, e.g., [17–Cor. 3.5.17])

$$\mathbb{E}\left[Z^{\mathrm{OPT}}\right] = \mathbb{E}\left[Z^{\mathrm{SEPT}}\right] = \sum_j \mathbb{E}\left[C_j^{\mathrm{SEPT}}\right] = m + \sum_{j=m+1}^{n} \frac{j}{m}.$$

The best fixed assignment policy assigns 2 jobs each to $\lceil m/2 \rceil$ of the machines, and 1 job each to $\lfloor m/2 \rfloor$ of the machines. Hence, there are m jobs with $\mathbb{E}\left[C_j\right]=1$, and $\lceil m/2 \rceil$ jobs with $\mathbb{E}\left[C_j\right]=2$. The expected performance for the best fixed assignment policy FIX is

$$\mathbb{E}\left[Z^{\mathrm{FIX}}\right] = \sum_j \mathbb{E}\left[C_j^{\mathrm{FIX}}\right] = m + 2 \cdot \lceil m/2 \rceil.$$

For small values $m = 2, 3, 4\ldots$, we calculate $\mathbb{E}\left[Z^{\mathrm{FIX}}\right]/\mathbb{E}\left[Z^{\mathrm{OPT}}\right] = 8/7$, $7/6, 32/27, \ldots$. It is easy to see that

$$\frac{\mathbb{E}\left[Z^{\mathrm{FIX}}\right]}{\mathbb{E}\left[Z^{\mathrm{OPT}}\right]} = \frac{16m^2}{13m^2 + \mathrm{o}(m^2)},$$

and for $m \to \infty$ we get a lower bound of $16/13 \approx 1.23$. Now the claim of the theorem follows along the same lines if we redefine the number of jobs as $n = m + \lceil \sqrt{2}m \rceil$. $\qquad\square$

4   Stochastic Online Scheduling

We next define a stochastic online scheduling policy for the problem without release dates, $\mathrm{P}|\,|\mathbb{E}\left[\sum w_j C_j\right]$. The basic idea is simple: any job j, once it appears, will be assigned to the machine where it causes the minimal increase in expected objective value (given that jobs $1, \ldots, j-1$ have been assigned already). In order to be able to do that, we first need to specify how the jobs will be scheduled on every single machine. We introduce a final bit of notation, letting M_i denote all jobs that are assigned to machine i, and letting $M_i(j) = \{k < j \mid k \in M_i\}$ denote the subset of jobs assigned to a machine i before some job j appears[4].

WSEPT (weighted shortest expected processing time first)
On each machine i, schedule the jobs M_i in non-decreasing order of their ratios of weight over expected processing time $w_j/\mathbb{E}\left[P_j\right]$.

This policy is known to be optimal for (stochastic) machine problems on a single machine, $1|\,|\mathbb{E}\left[\sum w_j C_j\right]$, by results of Smith and Rothkopf, respectively [15, 12]. Now we can define the MININCREASE policy as follows.

---

[4] Recall that we assume a numbering of the jobs in the order in which they appear in the online sequence; hence $k < j$ denotes jobs k that appeared earlier than job j.

MinIncrease

When a job j is presented to the scheduler, it is assigned to the machine i that minimizes the expression

$$\operatorname{incr}(j, i) = w_j \cdot \sum_{k \in H(j) \cap M_i(j)} \mathbb{E}[P_k] \;+\; \mathbb{E}[P_j] \cdot \sum_{k \in L(j) \cap M_i(j)} w_k \;+ w_j \mathbb{E}[P_j] \,.$$

In fact, given that WSEPT is used on each machine, MININCREASE just chooses the machine where job j causes the least increase in expected performance.

Theorem 2. *Consider the stochastic online scheduling problem on parallel machines, $\mathrm{P} \,|\, | \mathbb{E}\left[\sum w_j\, C_j\right]$. Given that $\operatorname{Var}[P_j]/\mathbb{E}[P_j]^2 \le \Delta$ for all jobs j and some constant $\Delta \ge 0$, the MININCREASE policy is a ρ–approximation, where*

$$\rho = 1 + \frac{(\Delta + 1)(m - 1)}{2m}.$$

Proof. Denote by $\mathbb{E}[\operatorname{incr}(j)]$ the increase in the expected objective value caused by fixing the assignment of a job j using MININCREASE. Since MININCREASE chooses the machine i on which j causes the least expected increase, the expected increase is

$$\mathbb{E}[\operatorname{incr}(j)] = \min_i \{\mathbb{E}[\operatorname{incr}(j, i)]\}$$

$$= \min_i \left\{ w_j \sum_{k \in H(j) \cap M_i(j)} \mathbb{E}[P_k] \;+\; \mathbb{E}[P_j] \sum_{k \in L(j) \cap M_i(j)} w_k \right\} + w_j \mathbb{E}[P_j]$$

$$\le \frac{1}{m} \left(w_j \sum_{k \in H(j), k < j} \mathbb{E}[P_k] \;+\; \mathbb{E}[P_j] \sum_{k \in L(j), k < j} w_k \right) + w_j \mathbb{E}[P_j] \,,$$

where the inequality holds because the least expected increase is not more than the average expected increase over all machines. By summing up these quantities over all jobs we obtain the expected performance $\mathbb{E}\left[Z^{\mathrm{MI}}\right]$ of the MININCREASE policy.

$$\mathbb{E}\left[Z^{\mathrm{MI}}\right] = \sum_j \mathbb{E}[\operatorname{incr}(j)]$$

$$\le \frac{1}{m} \sum_j \left(w_j \sum_{k \in H(j), k < j} \mathbb{E}[P_k] \;+\; \mathbb{E}[P_j] \sum_{k \in L(j), k < j} w_k \right) + \sum_j w_j \mathbb{E}[P_j]$$

$$= \frac{1}{m} \sum_j w_j \sum_{k \in H(j)} \mathbb{E}[P_k] \;+\; \frac{m - 1}{m} \sum_j w_j \mathbb{E}[P_j] \,,$$

where the last equality holds by index rearrangement, since

$$\sum_j \mathbb{E}[P_j] \sum_{k \in L(j), k < j} w_k \;=\; \sum_j w_j \sum_{k \in H(j), k > j} \mathbb{E}[P_k] \,.$$

Now, we plug in the inequality of Lemma 1, and using the trivial fact that $\sum_j w_j \mathbb{E}[P_j]$ is a lower bound for the expected performance $\mathbb{E}[Z^{\mathrm{OPT}}]$ of an optimal policy, we obtain

$$\mathbb{E}[Z^{\mathrm{MI}}] \leq \mathbb{E}[Z^{\mathrm{OPT}}] + \frac{(\Delta - 1)(m - 1)}{2m} \sum_j w_j \mathbb{E}[P_j] + \frac{m - 1}{m} \sum_j w_j \mathbb{E}[P_j]$$

$$\leq \left(1 + \frac{(\Delta + 1)(m - 1)}{2m}\right) \cdot \mathbb{E}[Z^{\mathrm{OPT}}]. \qquad \square$$

This performance guarantee matches the currently best known performance guarantee for the classical stochastic setting, which was derived for the performance of the WSEPT rule in [11]. The WSEPT rule, however, requires the knowledge of all jobs with their weights w_j and expected processing times $\mathbb{E}[P_j]$ at the outset. In contrast, the MinIncrease policy decides on machine assignments online, without any knowledge of the jobs to come. Finally, it is worthy to note that simple instances show that these two policies are indeed different.

Lower bounds for MinIncrease. The lower bound on the performance ratio for *any* fixed assignment policy given in Theorem 1 holds for the MinIncrease policy, too. Hence, in general, MinIncrease cannot be better than 1.24-approximative. We can strengthen the lower bound via more sophisticated instances, but the computations of the optimal values become unpleasant. We next give an instance for $m = 2$ machines.

Example 1. We are given 6 jobs with exponentially distributed processing times such that $\mathbb{E}[P_1] = \mathbb{E}[P_4] = 1, \mathbb{E}[P_2] = \mathbb{E}[P_5] = k$ and $\mathbb{E}[P_3] = \mathbb{E}[P_6] = 2k$, for some fixed k. The jobs appear in order of their indices in the online sequence.

Without going into further details, it turns out that the expected performance of the MinIncrease policy is $6 + 9k$, and the expected performance of an optimal scheduling policy is $5 + k(167/24 + 7/(1 + k) + 1/(2 + 4k))$. For $k \to \infty$, this yields a lower bound of $216/167 \approx 1.29$, whereas Theorem 2 yields a performance guarantee of 1.5.

For less restricted probability distributions, i.e., non-exponential and with larger coefficients of variation, we obtain a lower bound of $3/2$ on the expected performance of MinIncrease relative to an optimal scheduling policy. However, this is less meaningful compared to the performance bound of Theorem 2, which depends on an upper bound Δ on the squared coefficient of variation. We skip the details.

5   Stochastic Online Scheduling with Release Dates

In this section we consider the problem of stochastic online scheduling on parallel machines where jobs have release dates. As the optimal single machine scheduling policy is unknown to date for this problem, we analyze the expected performance of the MinIncrease policy which runs the following single machine scheduling policy.

α-Shift-WSEPT

Modify the release date r_j of each job j such that $r'_j = \max\{r_j, \alpha \, \mathbb{E}\,[P_j]\}$, for some fixed $0 < \alpha \le 1$. At any time t, when the machine is idle, start the job with the highest priority in the WSEPT order among all available jobs (respecting the modified release dates).

In the deterministic (online) setting, this policy was proposed for parallel machines in [9]. For the analysis of this policy, we restrict ourselves to random variables that we call δ-NBUE. This is a generalization of NBUE random variables.

Definition 1 (δ-NBUE). *A non-negative random variable X is δ-NBUE if, for $\delta \ge 1$,*

$$\mathbb{E}\,[X - t \,|\, X > t\,] \;\le\; \delta \,\mathbb{E}\,[X] \qquad \text{for all } t \ge 0.$$

Ordinary NBUE distributions are by definition 1-NBUE. For a NBUE random variable X, Hall and Wellner [7] showed that the (squared) coefficient of variation is bounded by 1, that is, $\mathrm{Var}[X]/\mathbb{E}\,[X]^2 \le 1$. From their work, it also follows that, if X is δ-NBUE, then $\mathrm{Var}[X]/\mathbb{E}\,[X]^2 \le 2\delta - 1$. Examples of ordinary NBUE (or 1-NBUE) distributions are exponential, Erlang, uniform, or Weibull distributions (with shape parameter at least 1). We next derive an upper bound on the expected completion time of a job, $\mathbb{E}\,[C_j]$, when scheduling jobs on a single machine according to the α-Shift-WSEPT policy. This bound is used later to analyze the expected performance of MinIncrease.

Lemma 2. *Let all processing times be δ-NBUE. Then the expected completion time of job j for α-Shift-WSEPT on a single machine can be bounded by*

$$\mathbb{E}\,[C_j] \le (1 + \delta/\alpha)\, r'_j + \sum_{k \in H(j)} \mathbb{E}\,[P_k]\,.$$

Proof. We consider some job j. Let us denote by $\mathcal{B}$ the event that the machine is busy processing some job at time r'_j, and let us denote by $\mathcal{I}$ the complement of $\mathcal{B}$, namely that the machine is idle (or just finished processing some job) at time r'_j. Under the condition $\mathcal{I}$ it could still be that there are higher priority jobs $k \in H(j) \setminus \{j\}$ available at time r'_j, but in any case the expected start time of job j can be postponed by at most

$$\sum_{k \in H(j) \setminus \{j\}} \mathbb{E}\,[P_k \,|\, \mathcal{I}\,]\,.$$

However, due to independence of processing times, we have that $\mathbb{E}\,[P_k \,|\, \mathcal{I}\,] = \mathbb{E}\,[P_k]$, and therefore

$$\mathbb{E}\,[S_j \,|\, \mathcal{I}\,] \;\le\; r'_j + \sum_{k \in H(j) \setminus \{j\}} \mathbb{E}\,[P_k]\,.$$

Consider condition $\mathcal{B}$ and let us denote by $\mathbb{E}\,[x(\mathcal{B})]$ the expected length of the time period until the machine becomes idle for the first time after r'_j. Under the condition $\mathcal{B}$, for any realization p of the processing times, conditioned on $\mathcal{B}$, some job $\ell(p)$ is in process at time r'_j (in fact, $\ell(p)$ might have lower or higher priority than j). Any such job ℓ was available at time $r'_\ell < r'_j$, and by definition of the modified release dates, we therefore know that $\mathbb{E}[P_\ell] \leq (1/\alpha)r'_\ell < (1/\alpha)r'_j$ for any such job ℓ. Moreover, letting $t = r'_j - S_\ell$, the expected remaining processing time of such job ℓ, conditioned on the fact that it is indeed in process at time r'_j, is $\mathbb{E}[P_\ell - t \mid P_\ell > t]$. Due to the assumption of δ-NBUE processing times, we thus know that

$$\mathbb{E}[P_\ell - t \mid P_\ell > t] \leq \delta\,\mathbb{E}[P_\ell] \leq (\delta/\alpha)r'_j\,.$$

Therefore, the expected remaining processing time of any job ℓ that might be in process at time r'_j is bounded by $(\delta/\alpha)r'_j$, and thus

$$\mathbb{E}[x(\mathcal{B})] \leq (\delta/\alpha)r'_j\,.$$

Repeating the same argument as above, we can now conclude that

$$\mathbb{E}[S_j \mid \mathcal{B}] \leq (1 + \delta/\alpha)\,r'_j + \sum_{k \in H(j)\setminus\{j\}} \mathbb{E}[P_k]\,. \tag{1}$$

As each of the two conditional expectations $\mathbb{E}[S_j \mid \mathcal{I}]$ and $\mathbb{E}[S_j \mid \mathcal{B}]$ is bounded by the right hand side of (1), we obtain that

$$\mathbb{E}[S_j] \leq (1 + \delta/\alpha)\,r'_j + \sum_{k \in H(j)\setminus\{j\}} \mathbb{E}[P_k]\,,$$

and the fact that $\mathbb{E}[C_j] = \mathbb{E}[S_j] + \mathbb{E}[P_j]$ concludes the proof. $\square$

In fact, it is quite straightforward to use Lemma 2 in order to show the following.

Corollary 1. *The α-Shift-WSEPT algorithm is a 3-approximation for the single machine problem $1|r_j|\mathbb{E}[\sum w_j C_j]$, for NBUE processing times.*

We just use that $\delta = 1$, and we choose $\alpha = 1$. We skip further details, and note that this matches the best known LP based performance bound derived in [11], which even holds for arbitrary processing time distributions.

The MinIncrease policy for the problem with release dates is now the following. In order to decide on which machine a job should go, we just ignore the release dates, and use the same policy for assigning jobs to machines that we used before in the setting without release dates.

Theorem 3. *Consider the stochastic online scheduling problem on parallel machines with release dates, $P|r_j|\mathbb{E}[\sum w_j C_j]$. Given that all processing times are δ-NBUE, the modified MinIncrease policy is a ρ-approximation, where*

$$\rho = 1 + \max\{1 + \delta/\alpha\,,\ \alpha + \delta + (m-1)(\Delta+1)/(2m)\}\,.$$

Here, Δ is such that $\mathrm{Var}[P_j]/\mathbb{E}[P_j]^2 \le \Delta$ for all jobs j. In particular, since all processing times are δ-NBUE, we know that $\Delta \le 2\delta - 1$ in the above performance bound.

Proof. Let i_j be the machine to which job j is assigned. Then, by Lemma 2 we know that

$$\mathbb{E}[C_j] \le (1 + \frac{\delta}{\alpha})r'_j + \sum_{k \in H(j) \cap M_{i_j}} \mathbb{E}[P_k], \tag{2}$$

and the expected value of MinIncrease can be bounded by

$$\mathbb{E}[Z^{\mathrm{MI}}] \le (1 + \frac{\delta}{\alpha}) \sum_j w_j r'_j + \sum_j w_j \sum_{k \in H(j) \cap M_{i_j}} \mathbb{E}[P_k] \tag{3}$$

Using an index rearrangement argument as in the proof of Theorem 2, we can write

$$\sum_j w_j \sum_{k \in H(j) \cap M_{i_j}} \mathbb{E}[P_k] = \sum_j \left(w_j \sum_{k \in H(j) \cap M_{i_j}(j)} \mathbb{E}[P_k] + \mathbb{E}[P_j] \sum_{k \in L(j) \cap M_{i_j}(j)} w_k + w_j \mathbb{E}[P_j] \right).$$

By definition of MinIncrease, we know that job j is assigned to the machine which minimizes the sums in parenthesis of the right hand side of this equation. Hence, by an averaging argument, we know that

$$\sum_j w_j \sum_{k \in H(j) \cap M_{i_j}} \mathbb{E}[P_k] \le \sum_j \left(w_j \sum_{k \in H(j), k<j} \frac{\mathbb{E}[P_k]}{m} + \mathbb{E}[P_j] \sum_{k \in L(j), k<j} \frac{w_k}{m} + w_j \mathbb{E}[P_j] \right)$$

$$= \sum_j w_j \sum_{k \in H(j)} \frac{\mathbb{E}[P_k]}{m} + \frac{m-1}{m} \sum_j w_j \mathbb{E}[P_j],$$

where the last equality follows from index rearrangement. Plugging this into (2), leads to the following bound on the expected performance of MinIncrease.

$$\mathbb{E}[Z^{\mathrm{MI}}] \le (1 + \frac{\delta}{\alpha}) \sum_j w_j r'_j + \sum_j w_j \sum_{k \in H(j)} \frac{\mathbb{E}[P_k]}{m} + \frac{m-1}{m} \sum_j w_j \mathbb{E}[P_j].$$

As mentioned before, the relaxed problem without release dates provides a lower bound on the expected optimum with release dates. We therefore can plug into the above inequality the bound of Lemma 1, and obtain

$$\mathbb{E}[Z^{\mathrm{MI}}] \le (1 + \frac{\delta}{\alpha}) \sum_j w_j r'_j + \mathbb{E}[Z^{\mathrm{OPT}}] + \frac{(m-1)(\Delta+1)}{2m} \sum_j w_j \mathbb{E}[P_j]$$

$$= \mathbb{E}[Z^{\mathrm{OPT}}] + \sum_j w_j \left((1 + \frac{\delta}{\alpha})r'_j + \frac{(m-1)(\Delta+1)}{2m} \mathbb{E}[P_j] \right). \tag{4}$$

By bounding r'_j by $r_j + \alpha \mathbb{E}[P_j]$, we obtain the following bound on the term in parenthesis of the sum in the right hand side of inequality (4).

$$\left(1 + \frac{\delta}{\alpha}\right)r'_j + \frac{(m-1)(\Delta+1)}{2m}\mathbb{E}[P_j]$$
$$\leq \left(1 + \frac{\delta}{\alpha}\right)r_j + \left(\alpha + \delta + \frac{(m-1)(\Delta+1)}{2m}\right)\mathbb{E}[P_j]$$
$$\leq (r_j + \mathbb{E}[P_j])\max\left\{1 + \frac{\delta}{\alpha}, \ \alpha + \delta + \frac{(m-1)(\Delta+1)}{2m}\right\}.$$

The proof is completed by using this inequality in equation (4), and applying the trivial lower bound $\sum_j w_j(r_j + \mathbb{E}[P_j]) \leq \mathbb{E}[Z^{\mathrm{OPT}}]$ on the expected optimum performance.   □

For NBUE processing times, where we can choose $\Delta = \delta = 1$, the approximation ratio is minimal for $\alpha = (\sqrt{5m^2 - 2m + 1} - m + 1)/(2m)$, obtaining a ratio of $2 + (\sqrt{5m^2 - 2m + 1} + m - 1)/(2m)$, which is less than $(5 + \sqrt{5})/2 - 1/(2m) \approx 3.62 - 1/(2m)$, improving upon the previously best know approximation ratio of $4 - 1/m$ from [11].

Acknowledgement. The authors would like to thank Rolf H. Möhring for helpful discussions, and Andreas S. Schulz for pointing out a misinterpretation in an earlier version of this paper.

References

1. J. L. Bruno, P. J. Downey, and G. N. Frederickson. Sequencing tasks with exponential service times to minimize the expected flowtime or makespan. *J. ACM*, 28:100–113, 1981.
2. S. Chakrabarti and S. Muthukrishnan. Resource scheduling for parallel database and scientific applications. In *Proc. 8th Ann. ACM Symp. on Parallel Algorithms and Architectures*, pages 329–335, Padua, Italy, 1996.
3. C. Chekuri, R. Johnson, R. Motwani, B. Natarajan, B. Rau, and M. Schlansker. An analysis of profile-driven instruction level parallel scheduling with application to super blocks. In *Proc. 29th IEEE/ACM Int. Symp. on Microarchitecture*, Paris, France, pages 58–69, 1996.
4. M. A. H. Dempster, J. K. Lenstra, and A. H. G. Rinnooy Kan, editors. *Deterministic and Stochastic Scheduling*. D. Reidel Publishing Company, Dordrecht, 1982.
5. W. Eastman, S. Even, and I. Isaacs. Bounds for the optimal scheduling of n jobs on m processors. *Mgmt. Sci.*, 11:268–279, 1964.
6. R. L. Graham, E. L. Lawler, J. K. Lenstra, and A. H. G. Rinnooy Kan. Optimization and approximation in deterministic sequencing and scheduling: A survey. *Ann. Discr. Math.*, 5:287–326, 1979.
7. W. J. Hall and J. A. Wellner. Mean residual life. In M. Csörgö, D. A. Dawson, J. N. K. Rao, and A. K. Md. E. Saleh, editors, *Proc. Int. Symp. on Statistics and Related Topics*, pages 169–184, Ottawa, ON, 1981. Amsterdam: North-Holland.

8. E. Koutsoupias and C. H. Papadimitriou. Beyond competitive analysis. *SIAM J. Comp.*, 30:300–317, 2000.

9. N. Megow and A. S. Schulz. On-line scheduling to minimize average completion time revisited. *Oper. Res. Lett.*, 32(5):485–490, 2004.

10. R. H. Möhring, F. J. Radermacher, and G. Weiss. Stochastic scheduling problems I: General strategies. *ZOR - Zeitschrift für Oper. Res.*, 28:193–260, 1984.

11. R. H. Möhring, A. S. Schulz, and M. Uetz. Approximation in stochastic scheduling: the power of LP-based priority policies. *J. ACM*, 46:924–942, 1999.

12. M.H. Rothkopf. Scheduling with random service times. *Mgmt. Sci.*, 12:703–713, 1966.

13. M. Scharbrodt, T. Schickinger, and A. Steger. A new average case analysis for completion time scheduling. In *Proc. 34th Ann. ACM Symp. on the Theory of Computing*, Montréal, QB, pages 170–178, 2002.

14. M. Skutella and M. Uetz. Scheduling precedence-constrained jobs with stochastic processing times on parallel machines. In *Proc. 12th Ann. ACM-SIAM Symp. on Discrete Algorithms*, pages 589–590, Washington, DC, 2001.

15. W. Smith. Various optimizers for single-stage production. *Naval Res. Log.*, 3:59–66, 1956.

16. A. Souza-Offermatt and A. Steger. The expected competitive ratio for weighted completion time scheduling. In *Proc. 21st Symp. on Theoretical Aspects of Computer Science*, Montpellier, France, 2004. To appear.

17. M. Uetz. *Algorithms for Deterministic and Stochastic Scheduling*. PhD thesis, Cuvillier Verlag, Göttingen, Germany, 2002.

18. G. Weiss. Approximation results in parallel machines stochastic scheduling. *Ann. Oper. Res.*, 26:195–242, 1990.

19. G. Weiss. Turnpike optimality of Smith's rule in parallel machines stochastic scheduling. *Math. Oper. Res.*, 17:255–270, 1992.

20. Gideon Weiss and Michael Pinedo. Scheduling tasks with exponential service times on non-identical processors to minimize various cost functions. *J. Appl. Prob.*, 17:187–202, 1980.

A $\frac{5}{4}$-Approximation Algorithm for Biconnecting a Graph with a Given Hamiltonian Path[*]

Davide Bilò[1] and Guido Proietti[1,2]

[1] Dipartimento di Informatica, Università di L'Aquila, Via Vetoio,
67010 L'Aquila, Italy
{davide.bilo, proietti}@di.univaq.it.

[2] Istituto di Analisi dei Sistemi ed Informatica "Antonio Ruberti", IASI-CNR,
Viale Manzoni 30, 00185 Roma, Italy

Abstract. Finding a minimum size 2-vertex connected spanning subgraph of a k-vertex connected graph $G = (V, E)$ with n vertices and m edges is known to be NP-hard and APX-hard, as well as approximable in $O(n^2 m)$ time within a factor of 4/3. Interestingly, the problem remains NP-hard even if a Hamiltonian path of G is given as part of the input. For this input-enriched version of the problem, we provide in this paper a linear time and space algorithm which approximates the optimal solution by a factor of no more than $\min\left\{ \frac{5}{4}, \frac{2k-1}{2(k-1)} \right\}$.

1 Introduction

The problem of finding a minimum size 2-vertex connected (simply *biconnected*, in the following) spanning subgraph (*MBSS problem*) of a biconnected, undirected graph $G = (V, E)$, with n vertices and m edges, is one of the classical problems in computer science and combinatorial optimization [9]. It is known to be NP-hard, since its decision version contains as a special case the *Hamiltonian cycle* problem (i.e., the problem of deciding whether a graph G contains a simple cycle that includes all the vertices), which is well-known to be NP-complete [5].

Due to its relevance and to the great number of applications it finds in different fields, several approximation algorithms for solving this problem have been devised in the past few years. Khuller and Vishkin [10] introduced the notions of *carving* of a graph to establish an approximation factor of no more than 5/3. Their algorithm has been firstly improved by Garg *et al.* [6], who obtained an approximation ratio of 3/2. After, this ratio was improved to 4/3 by Vempala and Vetta [13]. Concerning inapproximability results, the problem is known to be APX-hard [11].

The weighted version of the problem has been deeply investigated as well. In this case, the problem admits a $\left(2 + \frac{1}{n}\right)$-approximation algorithm [10], while if G satisfies the triangle inequality, then it is approximable within 3/2 [4]. Moreover, for any integer $d = o(\log n)$, if G is complete and Euclidean in $\mathbb{R}^d$

---

[*] This work has been partially supported by the Research Project GRID.IT, funded by the Italian Ministry of Education, University and Research.

(i.e., G is embedded in the Euclidean d-dimensional space and the edge weights correspond to the Euclidean d-dimensional distance between the corresponding endvertices), then the problem admits a PTAS [2]. Concerning inapproximability results, the problem is not approximable (unless P=NP) within $\frac{68569}{68564} - \varepsilon$, for any constant $\varepsilon > 0$ [1].

As far as the edge version of the problem is concerned, i.e., that of finding a minimum size 2-edge connected spanning subgraph of a 2-edge-connected, undirected graph, the best known approximation ratio is $5/4$ [7]. In the same paper, the authors claimed that their algorithm can be extended, by preserving the approximation ratio, to the vertex version of the problem, but unfortunately this seems not to be the case [8]. As a consequence, there is currently a gap between the approximability of the vertex and the edge version of the problem, i.e., $4/3$ versus $5/4$.

A question which naturally arises is that of studying whether the approximation guarantee can be improved once the input of the problem is enriched. In particular, Papadimitriou and Steiglitz [12] proved that the problem of determining whether a graph contains a Hamiltonian cycle remains NP-complete even if a Hamiltonian path is given as part of the input. It follows that the problem of determining whether a graph admits a biconnected spanning subgraph of size $k \geq n$, once a Hamiltonian path is given as part of the input, is NP-complete as well. In this paper we consider the optimization version of this latter problem (*MBSSHP problem* for short).

To the best of our knowledge, for the MBSSHP problem the same approximation factor as for the MBSS problem holds, also when the edge-version of the problem is considered. Hence, also in this case, there is a gap between the approximability of the vertex and the edge version of the problem, i.e., $4/3$ versus $5/4$.

In this paper, we get to the target of closing this gap. Indeed, we show that the MBSSHP problem can be approximated in linear time and space with a performance guarantee of $5/4$. Moreover, we show that if G is k-vertex connected, $k > 3$, then our algorithm can be enhanced to return a $\frac{2k-1}{2(k-1)}$-approximated solution. Our approach deviates significantly from that proposed for the MBSS problem by Vempala and Vetta [13], since their algorithm cannot guarantee an approximation factor better than $4/3$ when adapted to the MBSSHP problem.

From an application point of view, our algorithm has a practical impact on *chain communication networks*, where we have a set of vertices $v_1, v_2, \ldots, v_n$ which mutually exchange messages through a chain of links $(v_i, v_{i+1}), i = 1, \ldots, n-1$. Suppose now we have a set of potential additional links (v_i, v_j), $1 \leq i < j+1 \leq n$, such that the graph resulting from the chain enriched of the additional edges is biconnected. Then, one might be interested in making the communication between vertices immune to single vertex failures, by using a minimum number of links. In this case, our algorithm computes an approximated solution in linear time and space with a performance guarantee of $5/4$.

The paper is organized as follows: in Section 2 we introduce basic definitions and notations used in the paper. In Section 3 we present two simple algorithms for the MBSSHP problem. In Section 4, after analyzing the known lower bounds for the MBSS problem, we first present a new lower bound, then we refine the algorithms of Section 3.

2   Basic Definitions

Let $G = (V, E)$ be a simple undirected graph (i.e., without loops and parallel edges), where V is the set of vertices and $E \subseteq V \times V$ is the set of edges. Let $n \geq 3$ and m be the number of vertices and the number of edges, respectively. For all $v \in V$, $\delta_G(v)$ denotes the *degree* of v in G, i.e., the number of vertices adjacent to v with respect to the edge set of G. A graph $H = (V(H), E(H))$ is called a *subgraph* of G if $V(H) \subseteq V$ and $E(H) \subseteq E$. If $V(H) = V$ then H is called a *spanning subgraph* of G. For any $U \subseteq V$, the graph $G_U = (U, E_U)$ where $E_U = \{(v, v') \in E \mid v, v' \in U\}$ is said to be *induced* from U while the graph $G' = (U \cup \{x_U\}, E_U \cup E')$, where $x_U \notin V$ and $E' = \{(u, x_U) \mid (u, v) \in E$ with $u \in U \wedge v \in V \setminus U\}$, is said to be obtained from G by *shrinking* $V \setminus U$ into one vertex x_U.

A *simple path* Π (or a *path* for short) in G is a subgraph with $V(\Pi) = \{v_1, \ldots, v_k \mid v_i \neq v_j$ for $i \neq j\}$ and $E(\Pi) = \{e_i = (v_i, v_{i+1}) \mid 1 \leq i < k\}$, also denoted as $\Pi(v_1, v_k)$ or $\langle v_1, v_2, \ldots, v_k \rangle$. Path Π is said to go from v_1 to v_k, called the *endvertices* of Π, passing through the *internal* vertices $v_2, v_3, \ldots, v_{k-1}$. A *cycle* is a path whose endvertices coincide. G is said to be *Hamiltonian* if it has a spanning cycle. A spanning path $\Pi_G = \langle v_1, v_2, \ldots, v_n \rangle$ of G is called a *Hamiltonian path*. Edges in $E(\Pi_G)$ are called *path edges*, while edges in $F = E \setminus E(\Pi_G)$ are called *cycle edges*. By $E(e_i)$ we denote the set of all cycle edges forming a *fundamental cycle* with e_i, i.e., a cycle containing only one cycle edge. If $f \in E(e_i)$ then we say f *covers* e_i. Thus, $f = (v_j, v_h)$, with $j < h$, covers e_i iff $j \leq i < h$. For any cycle edge $f = (v_i, v_j)$, with $i < j$, we call v_i (resp., v_j) the *left* (resp., *right*) endvertex of f. We denote by $\mathcal{G}^n$ the class of graphs of n vertices having a Hamiltonian path.

A graph G is *connected* if, for any $u, v \in V$, there exists a path in G going from u to v. The *connected components* of a graph G are the maximal (w.r.t. vertex insertion) connected subgraphs of G. A graph G is *k-vertex connected* (or simply *k-connected*) if $n \geq k + 1$ and for any $V' \subsetneq V$ of $k - 1$ vertices, the graph induced by $V \setminus V'$ is connected. When $k = 2$, G is said to be *biconnected*. The maximum integer k such that G is k-connected is said to be the *connectivity number* of G and it is denoted by $\kappa(G)$.

A subset C of V, with $|C| = k$ is a *k-separator* (or simply *separator*) if $G_{V \setminus C}$ is not connected, that is, there exists a two partition V_ℓ, V_r of $V \setminus C$ such that G has no edges with one endvertex in V_ℓ and the other in V_r. We say that V_ℓ, C, V_r is a *k-separation* of G. A separator C is said to be *minimal* if no proper subset of C is a separator. Observe that $\kappa(C) \geq \kappa(G)$, for every separator C. Let $e_i \in E(\Pi_G)$. By $\kappa_\ell(e_i)$ (resp., $\kappa_r(e_i)$) we denote the cardinality of a minimal (unique) separation V_ℓ, C, V_r such that $C \cup V_\ell = \{v_1, v_2, \ldots, v_i\}$ (resp., $C \cup V_r = \{v_{i+1}, v_{i+2}, \ldots, v_n\}$). Then, $\kappa(e_i) = \min\{\kappa_\ell(e_i), \kappa_r(e_i)\}$.

In the rest of the paper, $\text{OPT}(G)$ will denote the size of a *minimum biconnected spanning subgraph* (MBSS) of a biconnected graph G. Clearly, $\text{OPT}(G) \geq n$.

3   Basic Algorithm

In this section we describe a very simple algorithm for finding a biconnected spanning subgraph of an undirected graph G, with the hypothesis that we have a Hamiltonian path Π_G of G and $\kappa(G) > 1$. We show that the approximation ratio depends on $\kappa(G)$. Henceforth, unless stated otherwise, $\Pi_G = \langle v_1, v_2, \dots, v_n \rangle$. Moreover, by f_i we denote a cycle edge of $E(e_i)$ with right endvertex of maximum index.

The basic idea of the algorithm is simple. Starting from a spanning subgraph H of G with no edges, the algorithm processes each vertex in order from v_1 to v_n. At each step it augments $E(H)$ by adding edges. The invariant property maintained by the algorithm is the following: if it is currently exploring Π_G from v_i to v_j, $i < j$, then the subgraph H' of H induced from $\{v_1, v_2, \dots, v_i\}$ is already biconnected. The set of edges to be added is determined by the function *Expand*. Thus, the more powerful the function *Expand*, the lower the size of the computed biconnected spanning subgraph. We propose a first simple version of the function *Expand*.

Algorithm $BSS(G, \Pi_G = \langle v_1, v_2, \dots v_n \rangle)$;
Input: A biconnected graph G and a Hamiltonian path Π_G of G;
Output: A biconnected spanning subgraph H of G.
  begin
      $H = (V, \emptyset)$;
      $\Pi_R = \Pi_G(v_2, v_n)$;
      while $\Pi_R \neq \langle v_n \rangle$ **do**
          $(F', E_\Pi) = Expand(G, \Pi_R)$; $\% F' \subseteq F = E \setminus E(\Pi_G), E_\Pi \subseteq E(\Pi_G)$
          $i = \max\{j \mid (v_h, v_j) \in F', h < j\}$;
          $E(H) = E(H) \cup F' \cup E_\Pi$;
          $\Pi_R = \Pi_G(v_i, v_n)$;
      end while
      return H;
  end.

Function $Expand(G, \Pi_R = \langle v_{i+1}, v_{i+2}, \dots, v_n \rangle)$;
Input: A graph G and a path Π_R of G;
Output: A set of cycle edges $F' \subseteq E \setminus E(\Pi_G)$ and a set of path edges $E_\Pi \subseteq E(\Pi_R)$.
  begin
      Let $E_\Pi \subseteq E(\Pi_R)$ be the set of path edges covered by f_i;
      return $(\{f_i\}, E_\Pi)$;
  end.

Theorem 1. *The algorithm BSS computes a $\frac{\kappa(G)}{\kappa(G)-1}$-approximated solution in* $\mathcal{O}(m)$ *time and space.*

Proof. At a given iteration, let $\Pi_R = \Pi_G(v_{i+1}, v_n)$. Since $\kappa(e_i) \geq \kappa(G)$, we have that $f_i = (v, v_{i+j+1})$, with $j \geq \min\{n - i - 1, \kappa(G) - 1\}$. So we add at most one cycle edge every $\kappa(G) - 1$ edges of Π_R plus one extra cycle edge if Π_R has a length less than $\kappa(G) - 1$. Since Π_G is a subgraph of H, we achieve the following approximation ratio:

$$\frac{|E(H)|}{\text{OPT}(G)} \leq \frac{n - 1 + \left\lceil \frac{n-1}{\kappa(G)-1} \right\rceil}{n} \leq 1 + \frac{\left\lceil \frac{n}{\kappa(G)-1} \right\rceil - 1}{n} \leq 1 + \frac{\frac{n}{\kappa(G)-1}}{n} = \frac{\kappa(G)}{\kappa(G) - 1}.$$

Moreover, at each iteration we only have to explore the cycle edges belonging to $E(e_i)$. As we take the one whose right endvertex has maximum index and since Π_R updates to $\Pi(v_{i+j+1}, v_n)$, then the algorithm does not need to explore these edges any more. Hence, the time and space complexity is $\mathcal{O}(m)$. $\qquad \square$

The function $Expand(G, \Pi_R)$ defined above uses little information implied by $\kappa(G)$. To improve the performance of the algorithm we introduce the relation of *semi-adjacency* between cycle edges. So we say that two cycle edges $f = (v_i, v_{j+1}), f' = (v_j, v_k)$, with $i < j < k - 1$, are *semi-adjacent*. Path edge e_j is said to be the *middle* of f and f'. We can now prove the following:

Lemma 1. *For every e_i, either f_i or a pair semi-adjacent edges f, f' (one of which belongs to $E(e_i)$) cover $\min\{n - i - 1, 2(\kappa(G) - 1)\}$ edges of $\Pi_G(v_{i+1}, v_n)$.*

Proof. Let v_{j+1} be the right endvertex of f_i. If f_i covers $\min\{n-i-1, 2(\kappa(G)-1)\}$ edges of $\Pi_G(v_{i+1}, v_n)$ then the claim is true. So we can assume that f_i covers $k < \min\{n - i - 1, 2(\kappa(G) - 1)\}$ edges of $\Pi_G(v_{i+1}, v_n)$. We say that v_h, with $i < h \leq j$ is a *potential semi-adjacent* vertex if v_{h+1} is an endvertex of some edges in $E(e_i)$. Indeed, if some edge in $E(e_j)$ has v_h as endvertex, then this edge is semi-adjacent to some edge in $E(e_i)$. Since $\kappa(e_i) \geq \kappa(G)$, then there are $k_1 \geq \kappa(G) - 1$ vertices of $\Pi_G(v_{i+2}, v_{j+1})$ that are endvertices of some edge in $E(e_i)$. Hence, k_1 vertices of $\Pi_G(v_{i+1}, v_j)$ are potential semi-adjacent vertices, while the remaining

$$k_2 = k - k_1 \leq k - \kappa(G) + 1$$

are not. Since $\kappa(e_j) \geq \kappa(G)$, then at least $\kappa(G) - 1$ vertices of $\Pi_G(v_{i+1}, v_j)$ are endvertices of some edges in $E(e_j)$. As $k_2 \leq k - \kappa(G) + 1 < \kappa(G) - 1$ this means that there exist edges in $E(e_j)$ semi-adjacent to some edges in $E(e_i)$. Among such edges, choose any one (say f) whose right endvertex has maximum index. We claim that f covers at least $\min\{n - j - 1, 2(\kappa(G) - 1) - k\}$ edges of $\Pi_G(v_{j+1}, v_n)$. To prove this, suppose it is not true, that is f covers $h < \min\{n - j - 1, 2(\kappa(G) - 1) - k\} \leq 2(\kappa(G) - 1) - k$ edges of $\Pi_G(v_{j+1}, v_n)$. In this case, if we remove the first $h + 1$ vertices of $\Pi_G(v_{j+1}, v_n)$ and all non potential semi-adjacent vertices in $\Pi_G(v_{i+1}, v_j)$, we break the graph into two connected components. But the number of vertices removed is

$$k_2 + h + 1 \leq k - \kappa(G) + 1 + h + 1 < k - \kappa(G) + 1 + 2\kappa(G) - 2 - k + 1 = \kappa(G)$$

and so G is not $\kappa(G)$-connected. We have obtained a contradiction. $\qquad \square$

Using Lemma 1 we can implement a new powerful function *Expand* such that:

Theorem 2. *The algorithm BSS computes a* $\frac{2\kappa(G)-1}{2(\kappa(G)-1)}$*-approximated solution in* $\mathcal{O}(m)$ *time and space.*

Proof. At a given iteration, let v_{j+1} in $\Pi_R = \Pi_G(v_{i+1}, v_n)$ be the right endvertex of f_i. If f_i covers $\min\{n - i - 1, 2(\kappa(G) - 1)\}$ edges of Π_R, set $E_C = \{f_i\}$ and $E_\Pi = \{e_{i+1}, e_{i+2}, \dots, e_j\}$, otherwise from Lemma 1 there are two semi-adjacent edges f, f', one of which belongs to $E(e_i)$, covering $\min\{n - i - 1, 2(\kappa(G) - 1)\}$ edges of Π_R. In this case E_Π it the set of path edges of Π_R covered by f, f' minus the middle one, while $E_C = \{f, f'\}$. So we add at most one cycle edge every $2(\kappa(G) - 1)$ edges of Π_R plus one extra cycle edge if Π_R has a length less than $2(\kappa(G) - 1)$. Thus, we achieve the following approximation ratio:

$$\frac{|E(H)|}{\text{OPT}(G)} \le \frac{n - 1 + \left\lceil \frac{n-1}{2(\kappa(G)-1)} \right\rceil}{n} \le 1 + \frac{\frac{n}{2(\kappa(G)-1)}}{n} = \frac{2\kappa(G) - 1}{2(\kappa(G) - 1)}.$$

From Lemma 1 it also follows that at each iteration we have to explore the edges belonging to $E(e_i)$ and the edges belonging to $E(e_j)$. Since the next iterations the algorithm will not explore edges in $E(e_i)$ any more, then the time and space complexity is $\mathcal{O}(m)$. □

The second function *Expand* we have just defined, uses $\kappa(G)$ as a lower bound for $\kappa(e_i), 1 \le i \le n-1$. Looking at the proof of Lemma 1 one may convince that

Remark 1. The practical approximation ratio we obtain is given by the minimum value $\kappa(e_i)$ of all sets $E(e_i)$ the algorithm considers.

The following lemma shows that the approximation ratio of Theorem 2 is tight when compared to the trivial lower bound n.

Lemma 2. $\forall k \ge 2$, *there exists a k-connected graph $G \in \mathcal{G}^n$ for which the ratio between* $\text{OPT}(G)$ *and n is equal to* $\frac{2k-1}{2(k-1)}$.

Proof. The proof is constructive. Let $n = 2j(k-1) + 4k - 1$, where j is a positive integer. We first build a bipartite graph $G' \in \mathcal{G}^n$. We number the vertices of G' from 1 to n. Let $V_C = \{u^i = v_{2k+2i(k-1)} | i = 0, \dots, j\}$. Let $V_0 = \{v_1, v_2, \dots, u^0\}$ and $V_{j+1} = \{u^j, \dots, n\}$, while

$$V_i = \{u^{i-1}, \dots, u^i\}, \text{ with } i = 1, \dots, j.$$

Notice that $V_i \cap V_{i+1} = \{u^i\}$. For every $i = 0, 1, \dots, j$, let V_i^e (resp., V_i^o) be the set of even (resp., odd) vertices of V_i. The set of edges of G' is defined as follows:

$$E(G') = \bigcup_{i=0}^{j+1} (V_i^e \times V_i^o).$$

Notice that $\langle v_1, v_2, \ldots, v_n \rangle$ is a Hamiltonian path of G'. Clearly $\kappa(G') = 1$, since every vertex in V_C is a *cut-vertex*, i.e., a vertex whose removal disconnects G'. From G' we build a new graph $G = (V, E)$ with $\kappa(G) = k$. Let $V = V(G')$ and $E = E(G') \cup E'$, where

$$E' = \{(u, v) | u \in V_i^e, v \in V_{i+1}^e, i = 0, 1, \ldots, j\}.$$

It is easy to see that G is k-connected. Let us consider the topological structure of a MBSS $\mathcal{H}$ of G. As every vertex $u^i \in V_C$ in G' is a cut-vertex, it follows that $E(\mathcal{H})$ must contain a cycle edge in E' *covering* u^i (i.e., an edge whose addition to G' makes u_i not to be a cut-vertex). By construction, the cycle edges covering u^i do not cover w^j, with $j \neq i$. So $\mathcal{H}$ has at least

$$|V_C| = j + 1 = \frac{n - 4k + 1}{2(k - 1)} + 1 = \frac{n - 2k - 1}{2(k - 1)}$$

edges of E'. Moreover, since there is no edge between pairs of odd vertices, this implies that $\mathcal{H}$ must have $2\left\lceil \frac{n}{2} \right\rceil = n + 1$ edges of $E(G')$, where the equality follows from the fact that n is odd. So the approximation ratio is

$$\lim_{n \to +\infty} \frac{n + 1 + |V_C|}{n} = 1 + \lim_{n \to +\infty} \frac{1 + \frac{n - 2k - 1}{2(k-1)}}{n} = 1 + \lim_{n \to +\infty} \frac{n - 3}{2n(k - 1)} = \frac{2k - 1}{2(k - 1)}.$$

$\square$

4   Improving the Algorithm

4.1   Considerations About Well-Known Lower Bounds

As seen in the previous section, the approximation ratio we can achieve is strictly related to the connectivity value of G. Since $\forall \kappa(G) \geq 3, \frac{2\kappa(G) - 1}{2(\kappa(G) - 1)} \leq \frac{5}{4}$, our aim now is to improve the approximation ratio for graphs G with $\kappa(G) = 2$. In [13], the authors give an improvement for the lower bound of $\text{OPT}(G)$ based on the following definitions:

Definition 1. *A vertex v is a* beta-vertex *if there exist two vertices u_1, u_2 such that the graph induced from $V \setminus \{u_1, u_2\}$ has at least three connected components, one of which only contains v.*

Definition 2. *Two vertices v_1, v_2 are a* beta-pair *if there exist two vertices u_1, u_2 such that the graph induced from $V \setminus \{u_1, u_2\}$ has at least three connected components, one of which only contains v_1, v_2.*

Definition 3. *A graph G is* beta-free *if it has no beta-structures, i.e., neither beta-vertices nor beta-pairs.*

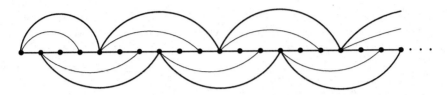

Fig. 1. A tight example for the lower bound given in [13] and in [7]. The edges of a MBSS are represented in bold

In [13], the authors only consider instances for which G is beta-free. They first show that the case of beta-pairs can be reduced to that of beta-vertices. Then, they consider the case of a beta-vertex v of G. Assume u_1, u_2 are the two vertices adjacent to v. Let $G' = G \setminus \{v\}$. Since the two edges incident to v are forced in any MBSS of G, an α-approximated solution for G can be obtained from an α-approximated solution for G' by adding the edges incident to v. In this case, it is easy to build a Hamiltonian path $\Pi_{G'}$ of G' from a Hamiltonian path Π_G of G: simply remove the beta-vertex v and add edge (u_1, u_2). If v is an endvertex of Π_G, then the graph induced from $V \setminus \{u_1, u_2\}$ has another beta-vertex u different from v. In this case remove u (as it cannot be an endvertex of Π_G) and add edge (u_1, u_2).

Now we describe a linear time algorithm that uses Π_G to remove all beta-vertices and all beta-pairs from G. For every vertex v_i, $1 < i \leq n-3$, if $\delta_G(v_{i+1}) = 2$ and f_{i-1} does not cover e_{i+2}, then v_{i+1} is a beta-vertex. Otherwise, if $v_{i+3} \neq v_n$, v_{i+1} is at most adjacent to v_i, v_{i+2}, v_{i+3}, while v_{i+2} is at most adjacent to v_{i+1}, v_{i+3}, v_i and f_{i-1} does not cover e_{i+3}, then v_{i+1}, v_{i+2} is a beta-pair.

In [13] it is shown that one can find a 4/3-approximated solution in $\mathcal{O}(n^2 m)$ time and linear space. The lower bound used there to estimate $\mathrm{OPT}(G)$ (the same lower bound was used in [7]) is given by the number of beta-structures removed from G plus the size of a minimum spanning subgraph H with $\delta_H(v) \geq 2, \forall v \in V(H)$, of the graph computed from the beta-free reduction of G. However, whenever $G \in \mathcal{G}^n$ is a beta-free graph, this lower bound is at most $n+1$. Indeed, a subgraph H of G made up by a Hamiltonian path Π_G, plus two extra edges $(v_1, u), (u', v_n) \in E$, is a spanning subgraph of G such that $\delta_H(v) \geq 2, \forall v \in V$. In Figure 1 we show a beta-free graph $G \in \mathcal{G}^n$ whose MBSS has size asymptotically equal to $\frac{4}{3}n$.

4.2   Our Lower Bound

As seen before, the core of the MBSS problem is not just the achievement of a better algorithm, but also the careful estimate of the size of an optimal solution. The purpose of this section is to present a new lower bound.

Let $\alpha(G)$ denote the *independence number* of G, i.e., the size of a largest set of vertices U (called *maximum independent set*) of G that induces an *empty graph*, i.e., a graph with no edges. The following two lemmas are well-known.

Lemma 3. *[3] Every graph G with $n \geq 3$ and $\kappa(G) \geq \alpha(G)$ is Hamiltonian.* $\square$

Lemma 4. *[6] For a biconnected graph G,* $\text{OPT}(G) \geq \max\{n, 2\alpha(G)\}$. $\qquad\square$

The *join* $G = G_1 + G_2$ of two graphs $G_1 = (V_1, E_1)$ and $G_2 = (V_2, E_2)$, where $V_1 \cap V_2 = \emptyset$ and $E_1 \cap E_2 = \emptyset$, is the *graph union* $G_1 \cup G_2$ (i.e., the graph $(V_1 \cup V_2, E_1 \cup E_2)$) together with all the edges joining V_1 and V_2.

Let $\mathcal{K}(|U|, \mathcal{K}_k)$ denote the *join* of the empty graph of $|U|$ vertices and the *complete graph* $\mathcal{K}_k$ (i.e., the graph having an edge between every pair of its k vertices). We can now prove the following:

Lemma 5. *For every integer $k \geq 2$, $\mathcal{K}(k+1, \mathcal{K}_k)$ is a maximum-size k-connected non-Hamiltonian graph with $2k + 1$ vertices. Moreover, any other k-connected non-Hamiltonian graph G with $2k + 1$ vertices is isomorphic to a subgraph of $\mathcal{K}(k + 1, \mathcal{K}_k)$.*

Proof. From Lemma 3, a non-Hamiltonian k-connected graph must have at least $2k + 1$ vertices. Since $\alpha(\mathcal{K}(k + 1, \mathcal{K}_k)) = k + 1$, from Lemma 4 it follows that $\mathcal{K}(k + 1, \mathcal{K}_k))$ is not Hamiltonian (notice that, in every biconnected spanning subgraph of $\mathcal{K}(k + 1, \mathcal{K}_k)$, at least 2 vertices of $V(\mathcal{K}(k + 1, \mathcal{K}_k)) \setminus U$ must have degree 3). The insertion of an edge in $\mathcal{K}(k + 1, \mathcal{K}_k)$ causes the independence number to decrease to k, and so from Lemma 3 the graph becomes Hamiltonian. Every non-Hamiltonian k-connected graph G with $2k + 1$ vertices must have $\alpha(G) = k + 1$, and so there exists a subgraph of $\mathcal{K}(k + 1, \mathcal{K}_k)$ isomorphic to G. The claim follows. $\qquad\square$

Let $G = (V, E)$ be a biconnected graph. For every k-partition $V_1, V_2, \ldots, V_k$ of V, there exist two edges with one endvertex in V_i and the other in $V \setminus V_i$, $i = 1, \ldots, k$. Looking at the proof of Lemma 5, we deduce the following:

Lemma 6. *Let C be a k-separator and let $V_1, V_2, \ldots V_{k+1}$ be a $(k + 1)$-partition of $V \setminus C$. If by shrinking each V_i into a node x_i we obtain a graph isomorphic to a subgraph of $\mathcal{K}(k + 1, \mathcal{K}_k)$ (with $x_1, x_2, \ldots, x_{k+1}$ mapped to U), then G is not Hamiltonian, i.e., $\text{OPT}(G) \geq n + 1$. Moreover, every biconnected spanning subgraph of G has two vertices of C with degree greater than 2.*

Now it becomes trivial to prove the following:

Corollary 1 (Lower Bound). *Let G be a biconnected graph and let $C_1, C_2, \ldots, C_p$ be disjoint separators. If $C_j, j = 1, \ldots, p$, satisfies conditions of Lemma 6, then $\text{OPT}(G) \geq n + p$.* $\qquad\square$

Before ending this subsection we introduce a new topological structure that let us allow to design a better algorithm w.r.t. the one described in Theorem 2.

Definition 4. *A path edge $e_{i+3} = (v_{i+3}, v_{i+4})$ in Π_G, generates a left hook (see Figure 2) if the following conditions hold:*

(i) $\delta_G(v_{i+2}) = 2$;
(ii) $E(e_i) \cap E(e_{i+1}) = \{(v_j, v_{i+3}), (v_j, v_{i+4})\}$, for a unique $j \leq i$ (v_j is the tip of the hook);

(iii) $E(e_{i+4})$ *contains at least one edge with v_{i+1} as left endvertex, plus (possibly) edges with v_{i+4} as left endvertex; these are the only admissible covering edges for e_{i+4};*

(iv) *there exists a vertex $v_t \neq v_j$, with $t \leq i$.*

Definition 5. *A path edge (v_i, v_{i+1}) in $\Pi_G = \langle v_1, v_2, \ldots, v_n \rangle$ generates a right hook if (u^{n-i}, u^{n-i+1}) generates a left hook in $\Pi_G^{-1} = \langle u^1 \equiv v_n, u^2 \equiv v_{n-1}, \ldots, u^{n-i+1} \equiv v_i, \ldots, u^n \equiv v_1 \rangle$.*

Definition 6. *Let $G \in \mathcal{G}^n$ and let Π_G a Hamiltonian path of G. G has a hook if there exists a path edges that generates either a left or a right hook.*

Looking at the definition of left hook, it is easy to devise an $\mathcal{O}(m)$ time algorithm that finds all the hooks of a graph G. In the next subsection we will prove that, if G has k hooks, then $\mathrm{OPT}(G) \geq n+k$. Moreover, we will prove that we can remove all hooks from G by creating a new graph and a Hamiltonian path for it, and without altering the size of MBSS.

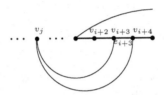

Fig. 2. A left hook generated by e_{i+3}

4.3   Graph Decomposition

Let G be a biconnected graph and let V_ℓ, C, V_r be a 2-separation. Henceforth, G_ℓ (resp., G_r) will denote the graph obtained from G by shrinking V_r (resp., V_ℓ) into one vertex x_ℓ (resp., x_r). Notice that G_ℓ, G_r are biconnected. Moreover, by $G_{\ell,r} = (V', E')$, where $V' = V_\ell \cup C \cup V_r$ and $E' = \{(u,v) \in E(G_\ell) \cup E(G_r) \mid u, v \in V'\}$, we denote the graph *built from G_ℓ and G_r w.r.t. x_ℓ, C, x_r*. Observe that $G = G_{\ell,r}$.

Lemma 7. *If H is a MBSS of G then H_ℓ, H_r are MBSS of G_ℓ, G_r, respectively. If H_ℓ, H_r are respectively MBSS of G_ℓ, G_r, then $H_{\ell,r}$ is a MBSS of G.*

Proof. Suppose H is a MBSS of G, but, w.l.o.g., H_ℓ is not a MBSS of G_ℓ. Let $\mathcal{H}_\ell^*$ be a biconnected spanning subgraph of G_ℓ and $|E(\mathcal{H}_\ell^*)| < |E(H_\ell)|$. It is easy to show that the graph H' built from $\mathcal{H}_\ell^*$ and H_r w.r.t. x_ℓ, C, x_r is biconnected. As x_ℓ, x_r have degree 2 in G_ℓ, G_r, respectively, it follows that

$$|E(H')| = |E(\mathcal{H}_\ell^*)| - 2 + |E(H_r)| - 2 < |E(H_\ell)| - 2 + |E(H_r)| - 2 = |E(H)|$$

and so H cannot be a MBSS of G, thus obtaining a contradiction.

Suppose now that H_ℓ, H_r are respectively MBSS of G_ℓ, G_r, but $H_{\ell,r}$ is not a MBSS of $G_{\ell,r}$. Let $\mathcal{H}$ be a biconnected spanning subgraph of G with $|E(\mathcal{H})| < |E(H)|$. Then, we have

$$|E(\mathcal{H})| = |E(\mathcal{H}_\ell)| - 2 + |E(\mathcal{H}_r)| - 2 < |E(H_\ell)| - 2 + |E(H_r)| - 2 = |E(H_{\ell,r})|,$$

from which we deduce

$$|E(\mathcal{H}_\ell)| + |E(\mathcal{H}_r)| < |E(H_\ell)| + |E(H_r)|.$$

Since all graphs used in the equation above are biconnected, we can claim that either H_ℓ or H_r is not a MBSS for the respective graph, thus obtaining a contradiction. $\qquad\square$

Let Π_G be a Hamiltonian path of a biconnected graph G. We say that G is *decomposable* if there exists a path edge e_i with $\kappa(e_i) = 2$ and the associated 2-separation V_ℓ, C, V_r is such that $|V_\ell|, |V_r| \neq 1$. The pair G_ℓ, G_r is a *decomposition* of G. A non decomposable graph is called *prime*. A *prime decomposition* $G_1, G_2, \ldots, G_k$ of a decomposable graph G is the repeated decomposition of non-prime graphs of a decomposition of G, until G_j is prime, $j = 1, 2, \ldots, k$.

Remark 2. How much does it cost to decompose G w.r.t. the 2-separation V_ℓ, C, V_r? Let $\delta_G(v) = \delta_\ell(v) + \delta_r(v)$, where $v \in C$ and $\delta_\ell(v)$ (resp., $\delta_r(v)$) is the number of edges incident to v and to a vertex in V_ℓ (resp., V_r). Then decomposing G into G_ℓ, G_r costs $\mathcal{O}\left(\max_{v\in C} \{\min\{\delta_\ell(v), \delta_r(v)\}\}\right)$ time. Moreover $\delta_{G_\ell}(v) = \delta_\ell(v) + 1$ and $\delta_{G_r}(v) = \delta_r(v) + 1$, $\forall v \in C$.

Remark 3. If $G \in \mathcal{G}^n$ is prime, then $\kappa(e_i) \geq 3$, $4 \leq i \leq n - 4$.

We can now prove the following lemma:

Lemma 8. *Let $G \in \mathcal{G}^n$ be a prime graph and let Π_G be a Hamiltonian path of G. If G has a left hook, then $\mathrm{OPT}(G) \geq n + 1$.*

Proof. First note that a prime graph cannot have more than one left hook. Indeed, if e_{i+3} generates a left hook, then $\{v_{i+1}, v_{i+4}\}$ is a 2-separator. Let v be the tip of the hook. It is not hard to see that every MBSS of G is such that either of v_{i+3}, v_{i+4} or v has degree at least 3, while v_{i+1} must always have degree at least 3. The claim follows. $\qquad\square$

The previous lemma naturally extends to right hooks. Looking at its proof, it is not hard to see that if G has a left hook generated by e_{i+3} with tip in v, then we can remove edge (v, v_{i+4}) without altering the size of a MBSS. Note that this process makes v_{i+2} become a beta-vertex.

Remark 4. From now on, we will consider biconnected beta-free graphs having no hooks.

Looking at Remarks 1 and 3, there is an advantage if the instance of our problem is a prime graph in $\mathcal{G}^n$. We can try to work with prime graphs if Lemma 7 applies in a nice way. Let us assume that there exists a path edge e_i such that $\kappa(e_i) = 2$. W.l.o.g. assume that $\kappa_\ell(e_i) = 2$, and let $V_\ell, C = \{v_j, v_i\}, V_r$ be the associated 2-separation such that $v_1 \in V_\ell$. Let G_ℓ (resp., G_r) be the graph obtained by shrinking the set V_r (resp., V_ℓ) into one vertex x_ℓ (resp., x_r). The pair G_ℓ, G_r is a decomposition of G. It is easy to see that $\langle v_1, v_2, \ldots, v_j, \ldots, v_i, x_\ell \rangle$ is a Hamiltonian path of G_ℓ, while $\langle v_j, x_r, v_i, v_{i+1}, v_{i+2}, \ldots, v_n \rangle$ is a Hamiltonian path of G_r. Moreover, note that x_r has degree 2 in G_r. Thus, by removing x_r from G_r and by adding (v_j, v_i) to G_r we obtain a new graph G'_r and $\langle v_j, v_i, v_{i+1}, v_{i+2}, \ldots, v_n \rangle$ is a Hamiltonian path of G'_r. It is easy to see that if $\mathcal{H}'_r$ is a MBSS of G'_r chosen among all biconnected spanning subgraphs of G'_r having edge (v_j, v_i), then Lemma 7 still holds. Indeed, the graph obtained from $\mathcal{H}'_r$ minus edge (v_j, v_i), plus vertex x_r and edges $(v_j, x_r), (x_r, v_i)$, is a MBSS of G_r. Moreover, notice that:

Remark 5. If H_ℓ, H'_r are biconnected spanning subgraphs of G_ℓ, G'_r, respectively, then the graph H built from H_ℓ and H'_r w.r.t. x_ℓ, C, x_r, is such that $|E(H)| = |E(H_\ell)| + |E(H'_r)| - 3$.

The pair G_ℓ, G'_r is said to be a *simplified decomposition* of G. A *simplified prime decomposition* $G_1, G_2, \ldots, G_k$ of a decomposable graph G is the repeated simplified decomposition of non-prime graphs of a simplified decomposition of G, until G_j is prime, $j = 1, \ldots, k$.

4.4   The Final Algorithm

In this section we improve the algorithms described in Section 3. Before showing the final algorithm, we first describe a linear time and space algorithm for decomposing a graph G into a collection of prime graphs. We assume that a Hamiltonian path Π_G of G is given in input. The algorithm begins by making a copy of G (say G') and by assuming that G' is the initial partial decomposition of G. Then, it decomposes the computed partial decomposition of G until each graph of the decomposition is prime. The algorithm explores path edges in order from e_1 to e_n. At a given iteration, let e_i be the path edge the algorithm must examine. Let v_h be the right endvertex of f_i and let v_{j+1} be the second right endvertex (different from h) of some edge in $E(e_i) \cup \{e_i\}$ (say f'_i) having maximum index. If $v_{j+1} = v_{i+1}$, then $\kappa_r(e_i) = 2$ and $C = \{v_{i+1}, v_h\}$ is a 2-separator. If a graph of the computed partial decomposition of G is decomposable w.r.t. C, then decompose it and skip to $e_j = e_{i+1}$. Otherwise, $\kappa_r(e_t) \geq 3, t = i, \ldots, j - 1$, and the algorithm can directly skip to e_j. Moreover, when examining e_j, it suffices to explore only cycle edges in $E(e_j)$ having the left endvertex with index greater than i. However, remember that f_i could be either f_j or f'_j. The same algorithm can be easily adapted to compute $\kappa_\ell(e_i)$. About the time and space complexity, notice that each cycle edge is explored a constant number of times. Moreover, as the number of vertices of a prime decomposition of G is $\mathcal{O}(n)$, then from Remark 2 it follows that the time and space complexity is $\mathcal{O}(m)$.

Lemma 9. *A biconnected beta-free graph $G \in \mathcal{G}^k$, with $k \leq 6$, is Hamiltonian.*

Proof. Since G is biconnected, from Lemma 3 it follows that graphs in $\mathcal{G}^3, \mathcal{G}^4$ are Hamiltonian. Let us assume that $G \in \mathcal{G}^6$ is not Hamiltonian and let $\mathcal{H}$ be a MBSS of G. In this case there are 2 vertices u_1, u_2 of $\mathcal{H}$ having degree 3. The graph obtained from $\mathcal{H}$ after the removal of u_1, u_2 has at least 3 connected components, two of these made by a single vertex v and v', respectively. Both v, v' cannot be adjacent to other vertices different from u_1, u_2 in G, otherwise G is Hamiltonian and $\mathcal{H}$ is not an optimal solution. So v is a beta-vertex in a beta-free graph. We have obtained a contradiction. The same technique can be used to prove that a beta-free graph $G \in \mathcal{G}^5$ is Hamiltonian.   □

Now we can implement a new more powerful function *Expand* such that

Lemma 10. *For a prime graph $G \in \mathcal{G}^n$ and a given Hamiltonian path Π_G of G, if $\mathrm{OPT}(G) \geq n + k$, then the algorithm BSS computes in $\mathcal{O}(m)$ time and space a solution H such that $|E(H)| \leq n + k + \lceil \frac{n-6}{4} \rceil < \frac{5}{4}\mathrm{OPT}(G)$.*

Proof. If $n \leq 6$ the claim follows from Lemma 9. Notice that as G is prime and has no beta-structures, it is not hard to see that we can cover the first (resp., last) 4 edges of Π_G by adding only one extra edge. So, looking at Remark 3 it is easy to prove the claim when $n \leq 9$. Hence, we can assume that $n \geq 10$. Moreover suppose we have added all edges of Π_G to $E(H)$. So we will remove all useless path edges and add cycle edges. We prove that we need to add one extra edge for the first (resp., last) 5 edges of Π_G, and one extra edge every 4 edges of $\Pi_G(v_6, v_n)$ (in an amortized sense). We prove this for the first path edges of Π_G, since for the last 5 path edges of Π_G the problem is symmetric. Let $\lambda = n$ be an initial lower bound for $|E(\mathcal{H})|$.

We first prove that we add only one extra edge for the first 5 path edges of Π_G. We must test sequentially the following (mutually exclusive) conditions.

(i) If f_1 covers 5 edges of $\Pi_G(v_1, v_n)$, then take it. Otherwise, if there are 2 semi-adjacent edges (one of which in $E(e_1)$) covering 5 edges of $\Pi_G(v_1, v_n)$, then take them and remove the middle one.

(ii) As G is prime, then $f_1 = (v_1, v_5)$ is the only cycle edge covering $E(e_1)$, and so it must be added to H. If there is an edge in $E(e_4)$ covering the first edges of $\Pi_G(v_5, v_n)$, then take it. Otherwise, if there are 2 semi-adjacent edges (one of which in $E(e_4)$) covering the first 5 edges of $\Pi_G(v_5, v_n)$, then take them and remove the middle one.

(iii) Add $f_1 = (v_1, v_5)$. In this case $E(e_5) \subseteq \{(v_2, v_j), (v_3, v_i) \mid 6 \leq i, j \leq 9\}$. Note that e_2 cannot generate a right hook (see Remark 4). Now the proof breaks into mutually exclusive cases that must be tested sequentially:

   (a) $(v_2, v_6) \in E(e_4)$ (resp., $(v_3, v_6) \in E(e_4)$). Add (v_2, v_6) (resp., (v_3, v_6)) plus one cycle edge in $E(e_5)$ with endvertex v_3 (resp., v_2), and remove e_2, e_5 (see Figure 3 (a)). In this case we add one extra edge for at least 6 path edges.

   (b) $(v_2, v_j), (v_3, v_{j+1}) \in E(e_4)$ (resp., $(v_3, v_j), (v_2, v_{j+1}) \in E(e_4)$). Add both edges and remove e_2, e_j (see Figure 3 (b)). In this case we add one extra edge for at least 7 path edges.

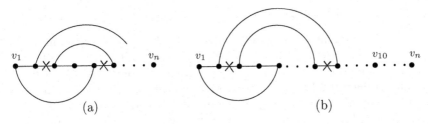

Fig. 3. Cases (a) and (b)

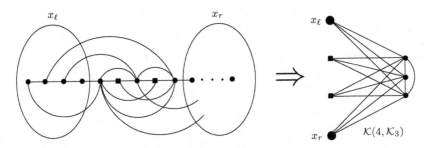

Fig. 4. Case (c)

(c) Otherwise, it must be $E(e_5) \subseteq \{(v_2, v_7), (v_2, v_9), (v_3, v_7), (v_3, v_9)\}$ and $E(e_9) \subseteq \{(v_5, v_j), (v_7, v_i), (v_9, v_k) \mid 10 \leq j, i, k\}$. Let G' be the graph obtained by shrinking the set $\{v_1, v_2, v_3, v_4\}$ (resp., $\{v_{10}, v_{11}, \ldots, v_n\}$) into one vertex x_ℓ (resp., x_r). Since $C = \{v_5, v_7, v_9\}$ is a separator and G is prime, it is not hard to see that G' is isomorphic to a subgraph of $\mathcal{K}(4, \mathcal{K}_3)$ (see Figure 4). In this case, add either $\{(v_2, v_7), (v_3, v_9)\}$ or $\{(v_3, v_7), (v_2, v_9)\}$, and remove e_2. Since from Corollary 1 we can increase λ by 1, it follows that in this case we add one extra edge for 9 path edges.

Since $\kappa(e_i) \geq 3$ for $4 \leq i \leq n - 4$, from Lemma 1 we have that we are able to add one cycle edge every 4 path edges. Let $\lambda = n + k \leq \text{OPT}(G)$ be our final lower bound. The computed solution H has size

$$|E(H)| \leq n - 1 + \left(1 + k + \left\lceil \frac{n-1-5}{4} \right\rceil\right) = n + k + \left\lceil \frac{n-6}{4} \right\rceil < \frac{5}{4}(n + k).$$

Comparing the size of H with $\text{OPT}(G)$ the approximation ratio follows. About the time and space complexity, as each edge is explored a constant number of times (see also Theorem 2), then the time and space complexity is $\mathcal{O}(m)$. □

We can finally prove the following:

Theorem 3. *The algorithm BSS returns a* $\min\left\{\frac{5}{4}, \frac{2\kappa(G)-1}{2(\kappa(G)-1)}\right\}$-*approximated solution for the MBSSHP-problem in* $\mathcal{O}(m)$ *time and space.*

Proof. If $\kappa(G) \geq 3$, the claim follows from Theorem 2. So, assume $\kappa(G) = 2$. Let $G_1, G_2, \ldots, G_i$ be the simplified prime decomposition of G and let Π_{G_j} be the given Hamiltonian path of G_j, $j = 1, 2, \ldots, i$. We first find a biconnected spanning subgraph H_j for each instance G_j, Π_{G_j}. Notice that we can build a biconnected spanning subgraph H of G from $H_1, H_2, \ldots, H_i$ in $\mathcal{O}(m)$ time and space. Let $p_j + 1$, with $j = 1, i$, be the length of path Π_{G_j}, and let $p_j + 1 + k_j$ be the lower bound for the size of a MBSS of G_j computed as in Lemma 10. Moreover, let $p_j + 2$, with $j = 2, \ldots, i - 1$, be the length of path Π_{G_j}, and let $p_j + 2 + k_j$ be the lower bound for the size of a MBSS of G_j computed as in Lemma 10. Then, let $k = \sum_{j=1}^{i} k_j$.

From Lemma 10, from Remark 5 and since $n \geq 1 + \sum_{j=1}^{i} p_j$, we have that

$$|E(H)| = \sum_{j=1,i} \left(p_j + 2 + k_j + \left\lceil \frac{p_j - 4}{4} \right\rceil \right) + \sum_{j=2}^{i-1} \left(p_j + 3 + k_j + \left\lceil \frac{p_j - 3}{4} \right\rceil \right) - 3(i - 1)$$

$$\leq k + n + \sum_{j=1}^{i} \left\lceil \frac{p_j - 3}{4} \right\rceil \leq k + n + \frac{1}{4} \sum_{j=1}^{i} p_j \leq \frac{5}{4}(n + k),$$

where the second inequality follows from the fact that, for every integer $1 \leq h \leq m$, being $m = hq + r$, where q and $r < h$ are positive integers, then:

$$\left\lceil \frac{m - (h - 1)}{h} \right\rceil = \left\lceil \frac{hq + (r + 1 - h)}{h} \right\rceil \leq q + \left\lceil \frac{r + 1 - h}{h} \right\rceil \leq q \leq \frac{m}{h}.$$

As Lemma 7 and Corollary 1 imply that $\text{OPT}(G) \geq n + k$, then the approximation ratio follows. The time and space complexity follows from Lemma 10 and from the fact we can find a prime decomposition of G in linear time and space. $\square$

References

1. H.-J. Böchenhauer, D. Bongartz, J. Hromkovič, R. Klasing, G. Proietti, S. Seibert, and W. Unger, On the hardness of constructing minimal 2-connected spanning subgraphs in complete graphs with sharpened triangle inequality, *Theoretical Computer Science*, 326:137–153, 2004. A preliminary version appeared on *Proc. of the 22nd Conf. on Foundations of Software Technology and Theoretical Computer Science (FSTTCS'02)*, Vol. 2556 of LNCS, Springer-Verlag, 59–70.
2. A. Czumaj and A. Lingas, On approximability of the minimum-cost k-connected spanning subgraph problem, *Proc. of the Tenth Annual ACM-SIAM Symposium on Discrete Algorithms (SODA'99)*, ACM Press, 281–290.
3. R. Diestel, *Graph theory*, Springer-Verlag electronic edition, 2000.
4. G.N. Frederickson and J. Jájá, On the relationship between the biconnectivity augmentation and traveling salesman problems, *Theoretical Computer Science* 19:189–201, 1982.
5. M.R. Garey and D.S. Johnson, *Computers and intractability: a guide to the theory of NP-Completeness*, W.H. Freeman and Company, New York, 1979.

6. N. Garg, V.S. Santosh, and A. Singla, Improved approximation algorithms for biconnected subgraphs via better lower bounding techniques, *Proc. of the 4th ACM-SIAM Symp. on Discrete Algorithms (SODA'93)*, ACM Press, 103–111.

7. R. Jothi, B. Raghavachari, and S. Varadarajan, A 5/4-approximation algorithm for minimum 2-edge-connectivity, *Proc. of the 14th ACM-SIAM Symp. on Discrete Algorithms (SODA'03)*, ACM Press, 725–734

8. R. Jothi, *Personal communication*, 2004.

9. S. Khuller, Approximation algorithms for finding highly connected subgraphs, in *Approximation Algorithms for NP-Hard Problems*, Dorit S. Hochbaum Eds., PWS Publishing Company, Boston, MA, 1996.

10. S. Khuller and U. Vishkin, Biconnectivity approximations and graph carvings, *J. ACM*, 41(2):214–235, 1994.

11. G. Kortsarz, R. Krauthgamer, and J.R. Lee, Hardness of approximation for vertex-connectivity network-design problems, *Proc. of the 5th Workshop on Approximation Algorithms (APPROX'02)*, Vol. 2462 of LNCS, Springer-Verlag, 185–199.

12. C.H. Papadimitriou and K. Steiglitz, Some complexity results for the Traveling Salesman Problem, *Proc. of the 8th ACM Symp. on Theory of Computing (STOC'76)*, ACM Press, 1–9.

13. S. Vempala and A. Vetta, Factor 4/3 approximations for minimum 2-connected subgraphs, *Proc. of the 3rd Workshop on Approximation Algorithms (APPROX 2000)*, Vol. 1913 of LNCS, Springer-Verlag, 262–273.

14. H. Whitney, Nonseparable and planar graphs, *Trans. Amer. Math. Soc.*, 34:339–362, 1932.

Order-Preserving Transformations and Greedy-Like Algorithms

Spyros Angelopoulos*

School of Computer Science, University of Waterloo,
Waterloo, Ontario, Canada N2L 3G1
sangelop@cs.uwaterloo.ca

Abstract. Borodin, Nielsen and Rackoff [5] proposed a framework for abstracting the main properties of greedy-like algorithms with emphasis on scheduling problems, and Davis and Impagliazzo [6] extended it so as to make it applicable to graph optimization problems. In this paper we propose a related model which places certain reasonable restrictions on the power of the greedy-like algorithm. Our goal is to define a model in which it is possible to filter out certain overly powerful algorithms, while still capturing a very rich class of greedy-like algorithms. We argue that this approach better motivates the lower-bound proofs and possibly yields better bounds. To illustrate the techniques involved we apply the model to the well-known problems of (complete) facility location and dominating set.

Keywords: Priority algorithms, inapproximability results, facility location, dominating set.

1 Introduction

Greedy algorithms have been a widely popular approach in combinatorial optimization and approximation algorithms. This is mainly due to their conceptual simplicity as well as their amenability to analysis. In fact, one reasonably expects a greedy algorithm to be one of the first approaches an algorithm designer employs when facing a specific optimization problem. It would therefore be desirable to know when such an approach is not likely to yield an efficient approximation. However, while it is relatively easy to identify a greedy algorithm based on intuition and personal experience, a precise definition of such a class of algorithms is needed so as to prove limitations on its power. Even more importantly, as argued in [5] it is expected that a rigorous framework for greedy algorithms can provide insight on how to develop better, more efficient algorithms.

Despite the popularity and importance of greedy algorithms as an algorithmic paradigm, it was only recently that a formal framework for their study emerged. In particular, Borodin, Nielsen and Rackoff introduced in [5] the class of *priority algorithms* as a model for abstracting the main properties of deterministic

---

* Research done while at the Department of Computer Science, University of Toronto.

greedy-like algorithms (we will hereafter refer to their model as the *BNR model*). In addition, Borodin, Nielsen and Rackoff showed how the framework can yield lower bounds on the approximation ratio achieved by priority algorithms for a variety of classical scheduling problems. In a follow-up paper, Regev [15] addressed one of the open questions in [5] related to scheduling in the subset model. The priority framework was subsequently applied in the context of the facility location and set cover problems by Angelopoulos and Borodin [2], and it has also been extended to capture greedy-like algorithms that allow randomization [1]. More recently, Davis and Impagliazzo [6] showed how to modify the BNR model in order to show limitations on the power of priority algorithms for *graph optimization problems*. To illustrate the technique, they applied the game to basic graph problems such as shortest path, metric Steiner tree, independent set and vertex cover. We refer to their model as the *DI model*.

In this paper we continue the study of greedy-like algorithms as formulated by priority algorithms. We first provide some key definitions. According to [5] priority algorithms are characterized by the following two properties:

1. The algorithm specifies an ordering of "the input items" and each input item is considered in this order.
2. As each input item is considered, the algorithm must make an "irrevocable decision" concerning the input input.

As one would expect, a precise definition of "input items" and "irrevocable decisions" pertains to the specific problem at hand.

Depending on whether the ordering changes throughout the execution of the algorithm, two classes of priority algorithms can be defined:

- Algorithms in the class FIXED PRIORITY decide the ordering before any input item is considered and this ordering does not change throughout the execution of the algorithm.
- Algorithms in the broader class ADAPTIVE PRIORITY are allowed to specify a new ordering after each input item is processed. The new ordering can thus depend on input items already considered.

In [5], **greedy** algorithms are defined as (fixed or adaptive) priority algorithms, which satisfy an additional property: the irrevocable decision is such that the objective function is *locally optimized*. More specifically, the objective function must be optimized as if the the input currently being considered is the last input. Note that in this context not every greedy-like (that is, priority) algorithm is greedy.

In both the work of Borodin *et al* and Davis and Impagliazzo, in order to show a lower bound on the approximation ratio one evaluates the performance of every priority algorithm for an appropriately constructed nemesis input. The construction of such a nemesis input can be seen as a *game* between *an adversary* and the algorithm. In both games, the adversary presents initially a (large) set of potential input items, and in each round removes certain input items according to the corresponding decisions made by the algorithm.

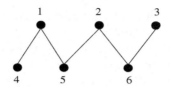

Fig. 1. An input for a graph problem

As noted in [5] it is expected that the study of priority algorithms will provide insights about how to develop better, more efficient greedy-like algorithms. To this end, it is essential that the adversary involved in the lower-bound arguments is "reasonably" powerful, or, from a different scope, that the priority algorithm does not have "unreasonable" power. Otherwise, it is expected that i) the arguments behind the lower bound proofs will be very elaborate (something which becomes even more critical in the context of graph problems, where the input items refer to each other) and it will be hard to get reasonably good lower bounds; and most importantly ii) the lower bounds will not necessarily reflect the limitations of "real" greedy-like algorithms, but rather those of artificial algorithms which use information that is conceptually difficult to generalize.

To illustrate the argument above, consider the graph shown in Figure 1, and suppose it is used as input to a FIXED PRIORITY algorithm for a certain unweighted graph optimization problem. In the DI model it is possible that the algorithm specifies the ordering 1,5,4,2,3,6 from highest to lowest priority. Note that in this ordering a vertex of degree one (namely vertex 3) receives both higher and lower priority than vertices of degree two (namely vertices 6 and 2, respectively). In other words, the algorithm has somehow the power to differentiate between two input items of the same degree such as vertices 6 and 2. However, it is very counterintuitive to think of a FIXED PRIORITY algorithm which can make such an unnatural distinction of seemingly identical input items.

In this paper we put forward a formal model which intends to capture the above observation, namely that the priority algorithm normally should give equal priority to input items which "look alike" (which is not necessarily the case in the DI and BNR models). We propose an adversary which applies to priority algorithms that are not necessarily as powerful as the algorithms assumed by the DI model, while still being able to capture a very wide class of natural greedy-like algorithms for graph problems. In a nutshell, we do not allow algorithms to acquire useful information from the id's of the input items (vertices). Interestingly, the proposed adversary does not remove input items, but instead can apply a more wide range of transformations over the potential input items. In particular, we allow transformations that do not affect the ordering in which the algorithm considers the input items, in the presence of an adversary. We believe this model reflects the fact that local information does not necessarily yield knowledge of the global structure of an instance, which is only self-evident in greedy-like algorithms.

To demonstrate our techniques, we apply our model to the *complete metric facility location* and *dominating set* problems. The former is a variant of the classic metric facility location problem in which every node is both a facility and a city (unlike the *disjoint* variant in which facilities and cities form disjoint sets); the latter can be seen as a variant of the set cover problem, where now each vertex can both cover and "be covered by" other vertices. We focus on these two graph optimization problems for two reasons: i) their corresponding variants, namely disjoint facility location and set cover have already been studied in [2] from the point of view of approximability by priority algorithms using the BNR game; ii) the lower bounds of [2] do not carry over to the problems we consider, since the input items are no longer "isolated", but refer to each other.

How do our results compare to the work of Davis and Impagliazzo? It is not clear to us whether the bounds as stated in Theorems 2 and 3 can be reproduced using the DI adversary. Similarly, we do not know whether Theorem 5 can be shown within the DI framework, but we believe that if so it would require a more elaborate proof. Theorems 1 and 4 can definitely be reproduced by the DI adversary. We nevertheless include them since not only they illustrate our transformations, but they are also interesting on their own (Theorem 1 uses an instance where the facilities have non-uniform opening costs).

We emphasize that as in deterministic priority algorithms (and similar to competitive analysis of online algorithms) the lower bounds are derived by exploiting the syntactic structure of the algorithms, and are orthogonal to any complexity considerations. In other words, we allow the algorithm unbounded time complexity.

Very recently and independently to this work, Borodin, Boyar and Larsen [4] addressed further the topic of priority algorithms for graph optimization problems. Their focus is primarily on the effect of memory on priority algorithms as defined in [5] and [6]; in particular they considered a model in which memoryless algorithms do not accept an input item once some other input item was rejected in a previous iteration (which they call the "acceptances-first" model) and presented lower bounds for problems such as vertex cover, independent set and vertex coloring. In addition, they showed that the "vertex-adjacency" model of representing input items (also assumed in our work) is more general than the "edge-adjacency" model. Finally, they proposed a formal definition of "greediness" in the context of graph problems; however, it is not clear whether their definition can lead to lower bounds for the class of "greedy" priority algorithms. The contributions of our paper are orthogonal to the work of Borodin, Boyar and Larsen even though the two papers address a similar topic.

2   The Model

2.1   Preliminaries

Input representation. An instance of a graph optimization problem Π can be described as an (undirected) graph $G = (V, E)$, with vertex and edge weights. A reasonable representation of an input item for Π is a pair $\langle d_v, w_v \rangle$ where

$v \in V$. Here, w_v is the *weight vector*[1] of v, and d_v is the *distance vector* of v. The weight vector represents the weight assignments for vertex v (e.g., the cost that is payed if v is included in the solution), while the distance vector is the vector of distances (edge weights) from v to every other vertex in V. In addition, each input item has a unique id. In what follows we use the notation $\langle v \rangle$ to denote the input item that corresponds to vertex v, although in some cases we will not make the distinction when it is clear from the context. Note that every time the priority algorithm considers an input item $\langle v \rangle$, all the information for $\langle v \rangle$ becomes available to the algorithm (including the id's of the neighbouring vertices in its distance vector).

Priority functions. As in [6] we will assume that the priority algorithm uses a *priority function* to assign real-valued priorities to the input items. Thus, in each iteration, the input item with the highest priority is the one that the priority function maps to the highest value among all remaining input items. To make the definition more precise, we must distinguish between the two classes of priority algorithms. Let $\langle V \rangle$ denote the infinite set of input items of the form described earlier. For FIXED PRIORITY algorithms, the priority function is $P : \langle V \rangle \to \mathbb{R}$, that is, the priority of an input item is determined solely by the input item itself. In contrast, in iteration k, an ADAPTIVE PRIORITY algorithm will take into consideration the *history* of the algorithm's execution, namely the set $H_{k-1} = \langle v_1 \rangle, \langle v_2 \rangle, \ldots, \langle v_{k-1} \rangle$ of input items considered in iterations 1 through $k - 1$. Hence for ADAPTIVE PRIORITY algorithms we can describe the priority function at iteration k as $P_k : \langle V \rangle \times H_{k-1} \to \mathbb{R}$.

We will assume that in the event two input items have the same highest priority, the algorithm cannot distinguish them. Equivalently, we will assume an adversary that dictates which input item should be considered next, in the event of a tie.

We also need to impose some natural, realistic restrictions on the capacity of the algorithm to assign priorities and differentiate between input items. Let us first introduce some notation. For a distance vector d_v denote by $M(d_v)$ the distance multiset of v, namely the multiset of all distances from v to every other vertex in G. Also, for a given set $S \subseteq V$, denote by $d_v(S)$ the vector of distances $d(v, u)$, for all $u \in S$.

Consider first algorithms in the class FIXED PRIORITY. Let $\langle v \rangle$, $\langle u \rangle$ be two input items. The priority function P must obey the following rule:

FP Rule: $P(\langle v \rangle) = P(\langle u \rangle)$ if $M(d_v) = M(d_u)$ and $w_v = w_u$. $(*)$

The interpretation of the above rule is that *id's do not carry information*. In this view the distance vectors degenerate to distance multisets, and thus two

---

[1] In general, the weight vector can represent more than one weights, according to the specific problem at hand. E.g., for the weighted complete facility location problem, each weight vector will store the weight of the facility as well as its opening cost.

input items with the same distance multisets and the same weight vectors will be indistinguishable to the algorithm.

A similar assumption will be made for ADAPTIVE PRIORITY algorithms; in this case, however, we must take into account the history. More specifically, let $\langle v \rangle$, $\langle u \rangle$ be two input items not in H_{k-1}. Also, let $\pi : V \setminus H_{k-1} \cup \{u\} \cup \{v\} \rightarrow V \setminus H_{k-1} \cup \{u\} \cup \{v\}$ be a permutation of vertices in $V \setminus H_{k-1} \cup \{u\} \cup \{v\}$. Then P_k must observe the following rule:

AP Rule: $P_k(\langle v \rangle, H_{k-1}) = P_k(\langle u \rangle, H_{k-1})$ if

- $d_v(H_{k-1}) = d_u(H_{k-1})$ and $w_u = w_v$,
- There exists a permutation π such that for every $x \in V \setminus H_{k-1} \{u\} \cup \{v\}$, $d(u, x) = d(v, \pi(x))$, $d_x(H_{k-1}) = d_{\pi(x)}(H_{k-1})$ and $w(x) = w(\pi(x))$.   (**)

The first constraint is related to the fact that the algorithm knows the distance vectors of all vertices in H_{k-1}. It implies that the history by itself is not sufficient to distinguish u from v. The second constraint is related to vertices not yet considered, and likewise signifies that such vertices cannot help in distinguishing u from v, assuming that id's do not carry information.

We conclude this section by mentioning that every time the priority algorithm considers an input item $\langle v \rangle$ it has to make an irrevocable decision concerning the input item. In several graph problems, the decision is whether to include the input item in the partial solution (we then say that the algorithm *accepts* the input item).

2.2   Order-Preserving Transformations

Similar to [5] and [6], in order to establish a lower bound on the approximability of a graph problem by a deterministic priority algorithm, we will use the concept of a game between the algorithm and an adversary. The game evolves in rounds, with each round corresponding to an iteration of the algorithm as described below (we focus on the more general case of ADAPTIVE PRIORITY algorithms).

The adversary presents, in the beginning of the first round of the game, a graph $G_1 = (V_1, E_1)$. Denote by $v_1^1, v_2^1, \ldots, v_n^1$, the input items for G_1, with the superscript "1" identifying the round of the game, and subscripts denoting the *labels* of vertices [2]. The algorithm considers the input item of highest priority, say v_1^1, and makes an irrevocable decision concerning it. More generally, suppose that by the end of the k-th round the algorithm has considered the input items with labels $1, \ldots, k$. At the end of the k-th round, the adversary applies a *transformation* $\phi_k : \langle V \rangle \rightarrow \langle V \rangle$. This transformation maps an input item $\langle v_i^k \rangle = \langle d_{v_i^k}, w_{v_i^k} \rangle$

---

[2] We distinguish between "labels" and "id's" intentionly, since the adversary is allowed to permute id's, as will become clear later. We need some invariant piece of information to refer to the input items in the order they are considered in the game, and the labels serve precisely this purpose.

to an input item $\langle v_i^{k+1} \rangle = \langle d'_{v_i^{k+1}}, w'_{v_i^{k+1}} \rangle$; in other words, the distance and weight vectors of the item may change. We emphasize that the transformation applies only to input items with labels $k + 1, \ldots, n$, i.e., the adversary cannot modify items considered in previous rounds. This gives rise to a new graph G_{k+1}, which can be uniquely described by the input items $v_1^{k+1}, v_2^{k+1}, \ldots, v_n^{k+1}$, for which $v_i^{k+1} \equiv v_i^k$, for all $i \leq k$. The game proceeds until the last round, namely round n. At that point the algorithm has considered all input items in the graph.

As one might expect, only limited types of transformations can be helpful for our purposes. We call ϕ_k an *order-preserving* transformation, if and only if for every $j \leq k$, and $i \geq k + 1$,

$$P_j(v_i^{k+1}, H_{j-1}) \leq P_j(v_j^j, H_{j-1}) \tag{1}$$

where H_{j-1} is the history at the end of round $j - 1$ of the game, namely $H_{j-1} = \{v_l^l \mid l \leq j - 1\}$.

Informally, the definition suggests that as v_i^k is "replaced" by v_i^{k+1}, the ordering of input items (in terms of their labels) considered by the algorithm up to round k will not be affected.

The following lemma[3] formalizes the use of the adversary/algorithm game as a tool for bounding the approximation ratio:

Lemma 1. *Suppose that the graph defined by the input items $v_1^n, \ldots v_n^n$ (as determined by the game between the adversary and the algorithm) is given as input to the algorithm. Then on this specific input, and in iteration i, the algorithm will consider input item v_i^n. In other words, the algorithm will consider input items in the order of their labels.*

The following transformations are implied in our model, but we emphasize them since they are of particular importance. First, if σ is the ordering of a set of input items, as produced by a FIXED PRIORITY algorithm, then the adversary can swap the positions in σ of two input items which have the same priority. Second, suppose that G_1 and G_2 are two **isomorphic** graphs, in the sense that there exists a permutation of the id's of vertices in G_1 which produces G_2. Suppose that on input G_1 the algorithm assigns label i to the vertex it considers in the i-th iteration. Then on input G_2, the algorithm will consider in the i-th iteration the vertex with label i, and will make the same decision as the decision made by the algorithm on input G_1 and in the i-th iteration.

2.3   What Is a "Greedy" Algorithm for a Graph Problem?

In the context of graph problems, a definition for greediness that treats the current input item as if it were the last one becomes problematic. First, note that throughout its execution the algorithm acquires local information which reveals

---

[3] In this preliminary version we omit certain proofs due to space restrictions. Full proofs will be provided in the journal version of the paper.

only part of the input graph. Hence, it is difficult to think of a specific input item as being the last one when the partial information suggests that there definitely exist input items that should follow. Second, and more importantly, it is not easy to identify a unique "locally optimal" decision at each iteration, precisely because the partial information does not represent a valid graph instance (some distances and weights have not yet been revealed to the algorithm). For the above reasons we will employ only intuitive and in a sense ad-hoc definitions of **greedy** (as opposed to greedy-like) algorithms which are specific to the problems we consider. The definitions which we propose are still broad enough to capture known algorithms, and provide some flavour of what a "locally optimal" decision is meant to be. We insist on providing some meaningful definitions not only because of the historical interest in this concept, but mainly because the concept itself is widely used in practice. As one would expect, it is possible to show much better lower bounds for greedy priority algorithms (for instance the bounds of Theorem 3 and Theorem 5 are tight for the corresponding classes of algorithms). In such cases, the bounds suggest some directions towards the design of better algorithms: namely, that in order to improve the approximation ratio it is essential that the algorithm is not greedy.

We first provide a definition of what we consider to be a greedy priority algorithm for complete facility location. Let v_i be the node considered by the algorithm at iteration i. We capture the greedy behaviour of the algorithm (which one would informally describe by the motto "live for today"), by requiring that it **always opens v_i if this results in lowering the cost of the "current solution"**. Of course, we must clearly define the intuitive term "current solution". Note that the algorithm has only limited information (i.e., what can be deduced by the triangle inequality) about the distance $d(u, v)$ of every two nodes u, v which have not been considered yet (prior to iteration i), as well as the facility cost of such nodes. The current solution is then the optimal solution with the constraint that the algorithm cannot open a yet unconsidered node (which implies that every unconsidered node has to be connected to a node which was opened by iteration i), and cannot revoke the decision about nodes which were considered before the current iteration (i.e., cannot open a facility it did not open in a previous iteration, and cannot close a facility that it opened in a previous iteration).

For dominating set, we propose the following definition of a greedy priority algorithm. Let v be the vertex considered in the current iteration. Then v will be accepted **if it is adjacent to at least two vertices which have not been considered yet and which are not dominated by vertices accepted thus far**. We emphasize that in the case where the above condition does not hold the algorithm may or may not open v; in other words no restrictions are placed on the algorithm in situations other than the one we described earlier. This is important since we do not want a definition that forces the algorithm to open a large number of vertices, because the lower bounds then become artificial. For instance, consider the situation where v is adjacent to only one yet unconsidered and undominated vertex u, and furthermore v is adjacent to no other undominated vertices. Then it would make sense for the algorithm to reject v, wait until u is considered in a subsequent iteration, and then accept u; this would not increase the total cost of the algorithm.

3   Applications: Complete Facility Location and Dominating Set

Problem definitions. In the *(uncapacitated, unweighted) facility location* problem, the input consists of a set $\mathcal{F}$ of facilities and a set $\mathcal{C}$ of cities. The set $\mathcal{N} = \mathcal{F} \cup \mathcal{C}$ corresponds to the set of *nodes* in the graph, i.e., a node can be either a facility or a city, or both. Each facility $i \in \mathcal{F}$ is associated with an *opening cost* f_i which reflects the cost that must be paid to utilize the facility. Furthermore, for every facility $i \in \mathcal{F}$ and city $j \in \mathcal{C}$, the non-negative *distance* or *connection cost* c_{ij} is the cost that must be paid to connect city j to facility i. In the version of the problem that is known as the *complete facility location* problem, we have $\mathcal{F} = \mathcal{C} = \mathcal{N}$, i.e., every node is both a facility and a city. The objective is to open the facilities at a subset of the nodes and connect every other node to some open facility so that the total cost incurred, namely the sum of the cost of open facilities and the total connection cost is minimized[4]. We shall focus exclusively on the *metric* version of the problem, in which the connection costs satisfy the triangle inequality.

Observe that from the point of view of algorithms (upper bounds), the disjoint version (namely the version in which $\mathcal{F} \cap \mathcal{C} = \emptyset$) subsumes the complete version. This is because we can always "split" a node in the complete version of the problem to a corresponding facility and a corresponding city at zero distance from each other. However, we emphasize that the lower bounds for priority algorithms for the disjoint version (see [2]) do not carry over to the complete version.

In the *dominating set* problem, the input is an undirected, unweighted graph $G = (V, E)$. We seek a set $V' \subseteq V$ of smallest cardinality such that every vertex $u \in V \setminus V'$ is adjacent to at least one vertex in V'.

3.1   Complete (Metric) Facility Location

The first constant-factor polynomial-time approximation algorithm for (metric) facility location was given by Shmoys, Tardos and Aardal [16]. Interestingly, the best-known approximation ratio (1.52) is due to Mahdian, Ye and Zhang [13], and is achieved by a priority algorithm. Other algorithms that follow the priority framework include the ADAPTIVE PRIORITY greedy algorithm of Mahdian, Markakis, Saberi and Vazirani [12], which is a 1.861-approximation algorithms, and the ADAPTIVE PRIORITY algorithm of Jain, Mahdian and Saberi [9]. On the other hand, Mettu and Plaxton [14] showed that an algorithm which belongs in the class FIXED PRIORITY greedy yields a 3-approximation for the

---

[4] We say that a node is *opened*, when the facility on the said node is opened.

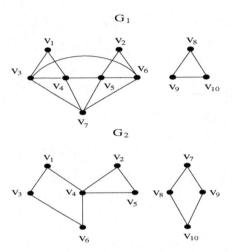

Fig. 2. The graphs G_1 and G_2 for the proof of Theorem 1. The edges shown indicate edges of distance 1, while all other distances are equal to 2

problem. It should be mentioned that their algorithm is in fact a re-statement of a primal-dual algorithm due to Jain and Vazirani [10].

On the negative side, Guha and Khuller [8] have shown that the *disjoint* facility location problem is not approximable within a factor better than 1.463, unless $NP \subseteq DTIME(n^{O(\log \log n)})$. They also showed, that under the same complexity assumption *complete* facility location is not approximable within a factor better than 1.278. Negative results for facility location priority algorithms in the disjoint model were given by Angelopoulos and Borodin in [2]. In particular, they showed lower bounds of 4/3 and 1.463 for general ADAPTIVE PRIORITY algorithms and *memoryless* ADAPTIVE PRIORITY algorithms respectively, as well as a lower bound of 3-ϵ for FIXED PRIORITY *greedy* algorithms.

In the lower-bound constructions shown in this Section we say that a node v *covers* a set $U \subset \mathcal{N}$ of nodes if and only if the distance from v to every node in U is equal to 1. The vertices of the graphs in our constructions will correspond to the nodes of the facility location instance.

Theorem 1. *No* ADAPTIVE PRIORITY *(not necessarily greedy) algorithm is better than an α-approximation, where α is a constant slightly greater than 36/35.*

Proof. We will use instances that consist of 10 nodes. Every node will cover either 2 or 4 facilities, and its facility cost will be denoted by f_2 and f_4, for the two cases, respectively. Here, f_2 and f_4 are suitably chosen constants (in particular, the 36/35 bound is obtained for $(f_2, f_4) = (3, 4.5)$).

Initially the adversary presents graph G_1, shown in Figure 2. We consider the following cases:

Case 1: The algorithm gives highest priority to an f_2 node, which the adversary specifies to be v_1, and the algorithm does not open it. In this case the input to the algorithm is G_1 itself (no transformation takes place).

Case 2: The algorithm gives highest priority to an f_2 node, (again, assume it to be v_1) which it opens. The adversary will perform an order-preserving transformation to derive graph G_2 also shown in Figure 2.

The remaining two cases, namely when the algorithm considers first a f_4 node, and either opens it or does not open it, are symmetric to Cases (1) and (2). In particular, in the case where the algorithm opens the f_4 node, the input to the algorithm will be graph G_1, while in the event it does not open it, the input is graph G_2. To complete the proof, it suffices to optimize with respect to f_2 and f_4.   □

Using ideas similar to Theorem 1, we can show the following:

Theorem 2. *There exists $\beta > \alpha$, where α is the lower bound of Theorem 1 such that no* FIXED PRIORITY *(but not necessarily greedy) algorithm achieves an approximation ratio better than β. In particular, β is slightly greater than 34/33.*

The lower bound we show in the following theorem matches the upper bound of Jain and Vazirani [10] and Mettu and Plaxton [14] (which it is easy to show that they belong in the class FIXED PRIORITY greedy).

Theorem 3. *No* FIXED PRIORITY greedy *algorithm has an approximation ratio better than $3 - \epsilon$ for arbitrarily small ϵ.*

Proof. The adversary presents to the algorithm a graph $G = (V, E)$, with $V = \{v_1, v_2, \ldots v_k, u_1, u_2, \ldots u_l\}$ and $l = c + d(k-1)$. Here, c and d are large constants, such that $c >> d^2$ (and whose importance will become evident later). Note that the total number of nodes in the graph is $n = k + c + d(k - 1) = k(d + 1) + c - d$; that is, n is a linear function of k. Each v_i, with $i \in [k]$ is at distance 1 from nodes $u_{d(i-1)+1} \cdots, u_{c+d(i-1)}$, and has a facility cost equal to $2d - \epsilon'$, where ϵ' is infinitesimally small. The distance between any two nodes v_i, v_j as well as the distance between any two nodes u_i, u_j is equal to 2. Every other distance is equal to 3. The facility cost of every u_i node is infinite (arbitrarily large). Figure 3 illustrates G for $k = 5$, $d = 2$, $c = 4$, with edges denoting distances equal to 1.

Lemma 2. $cost(OPT) \leq (2d^2/c + d + 3)k + O(1)$ and $cost(ALG) \geq k(3d - \epsilon')$ *for arbitrarily small ϵ'.*

The lower bound on the approximation ratio follows directly from Lemma 2. As k grows to infinity, and for large constants c, d, with $c >> d^2$, it is easy to verify that $cost(ALG)/cost(OPT) \geq 3 - \epsilon$, for arbitrarily small ϵ.   □
For completeness we mention that a simple argument can be used to show the following theorem.

Theorem 4. *No* ADAPTIVE PRIORITY greedy *algorithm is better than a 10/9-approximation.*

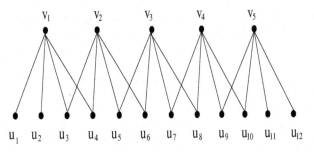

Fig. 3. Input for the proof of Theorem 3, for the case $k = 5$, $c = 4$, $d = 2$. The edges shown correspond to edges of distance 1 only

3.2   Dominating Set

Dominating set is known to be equivalent to Set Cover under L-reductions [3]. This result implies that dominating set is not approximable within a factor better than $(1 - \epsilon) \ln n$ unless $NP \subseteq DTIME(n^{O(\log \log n)})$, by a result due to Feige [7]. Here, n is the number of vertices in the graph. The well known greedy algorithm for set cover [11] can in fact be applied to yield a $\Theta(\log(n))$-approximation priority algorithm for dominating set, which is also greedy (as defined in Section 2.3, since the algorithm will always accept vertices as long as undominated vertices remain). Note that in [2] it was shown that no priority algorithm for set cover is better than $\ln n - o(\ln n)$-approximation, however the result does not carry over to dominating set.

Theorem 5. *Every* ADAPTIVE PRIORITY *greedy algorithm has approximation ratio* $\Omega(\log n)$ *where n is the number of vertices in the graph.*

Proof. The adversary presents the graph G, defined as follows. There is a set V of $k = 2^m$ triangles (3-cliques) of vertices v_l^1, v_l^2, v_l^3, with $l \in \{0, \dots, k - 1\}$, for some integer m. We call triangle $\{v_l^1, v_l^2, v_l^3\}$ *triangle l*, and we say that a vertex *is adjacent to triangle l* if and only if it is adjacent to all three vertices of triangle l. In addition, there is a set U of $m + 1 = \log k + 1$ vertices $u_1, \dots, u_{m+1}$ with the following property: vertex u_j, with $j \in [m]$ is adjacent to all triangles i for which the binary representation of i has the j-th most significant bit equal to 0. Vertex v_{m+1}, on the other hand, is adjacent to all triangles whose least significant bit is equal to 1. We call vertices $u_i, u_j \in U$ *complementary* if and only if every triangle is adjacent to one of u_i, u_j. Note that u_m and u_{m+1} are the only complementary vertices in G. Figure 4 illustrates G for the case $m = 3$ (for the sake of clarity we substituted the triangles by filled-in nodes). We remind the reader that the u_i's and v_j's play the role of the id's. Note that the total number of vertices in G is $\Theta(k)$. Then OPT has a cost of at most 3, since it suffices to accept vertices u_m, u_{m+1}, v_0^1 in order to dominate every vertex in G.

Consider the class $\mathcal{A}$ of priority algorithms with the following statement. On input G, every algorithm $A \in \mathcal{A}$ works in rounds, with a round consisting of several iterations. In particular, in the beginning of round j, A considers and

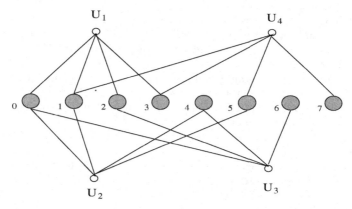

Fig. 4. The graph G for the proof of Theorem 5. Each filled-in node j corresponds to a triangle, and an edge between a vertex u_i and a triangle j indicates that all three vertices v_j^1, v_j^2, v_j^3 that comprise triangle j are adjacent to u_i

accepts a vertex in U. In subsequent iterations during round j, A considers (and possibly accepts) certain vertices in V, all of which are dominated by the set of vertices in U which were accepted by A in rounds $1, \ldots, j$. Round j ends (and round $j + 1$ begins) when A considers and accepts a new vertex in U.

We shall focus on $\mathcal{A}$ since it is easier to lower-bound the cost of algorithms in this class. First, we show that the cost of every greedy priority algorithm is lower-bounded (within a factor of $1/2$) by the cost of some algorithm in $\mathcal{A}$.

Lemma 3. *Let A' be a greedy priority algorithm. Then there is an algorithm $A \in \mathcal{A}$ such that on input G $cost(A') \geq (1/2) \cdot cost(A)$.*

It now suffices to prove the following lemma. Interestingly, we will show that permuting the id's of the vertices is sufficient for the adversary to force a logarithmic bound on the approximation ratio.

Lemma 4. *Let A be an algorithm in class $\mathcal{A}$. For every $j \leq m$ there exists a graph G_j isomorphic to G such that on input G_j the adversary can force A to consider and accept j vertices in U no two of which are complementary in the first j rounds of A.*

The theorem follows from Lemma 3 and Lemma 4, and the observation that on input G (or any graph isomorphic to G) no algorithm in $\mathcal{A}$ is correct unless it accepts a pair of complementary vertices. $\qquad\square$

Acknowledgements. I would like to thank the authors of [4] for providing an early copy of their paper as well as for comments on this paper.

References

1. S. Angelopoulos. Randomized priority algorithms. In *Proceedings of the 1st International Workshop on Approximation and Online Algorithms*, pages 27–40, 2003.

2. S. Angelopoulos and A. Borodin. On the power of priority algorithms for facility location and set cover. In *Proceedings of the 5th International Workshop on Approximation Algorithms for Combinatorial Optimization Problems (APPROX)*, pages 26–39, 2002.

3. R. Bar-Yehuda and S. Moran. On approximation problems related to the independent set and vertex cover problems. *Disc. Appl. Math*, 9:1–10, 1984.

4. A. Borodin, J. Boyar, and K. Larsen. Priority algorithms for graph optimization problems. These proceedings.

5. A. Borodin, M. Nielsen, and C. Rackoff. (Incremental) priority algorithms. *Algorithmica*, 37:295–326, 2003.

6. S. Davis and R. Impagliazzo. Models of greedy algorithms for graph problems. In *Proceedings of the 15th ACM-SIAM Symposium on Discrete Algorithms*, 2004.

7. U. Feige. A threshold of ln n for approximating set cover. *Journal of the ACM*, 45(4):634–652, 1998.

8. S. Guha and S. Khuller. Greedy strikes back: Improved facility location algorithms. In *Proceedings of the 9th ACM-SIAM Symposium on Discrete Algorithms*, pages 649–657, 1998.

9. K. Jain, M. Mahdian, and A. Saberi. A new greedy approach for facility location problems. In *Proceedings of the 34th Annual ACM Symposium on Theory of Computation*, pages 731–740, 2002.

10. K. Jain and V.V. Vazirani. Approximation algorithms for metric facility location and k-median problems using the primal-dual schema and lagrangian relaxation. *Journal of the ACM*, 48(2):274–296, 2001.

11. D.S. Johnson. Approximation algorithms for combinatorial problems. *JCSS*, 9(3):256–278, 1974.

12. M. Mahdian, E. Markakis, A. Saberi, and V. V. Vazirani. A greedy facility location algorithm analyzed using dual fitting. In *Proceedings of the 4th International Workshop on Approximation Algorithms for Combinatorial Optimization Problems (APPROX)*, pages 127–137, 2001.

13. M. Mahdian, J. Ye, and J. Zhang. Improved approximation algorithms for metric facility location problems. In *Proceedings of the 5th International Workshop on Approximation Algorithms for Combinatorial Optimization Problems (APPROX)*, pages 229–242, 2002.

14. R. R. Mettu and C. G. Plaxton. The online median problem. In *Proceedings of the 41st Annual IEEE Symposium on Foundations of Computer Science*, pages 339–348, 2000.

15. O. Regev. Priority algorithms for makespan minimization in the subset model. *IPL*, 84(3):153–157, 2002.

16. D.B. Shmoys, E. Tardos, and K. Aardal. Approximation algorithms for facility location problems (extended abstract). In *Proceedings of the 29th Annual ACM Symposium on Theory of Computing*, pages 265–274, 1997.

Off-line Admission Control for Advance Reservations in Star Networks

Udo Adamy[1], Thomas Erlebach[2,*], Dieter Mitsche[1], Ingo Schurr[1,**],
Bettina Speckmann[3], and Emo Welzl[1]

[1] Institute for Theoretical Computer Science, ETH Zürich, 8092 Zürich, Switzerland
{adamy, dmitsche, schurr, welzl}@inf.ethz.ch
[2] Department of Computer Science, University of Leicester, Leicester LE1 7RH, U.K
t.erlebach@mcs.le.ac.uk
[3] Department of Mathematics and Computer Science, TU Eindhoven,
5600 MB Eindhoven, The Netherlands
speckman@win.tue.nl

Abstract. Given a network together with a set of connection requests,
call admission control is the problem of deciding which calls to accept
and which ones to reject in order to maximize the total profit of the
accepted requests. We consider call admission control problems with advance reservations in star networks. For the most general variant we
present a constant-factor approximation algorithm resolving an open
problem due to Erlebach. Our method is randomized and achieves an
approximation ratio of 1/18. It can be generalized to accommodate call
alternatives, in which case the approximation ratio is 1/24. We show
how our method can be derandomized. In addition we prove that call
admission control in star networks is $\mathcal{APX}$-hard even for very restricted
variants of the problem.

1 Introduction

Call admission control (CAC) is a fundamental problem in the operation of
communication networks. In its general form each connection request (call) has a
certain bandwidth requirement and some time specification given by its starting
time and its duration. If the network establishes a call, it first decides on a path
from the sender to the receiver through which the call is being routed. Then it
allocates the requested amount of bandwidth on all links along that path during
the time period in which the call is active. In addition, each call is associated

* Work mainly done while the author was with ETH Zürich, supported by the Swiss
National Science Foundation under Contract No. 21-63563.00 (Project AAPCN) and
the EU Thematic Network APPOL II (IST-2001-32007), with funding provided by
the Swiss Federal Office for Education and Science.
** Supported by the joint Berlin/Zürich graduate program Combinatorics, Geometry,
and Computation (CGC), financed by the German Science Foundation (DFG) and
ETH Zürich.

G. Persiano and R. Solis-Oba (Eds.): WAOA 2004, LNCS 3351, pp. 211–224, 2005.

with some profit, which is gained by the network provider only if the desired connection is established. CAC is the problem of deciding which calls to accept and which to reject with the goal of maximizing the total profit accrued by the accepted requests.

In this paper we consider CAC problems in star networks with advance reservations. In star networks there is always a unique path for each sender–receiver pair, and the routing issue mentioned above is trivial. Advance reservation means that resources are requested in advance of when they are really needed. In this setting the decision of acceptance or rejection of calls does not have to be made on-line, i.e., immediately when a new call arrives, but can be made off-line. The network can collect incoming requests over some period of time and then decide for the set of collected calls which of them to accept and which to reject. Advance reservations are therefore helpful for the network in optimizing its CAC decisions.

We also deal with call alternatives. In this scenario the user can specify several alternative requests for a connection to be established. The network can either accept exactly one of the call alternatives, or reject a call completely by rejecting all its alternatives. By specifying several alternatives the user can increase the chances that the network accepts the call.

Problem Definition. A star network consists of a set of outer nodes, each of which is connected exclusively to a unique center node. It is modeled by a simple, undirected graph $G = (V, E)$, whose vertex set $V = \{0, 1, \ldots, n\}$ represents the nodes of the network with vertex 0 being the center node. The edge set E consists of the edges $e_i = \{i, 0\}$ for $i = 1, \ldots, n$ corresponding to the links in the star network that connect an outer vertex i to the center node 0. The capacities of the links are given as a capacity function $c : E \mapsto \mathbf{R}^+$ that maps each edge $e \in E$ to a positive capacity $c(e)$.

A connection request or *call i* is specified by a tuple $(u_i, v_i, t_i, d_i, b_i, p_i)$ consisting of a source node $u_i \in V$, a destination node $v_i \in V$, a starting time $t_i \in \mathbf{N}$, a duration $d_i \in \mathbf{N}$, a positive bandwidth requirement $b_i \in \mathbf{R}^+$, and a profit $p_i \in \mathbf{R}$. For a set $\mathcal{R}$ of connection requests, a solution is a subset $Q \subseteq \mathcal{R}$ of accepted calls. It is *feasible* if the sum of bandwidth requirements of simultaneously active calls using the same edge does not exceed the capacity of that edge:

$$\forall e \in E : \forall t \in \mathbf{N} : \sum_{i \in Q(e,t)} b_i \leq c(e),$$

where $Q(e, t)$ is the set of accepted calls that use the edge e at time t. Our goal is to compute a feasible solution that maximizes the sum $\sum_{i \in Q} p_i$ of profits of the accepted calls. We refer to this problem as the general call admission control problem in star networks, or GCA in stars for short.

GCA in stars is an $\mathcal{NP}$-hard problem. If we drop the time specifications of the calls and restrict the network to consist of a single edge only, it is equivalent to the KNAPSACK problem, which is known to be $\mathcal{NP}$-hard [12]. Therefore, we are interested in finding good approximate solutions. A feasible solution for GCA is called a ρ-approximation, $\rho \leq 1$, if its total profit is at least a ρ-fraction of the profit of an optimal solution. An algorithm is called a ρ-approximation algorithm if it runs in polynomial time and always outputs a ρ-approximation. The parameter ρ is called the approximation ratio of such an algorithm.

Motivation. A lot of work on CAC considered the scenario that a connection request is presented to the network only at the time when the connection should be established. This model ignores the possibility that resources might be requested ahead of time when they are needed. A natural concept to incorporate this possibility is advance reservation [13, 18], which is attractive for both the network provider and the users. The network provider has the advantage of a higher flexibility and can use better algorithms to maximize the profit gained from the accepted calls. The user benefits from a guaranteed quality of service (QoS) once his/her call is accepted.

If we omit the time specification of the calls, GCA contains the maximum edge-disjoint paths (MEDP) problem as a special case (see [16] for more on this problem). Since already the MEDP problem is hard to approximate in general directed graphs [14], it is natural to restrict the network topology to simpler classes. In this paper we restrict ourselves to star networks. Although star networks are a very restricted class of networks, we note that with the results we present in this paper, star networks are now the most general class of networks for which a constant-factor approximation algorithm for GCA is known; even for line networks, i.e., networks consisting of a single path, the existence of a constant-factor approximation algorithm is an open problem (GCA in lines includes the maximum independent set problem in rectangle intersection graphs as a special case; see [5] for the best known approximation results for the latter problem).

Interestingly, CAC algorithms for star networks apply to general networks if the provider has rented the capacity of his/her network from another provider according to the hose model of bandwidth reservations. In the hose model [9], one requests a logical network connecting a set of terminal nodes and specifies for each terminal node the maximum rates at which traffic will ever be transmitted or received by that node. The provider has to reserve sufficient capacity for the logical network to ensure that any traffic matrix consistent with the specifications can be accommodated. Therefore, with respect to the available bandwidth, the logical network behaves like a star network with the terminal nodes as outer nodes. Thus, if a provider of a video-on-demand system, say, has rented capacity for the distribution network according to the hose model, the handling of advance reservations for video transmissions leads naturally to the problem of GCA in star networks.

It is an important task to investigate CAC problems with arbitrary capacities on the links of the network. For line networks, for example, there is a constant-factor approximation algorithm in the case of uniform capacities on the edges and no time specifications for the calls [3]. However, if we allow arbitrary edge capacities and assume no restrictions on the bandwidth requirements, no constant-factor approximation algorithm is known so far. Constant-factor approximations for this problem are known only under the "no-bottleneck" assumption, which requires that the maximum bandwidth requirement is not larger than the minimum edge capacity [7, 8].

Similarly, for star networks there is a constant-factor approximation algorithm for GCA with unit edge capacities, and only a logarithmic approximation for the case of arbitrary edge capacities. These algorithms were presented in a paper of Erlebach [11], who raised the open question whether there is a constant-factor approximation algorithm for GCA in stars.

Our Results. We resolve this open question by presenting the first constant-factor approximation algorithm for GCA in star networks. Our method is randomized and achieves an approximation ratio of $1/18$. It is fast, because it does not involve linear programming. Instead, our algorithm is based on the local ratio technique of [3] and can easily be derandomized. We can also modify our algorithm to handle call alternatives and obtain a $1/24$-approximation algorithm for this case. On the other hand, we prove that GCA in stars is $\mathcal{APX}$-hard even for very restricted versions of the problem.

Related Work. Many authors have investigated CAC problems for various network topologies in the on-line and off-line setting. In the on-line scenario each request must be accepted or rejected immediately without knowing the future events. We refer to [17, 6] for surveys about on-line CAC algorithms.

GCA for various network topologies is studied in a paper by Erlebach [11]. For trees and trees of rings with unit edge capacities he presents constant-factor approximation algorithms for CAC without time specifications for the calls. For star networks he obtains a $1/10$-approximation algorithm both for GCA without time specifications and for GCA with unit edge capacities. For GCA in stars with arbitrary edge capacities he achieves a $1/O(\log R)$-approximation algorithm, where R is the ratio of the maximum edge capacity to the minimum edge capacity. All these algorithms are obtained by first solving a linear programming relaxation and then decomposing the optimal fractional solution into a convex combination of integral solutions. The output of the algorithm is the best of the integral solutions obtained this way.

Bar-Noy et al. [3] extend the so-called local ratio technique [4] to resource allocation and scheduling problems. They obtain a $\frac{1}{5}$-approximation algorithm for GCA on a single link. Their approach is also used by Lewin-Eytan et al. [18] to obtain results for lines, rings, and trees. For trees with unit edge capacities, they present a $\frac{1}{5}$-approximation for GCA without time specifications and a $1/O(\log m)$-approximation algorithm for GCA, where m is the number of requests.

2   Preliminaries: The Local Ratio Technique

Our approximation algorithm is based on the local ratio technique, which was developed by Bar-Yehuda and Even [4] and first used in our context by Bar-Noy et al. [3]. The general framework can be described as follows. Assume that we are given a profit vector $\boldsymbol{p} \in \mathbf{R}^n$ and a set $\mathcal{F}$ of feasibility constraints on vectors $x \in \mathbf{R}^n$. A vector $x \in \mathbf{R}^n$ is a *feasible* solution for the problem $(\mathcal{F}, p)$ if it satisfies all the constraints in $\mathcal{F}$. The *value* of a feasible solution x is the inner product $p \cdot x$. For a maximization problem, a feasible solution x is an r-approximation if $p \cdot x \geq r \cdot p \cdot x^*$, where x^* is an *optimal* solution, i.e., a solution whose value is maximal among all feasible solutions. The power of the technique is based on the following theorem.

Theorem 1. (Local Ratio [4, 3]) *Let $\mathcal{F}$ be a set of constraints and let p, p' and p'' be profit vectors such that $p = p' + p''$. Then, if x is an r-approximation with respect to $(\mathcal{F}, p')$ and with respect to $(\mathcal{F}, p'')$, then x is an r-approximation with respect to $(\mathcal{F}, p)$.*

For computing approximate solutions we will use the so-called *unified algorithm* proposed in [3]. We generalize our problem by allowing negative profits as well.

The Unified Algorithm:

1. Delete all calls with non-positive profit.
2. If no calls remain, return $Q = \emptyset$.
3. Otherwise, select a call i and decompose $p = p' + p''$. The choice of i and the decomposition depends on the problem at hand. For the GCA problem, we will always select i as a call with minimum end-time $t_i + d_i$, breaking ties arbitrarily.
4. Solve the problem recursively using p'' as the profit vector. Let Q'' be the set returned.
5. If $Q'' \cup \{i\}$ is a feasible solution, return $Q = Q'' \cup \{i\}$. Otherwise, return $Q = Q''$.

The decomposition of p into p' and p'' in step 3 will be specified separately for each problem to which we apply the algorithm.

We call a feasible solution i-*maximal* if it either contains the call i, or it does not contain the call i, but adding i to the solution will render it infeasible. The following lemma analyzes the quality of the solution produced by the algorithm.

Lemma 1. ([3]) *Let r be a constant. Suppose that in the algorithm above the choice of i and the decomposition $p = p' + p''$ is always such that: (1) $p''_i = 0$, and (2) every i-maximal solution is an r-approximation with respect to p'. Then, the algorithm terminates and the solution Q produced is an r-approximation with respect to p.*

Proof. First note that since $p''_i = 0$, the call i will be deleted in the recursive call, and the algorithm will eventually terminate. Because the deletion of calls with non-positive profit in step 1 does not change the optimum value, it is sufficient to show that Q is an r-approximation with respect to the remaining calls. Since Q is i-maximal, the condition (2) implies that Q is an r-approximation with respect to p'. It remains to show that Q is also an r-approximation with respect to p'' (and then apply the local ratio theorem). We proceed by induction on the number of recursive calls of the algorithm. At the basis of the recursion, the algorithm returns the empty set, since no calls remain. This is clearly an optimal solution and, hence, also an r-approximation. For the induction step, assume that Q'' is an r-approximation with respect to p''. Since $p_i'' = 0$ and Q is either Q'' or $Q'' \cup \{i\}$, it follows that Q is an r-approximation with respect to p'' as well. By the local ratio theorem (Theorem 1), Q is also an r-approximation with respect to p. □

By Lemma 1 we only need to find a constant $r > 0$ such that every i-maximal solution is an r-approximation with respect to p'. To do so, we will derive an upper bound p_{opt} on the optimum p'-profit and a lower bound p_{max} on the p'-profit of every i-maximal solution. The ratio $r = p_{max}/p_{opt}$ is then a lower bound on the approximation ratio of the algorithm.

3   Constant-Factor Approximation for GCA in Stars

In this section we present a $1/18$-approximation algorithm for GCA in star networks. Without loss of generality, we can assume that every call in the set $\mathcal{R}$ uses exactly two edges. We partition the calls in $\mathcal{R}$ into three classes according to their bandwidth requirements. Consider a call $i \in \mathcal{R}$ and denote by e and f the edges that it uses. We classify the call i to be

- a *small* call, if it uses at most half of the capacity on both of its edges, i.e., $b_i \leq c(e)/2$ and $b_i \leq c(f)/2$,
- a *big* call, if it uses more than half of the capacity on both of its edges, i.e., $b_i > c(e)/2$ and $b_i > c(f)/2$,
- a *mixed* call, otherwise.

We denote the set of small, big, and mixed calls by $\mathcal{R}_{small}$, $\mathcal{R}_{big}$, and $\mathcal{R}_{mixed}$, respectively. We give a constant-factor approximation for each of the three classes. The algorithm for the original problem is then as follows. Partition the calls in $\mathcal{R}$ into the three classes and solve the problem for each of the classes separately. Output the solution with the largest profit among the three solutions for the classes.

Small Calls. We approximate this set using the unified algorithm from above. Let i be the call with minimum end-time $t_i + d_i$ selected by the algorithm in step 3 and denote by e and f the edges that call i uses, such that $c(e) \leq c(f)$. We decompose the profit function $p = p' + p''$ by defining p' according to

$$
p_j' = p_i \cdot \begin{cases} 1 & \text{if } j = i, \\ \alpha \cdot (c(f) - b_i) \cdot b_j & \text{if } i \text{ and } j \text{ intersect on edge } e \text{ at time } t_i + d_i, \\ \alpha \cdot (c(e) - b_i) \cdot b_j & \text{if } i \text{ and } j \text{ intersect only on edge } f \text{ at time } t_i + d_i, \\ 0 & \text{otherwise,} \end{cases}
$$

where α is a parameter that will be determined later. Note that $p_i'' = 0$. The values of the profit function $p'' = p - p'$ may be non-positive.

Lemma 2. *The set of small calls admits an approximation ratio of $1/4$.*

Proof. Due to Lemma 1 it suffices to show that every i-maximal solution is a $1/4$-approximation with respect to $\boldsymbol{p}'$. Let Q^* be a $\boldsymbol{p}'$-optimal solution. In case i is not in Q^*, the contribution of other calls in Q^* using edge e is at most $p_i \cdot \alpha \cdot (c(f) - b_i) \cdot c(e)$, since the total bandwidth of all these calls intersecting at time $t_i + d_i$ is at most $c(e)$. Likewise the contribution on edge f is at most $p_i \cdot \alpha \cdot (c(e) - b_i) \cdot c(f)$. In case i belongs to Q^*, the call i adds p_i to the $\boldsymbol{p}'$-profit of Q^*, and uses b_i of the capacity of the edges e and f. The contribution of other calls in Q^* to the $\boldsymbol{p}'$-profit is then at most $p_i \cdot \alpha \cdot (c(f) - b_i) \cdot (c(e) - b_i)$ on edge e, and $p_i \cdot \alpha \cdot (c(e) - b_i) \cdot (c(f) - b_i)$ on edge f.

If we set $X = (c(f) - b_i) \cdot (c(e) - b_i)$, then the $\boldsymbol{p}'$-profit of Q^* is at most

$$p_i \cdot \max\{\alpha \cdot (c(f) - b_i) \cdot c(e) + \alpha \cdot (c(e) - b_i) \cdot c(f), 1 + 2 \cdot \alpha \cdot X\}.$$

To derive a lower bound on the $\boldsymbol{p}'$-value of an i-maximal solution Q, we distinguish two cases. If the call i is in Q, it contributes p_i itself. Thus the $\boldsymbol{p}'$-profit of Q is at least p_i. Otherwise, the call i is blocked by other calls in Q. This means that the total bandwidth of the calls preventing call i from being accepted must be greater than $c(e) - b_i$ on edge e or greater than $c(f) - b_i$ on edge f, respectively. Hence, the $\boldsymbol{p}'$-profit of the blocking calls is at least $p_i \cdot \alpha \cdot X$ (no matter on which of the edges e or f the call i is blocked). The minimum $p_i \cdot \min\{1, \alpha \cdot X\}$ of both expressions is a lower bound on the $\boldsymbol{p}'$-value of every i-maximal solution. Altogether, the approximation ratio r is given by

$$r = \frac{\min\{1, \alpha \cdot X\}}{\max\{\alpha \cdot (c(f) - b_i) \cdot c(e) + \alpha \cdot (c(e) - b_i) \cdot c(f), 1 + 2 \cdot \alpha \cdot X\}},$$

which for $\alpha = \frac{1}{X} = \frac{1}{(c(f) - b_i) \cdot (c(e) - b_i)}$ gives $r = 1/\max\left\{\frac{c(e)}{c(e) - b_i} + \frac{c(f)}{c(f) - b_i}, 3\right\}$. Finally, the fact that i is a small call, i.e., $b_i \leq c(e)/2$ and $b_i \leq c(f)/2$, implies that r is at least $1/4$. $\qquad\square$

Big Calls. In the case of big calls, no two calls using the same edge simultaneously can be in a feasible solution. Therefore we may assume that $b_j = 1$ for all big calls j and that $c(e) = 1$ for all edges $e \in E$. To define the decomposition $\boldsymbol{p} = \boldsymbol{p}' + \boldsymbol{p}''$, we set $\boldsymbol{p}'$ to be

$$p'_j = p_i \cdot \begin{cases} 1 & \text{if } j = i \text{ or } i \text{ and } j \text{ intersect at time } t_i + d_i, \\ 0 & \text{otherwise.} \end{cases}$$

The proof of the following lemma is similar to (but easier than) the proof of Lemma 2.

Lemma 3. *The set of big calls admits an approximation ratio of $1/2$.*

Mixed Calls. Mixed calls use at most half of the capacity of one of their two edges, and need more than half of the capacity of the other edge. They cause the logarithmic factor in the approximation ratio shown by Erlebach in [11]. We briefly sketch the reason for this. Erlebach defines mixed calls as calls using at most $1/3$ of the capacity of one of their edges and more than $1/2$ of the capacity of the other. For an edge e, some mixed calls using e may use at most $1/3$ of the capacity of e, while others may use more than $1/2$. For a call using at most $1/3$ of the capacity of edge e, Erlebach [11] has to argue about the second edge f used by this call; on that edge f, the call uses more than half of the capacity. If there is another call using at most $1/3$ of the capacity of that edge f, he has to argue about the second edge of that call, and so on. The capacity of the edges considered in this chain of arguments decreases by a factor of $2/3$ in each step; thus the number of steps is $O(\log R)$, where R is the ratio of the maximum to the minimum edge capacity, and this factor enters into the approximation ratio.

Here, we present a randomized procedure to approximate the mixed calls within a constant factor of the optimum. Our method can easily be derandomized, as we will discuss later on. We perform the following random experiment on the star network. Every outer node $v \in \{1, \ldots, n\}$ is put independently with probability $1/2$ in a set A and with probability $1/2$ in a set B. Then we consider only those mixed calls that have one endpoint in the set A and the other endpoint in the set B. We denote this set of calls by $\mathcal{R}_{A,B}$. The probability that a mixed call belongs to $\mathcal{R}_{A,B}$ is exactly $1/2$. This implies that the expected profit of an optimal solution for the set $\mathcal{R}_{A,B}$ is at least half the profit of an optimal solution for all mixed calls. Hence, we lose only a factor of 2 in expectation.

After this random experiment, all calls in $\mathcal{R}_{A,B}$ connect a vertex in A with a vertex in B. For a call $i \in \mathcal{R}_{A,B}$, let $e(i)$ denote the edge of the path of i connecting its endpoint in A with the center node 0, and let $f(i)$ denote the edge of the path between the center node and its endpoint in B. We further partition the calls in $\mathcal{R}_{A,B}$ according to their bandwidths. If a mixed call $i \in \mathcal{R}_{A,B}$ uses more than half of the capacity of the edge $e(i)$ (and at most half of the capacity of the edge $f(i)$), we put the call i in the set $\mathcal{R}_A$. Otherwise, we put the call i in the set $\mathcal{R}_B$. By considering $\mathcal{R}_A$ and $\mathcal{R}_B$ separately, we avoid the problem encountered in the analysis in [11], i.e., we do no longer have to deal with mixed calls occupying a large fraction and those occupying a small fraction of the capacity of the same edge at the same time.

Lemma 4. *Each of the sets $\mathcal{R}_A$ and $\mathcal{R}_B$ admits an approximation ratio of $1/3$.*

Proof. To approximate the sets $\mathcal{R}_A$ and $\mathcal{R}_B$ we again use the unified algorithm. For the set $\mathcal{R}_A$ we decompose $\boldsymbol{p} = \boldsymbol{p}' + \boldsymbol{p}''$ by specifying $\boldsymbol{p}'$ to be

$$
p'_j = p_i \cdot \begin{cases} 1 & \text{if } j = i, \\ \alpha/2 & \text{if } i \text{ and } j \text{ intersect on edge } e(i) \text{ at time } t_i + d_i, \\ \alpha \cdot \frac{b_j}{c(f(i))} & \text{if } i \text{ and } j \text{ intersect only on edge } f(i) \text{ at time } t_i + d_i, \\ 0 & \text{otherwise.} \end{cases}
$$

A $\boldsymbol{p}'$-optimal solution accepts on edge $e(i)$ either the call i with $\boldsymbol{p}'$-profit p_i or one other call that intersects call i on the edge $e(i)$ and gives profit $p_i \cdot \alpha/2$. On the edge $f(i)$ the total bandwidth of accepted calls is bounded by $c(f(i))$, yielding a bound of $p_i \cdot \alpha$ on the optimal $\boldsymbol{p}'$-profit on that edge. An i-maximal solution contains either the call i with $\boldsymbol{p}'$-profit p_i or is blocked by calls with profit at least $p_i \cdot \alpha/2$. Hence, the approximation ratio r is given by

$$r = \frac{\min\{1, \alpha/2\}}{\max\{1, \alpha/2\} + \alpha},$$

which, for $\alpha = 2$, gives $r = 1/3$.

By symmetry, the set $\mathcal{R}_B$ can be approximated with the same ratio.   □

After approximating both sets $\mathcal{R}_A$ and $\mathcal{R}_B$ by the unified algorithm, we output the solution that has a larger profit.

Corollary 1. *The set of mixed calls admits an approximation ratio of* $1/12$.

Proof. The larger solution for the sets $\mathcal{R}_A$ and $\mathcal{R}_B$ has profit at least $1/6$ times the profit of an optimal solution for the set $\mathcal{R}_{A,B}$, which, in expectation, is a factor $1/2$ away from the optimal solution for all mixed calls.   □

We have approximation ratio $1/4$, $1/2$, and $1/12$ for the small calls, big calls, and mixed calls, respectively. As our algorithm outputs the solution with largest profit among the three individual solutions, we obtain the following theorem.

Theorem 2. *The above algorithm is a* $1/18$-*approximation algorithm for GCA in stars.*

3.1   Derandomization

The random process for filtering the mixed calls can be derandomized by reducing the size of the sample space. We refer to [1–Chapter 15] for an overview. In the analysis of our randomized algorithm, we used two properties of our random assignment. Firstly, the probability that a node v is assigned to the set A is $1/2$, and secondly the pairwise independence of the events, which guarantees that each call "survives" the experiment (i.e., its two endpoints are put into different sets) with probability $1/2$.

We employ a linear-size sample space that preserves these two properties. If we choose assignments uniformly at random from this sample space, our previous analysis remains valid. Thus, if we exhaustively search the sample space, we are guaranteed to find an assignment of nodes to the sets A and B such that the profit of an optimal solution for calls in $\mathcal{R}_{A,B}$ is at least half the profit of an optimal solution for all mixed calls. The construction of the linear-size sample space is omitted due to space limitations.

3.2   Call Alternatives

Our approximation algorithm can be generalized to accommodate alternatives for the calls with only a slight decrease of the approximation ratio. Call alternatives allow the specification of several alternatives for establishing a connection request. For example, a connection can be established either from 8:00 a.m. to

11:00 a.m. gaining profit p_1 or from 2:00 p.m. to 4:00 p.m. with profit p_2. We allow all parameters (including source node, destination node, and bandwidth requirement) of different alternatives of a call to be different. A solution obeying the capacity constraints is feasible if it contains at most one of the alternatives per call. The goal is to find the feasible solution with the largest profit. For this generalized problem, we use exactly the same algorithm as above and change only the decomposition of the profit function in the unified algorithm for each class of calls. In this case, the approximation ratios for the set of small calls, big calls, and mixed calls become 1/5, 1/3, and 1/16, respectively. This leads to the following theorem.

Theorem 3. *GCA in stars with call alternatives admits an approximation ratio of 1/24.*

4 $\mathcal{APX}$-Hardness

In this section we prove that already a very restricted variant of the general call admission control problem in stars is $\mathcal{APX}$-hard. The problem variant we consider is the following. We are given a star network with unit capacity on every edge, and a set $\mathcal{R}$ of calls. Each call i is associated with a starting time $t_i \in \{0, 1, 2\}$, has unit profit, and needs one unit of bandwidth on its edges. Every call i has duration $d_i = 2$. The goal is to compute a feasible subset $Q \subseteq \mathcal{R}$ of maximum cardinality. In the following we refer to this problem variant as STAR-GCA-SIMPLE, because many parameters of the general version are simplified by fixing them to be small constants. A set $Q \subseteq \mathcal{R}$ is a feasible solution for STAR-GCA-SIMPLE if at most one path per edge is active at any time.

We remark that the cases with one or two different starting times can be solved optimally in polynomial time (still assuming that all calls have the same duration). To see this, note that a maximum cardinality subset among a given set of calls that overlap in time can be obtained using a maximum matching computation [10, 19] in the graph with an edge $\{u, v\}$ for every request with endpoints u and v. This settles the case of one starting time. If there are two different starting times, either all calls overlap in time or the problem decomposes into two instances on disjoint time intervals.

If we allow three or more different starting times, the restricted variant of general call admission in star networks becomes difficult to solve, which is expressed in the next theorem.

Theorem 4. *The problem* STAR-GCA-SIMPLE *is $\mathcal{APX}$-hard.*

Corollary 2. *GCA in stars is $\mathcal{APX}$-hard.*

In particular, Theorem 4 implies that there is no polynomial-time approximation scheme for the restricted and its more general variants unless $\mathcal{P} = \mathcal{NP}$. Thus, the constant-factor approximations we have presented are best possible (except possibly for the constants) in this sense.

We will prove the theorem by an approximation preserving reduction, which is defined as follows [2].

Definition 1 (AP-reduction). *Let P_1 and P_2 be two optimization problems in NPO. For a solution y to an instance x of P_i, let ratio$_{P_i}(x, y)$ denote the ratio between the value of an optimal solution to x and the value of y (or the reciprocal of this ratio, whichever is larger than 1). The problem P_1 is called AP-reducible to P_2 if two functions f and g and a positive constant $\alpha > 1$ with the following properties exist:*

(i) *For any instance x of P_1 and for any rational $r > 1$, $f(x, r)$ is an instance of P_2.*

(ii) *For any instance x of P_1 and for any rational $r > 1$, if there is a solution to x, then there is also a solution to $f(x, r)$.*

(iii) *For any instance x of P_1, for any rational $r > 1$, and for any solution y to $f(x, r)$, $g(x, y, r)$ is a solution to x.*

(iv) *f and g are computable in polynomial time for any fixed rational r.*

(v) *For any instance x of P_1, for any rational $r > 1$, and for any solution y to $f(x, r)$, ratio$_{P_2}(f(x, r), y) \leq r$ implies ratio$_{P_1}(x, g(x, y, r)) \leq 1 + \alpha(r - 1)$.*

The properties (i)-(iv) ensure that there are polynomial time transformations f and g that map instances of P_1 to instances of P_2 and solutions for instances of P_2 back to solutions for the original instance of P_1, respectively. The heart of the AP-reduction is given by property (v), which intuitively says that an r-approximation algorithm for P_2 implies the existence of a $(1 + \alpha(r - 1))$-approximation algorithm for P_1.

In the sequel we present an AP-reduction from the maximum 3-dimensional matching problem, which is defined as follows: Given a set $D \subseteq X \times Y \times Z$, where X, Y and Z are disjoint sets, the goal is to find a matching $M \subseteq D$ for D of maximum cardinality, i.e., a largest set $M \subseteq D$ such that no two elements in M agree in any coordinate. The maximum 3-dimensional matching problem is known to be $\mathcal{APX}$-complete even if each of the elements in X, Y and Z occurs in at most three triples in D [15]. We refer to this problem as the bounded maximum 3-dimensional matching problem.

In this bounded version of the problem, each triple can intersect at most six other triples, which implies that the maximum matching contains at least $|D|/7$ triples. Moreover, the following lemma is easy to prove.

Lemma 5. *There is a greedy procedure that computes a 1/3-approximation for the bounded maximum 3-dimensional matching problem.*

Let $D \subseteq X \times Y \times Z$ be an instance of the maximum 3-dimensional matching problem. The function f of the AP-reduction is given by the following construction of an instance of STAR-GCA-SIMPLE. It does not depend on the parameter r.

Let vertex 0 be the center vertex of the star. For every element $x_i \in X$, we add the vertex x_i to the star and connect it to vertex 0 by an edge $\{x_i, 0\}$. We do the same for every $y_i \in Y$ and for every $z_i \in Z$. For each triple $d_j = (x_j, y_j, z_j) \in D$,

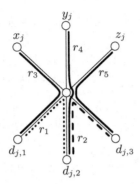

Fig. 1. The building block for a triple $d_j = (x_j, y_j, z_j)$

we add three more vertices $d_{j,1}$, $d_{j,2}$, and $d_{j,3}$ to the star and connect them to the center by the three edges $\{d_{j,1}, 0\}$, $\{d_{j,2}, 0\}$, and $\{d_{j,3}, 0\}$. In addition, we add the following 5 requests to $\mathcal{R}$ (see Figure 1):

- $r_1 = (d_{j,1}, d_{j,2})$ with interval $[0, 2)$ - $r_2 = (d_{j,2}, d_{j,3})$ with interval $[2, 4)$
- $r_3 = (d_{j,1}, x_j)$ with interval $[1, 3)$ - $r_4 = (d_{j,2}, y_j)$ with interval $[1, 3)$
- $r_5 = (d_{j,3}, z_j)$ with interval $[1, 3)$

Note that the dotted request r_1 and the dashed request r_2 have disjoint time intervals, whereas the solid requests r_3, r_4, and r_5 do not share any edge. The idea behind this is the following. For every triple $d_i \in D$, a solution either accepts the two request r_1 and r_2 without affecting any edge connecting a vertex from the sets X, Y, Z to the center, but blocking the requests r_3, r_4 and r_5, or it accepts the three requests r_3, r_4 and r_5 at the price of blocking all three edges connecting the center to the elements of the triple d_i. More than three requests per triple are not feasible.

Lemma 6. *Let $D \subseteq X \times Y \times Z$ be an instance of the maximum 3-dimensional matching problem, and let $(G, \mathcal{R})$ be the corresponding instance of* STAR-GCA-SIMPLE *defined above. There is a feasible solution for $(G, \mathcal{R})$ that accepts $2|D| + k$ requests if and only if D has a matching of size k.*

Proof. Suppose there is a feasible solution Q for the instance $(G, \mathcal{R})$ of size $2|D| + k$. Since no more than three requests in Q belong to the same triple, there are at least k triples for which three requests are in Q. The only possibility for one triple $d_i = (x_i, y_i, z_i)$ to have three of its requests accepted is the choice that accepts the three requests containing the vertices x_i, y_i and z_i. But then these vertices are blocked for the requests of all other triples. The feasibility of Q implies that all k triples are disjoint. Hence, they form a matching of size k.

Conversely, if there is a matching $M \subseteq D$ of size k, we can construct a feasible solution Q for the instance $(G, \mathcal{R})$ as follows. For every triple $d_i \in M$, put the three requests r_3, r_4 and r_5 into Q. Since the triples in M are disjoint, Q is feasible so far. For each of the remaining triples in $D \setminus M$, we can safely add the two requests r_1 and r_2 to Q without creating any conflict. Thus, Q is feasible by construction, and consists of $2|D| + k$ requests. □

The function g of the AP-reduction takes as arguments the instance D of the bounded maximum 3-dimensional matching problem, a solution Q of the instance $(G, \mathcal{R})$ and the parameter r, which will not be used. It first computes a solution M_1 for the instance D using the greedy procedure of Lemma 5. Secondly, it composes a matching M_2 out of the accepted requests in the solution Q. Whenever all three requests r_3, r_4 and r_5 corresponding to some triple $d_i \in D$ are in Q, it adds the triple d_i to the matching M_2. The value of $g(D, Q, r)$ is given by the larger of the two matchings M_1 and M_2. Thus, $|g(D, Q, r)| = \max\{|M_1|, |M_2|\}$. In addition, we have $|M_1| \geq |M^*|/3$ by Lemma 5, where M^* is a maximum matching for D, and $|M_2| \geq |Q| - 2|D|$ by Lemma 6.

So far the properties (i) – (iv) of the AP-reduction have been shown to be satisfied, and we will now show that property (v) holds with $\alpha = 43$. Therefore, let M^* be a maximum matching for the instance D. Then by Lemma 6 an optimal solution Q for $(G, \mathcal{R})$ consists of $|M^*| + 2|D|$ requests. Assume that we have an r-approximation for $(G, \mathcal{R})$, that is a solution Q that contains at least $(|M^*| + 2|D|)/r$ requests.

If $r \geq 45/43$, the inequality $|g(D, Q, r)| \geq |M_1| \geq |M^*|/3$ shows that g computes a $1/3$-approximation. Since $3 = 1 + 43(\frac{45}{43} - 1) \leq 1 + 43(r - 1)$, property (v) with $\alpha = 43$ holds in this case.

Otherwise $r < 45/43$. From $|Q| \geq (|M^*| + 2|D|)/r$, we get

$$|Q| \geq \frac{2r|D| + |M^*| - 2(r - 1)|D|}{r} = 2|D| + \frac{|M^*| - 2(r - 1)|D|}{r}$$
$$\geq 2|D| + \frac{|M^*|(1 - 14(r - 1))}{r},$$

where we used $|D| \leq 7|M^*|$ (which holds in the bounded version) in the last inequality. As $|g(D, Q, r)| \geq |M_2| \geq |Q| - 2|D| \geq (1 - 14(r - 1))|M^*|/r$, we get that ratio$_{P_1}(D, g(D, Q, r))$ is at most

$$\frac{|M^*|}{(1 - 14(r - 1))|M^*|/r} = 1 + \frac{15}{15 - 14r}(r - 1) \leq 1 + 43(r - 1),$$

where the last inequality holds for $1 < r < 45/43$. Again, property (v) is fulfilled with $\alpha = 43$, which completes the proof of the theorem.

References

1. N. Alon and J. H. Spencer. *The Probabilistic Method*. Wiley, New York, Second edition, 1992.
2. G. Ausiello, P. Crescenzi, G. Gambosi, V. Kann, A. Marchetti-Spaccamela, and M. Protasi. *Complexity and Approximation. Combinatorial Optimization Problems and their Approximability Properties*. Springer-Verlag, Berlin, 1999.
3. A. Bar-Noy, R. Bar-Yehuda, A. Freund, J. S. Naor, and B. Schieber. A unified approach to approximating resource allocation and scheduling. *Journal of the ACM (JACM)*, 48(5):1069–1090, 2001.

4. R. Bar-Yehuda and S. Even. A local-ratio theorem for approximating the weighted vertex cover problem. *Annals of Discrete Mathematics*, 25:27–46, 1985.
5. P. Berman, B. DasGupta, S. Muthukrishnan, and S. Ramaswami. Improved approximation algorithms for rectangle tiling and packing. In *Proceedings of the 12th Annual ACM–SIAM Symposium on Discrete Algorithms (SODA)*, pages 427–436, 2001.
6. A. Borodin and R. El-Yaniv. *Online Computation and Competitive Analysis*. Cambridge University Press, 1998.
7. A. Chakrabarti, C. Chekuri, A. Gupta, and A. Kumar. Approximation algorithms for the unsplittable flow problem. In *Proceedings of the 5th International Workshop on Approximation Algorithms for Combinatorial Optimization (APPROX)*, LNCS 2462, pages 51–66. Springer-Verlag, 2002.
8. C. Chekuri, M. Mydlarz, and F. B. Shepherd. Multicommodity demand flow in a tree. In *Proceedings of the 30th International Colloquium on Automata, Languages, and Programming (ICALP)*, LNCS 2719, pages 410–425. Springer-Verlag, 2003.
9. N. G. Duffield, P. Goyal, A. Greenberg, P. Mishra, K. K. Ramakrishnan, and J. E. van der Merwe. Resource management with hoses: point-to-cloud services for virtual private networks. *IEEE/ACM Transactions on Networking (TON)*, 10(5):679–692, 2002.
10. J. Edmonds. Paths, trees and flowers. *Canadian Journal of Mathematics*, 17:449–467, 1965.
11. T. Erlebach. Call admission control for advance reservation requests with alternatives. In *Proceedings of the 3rd Workshop on Approximation and Randomization Algorithms in Communication Networks (ARACNE)*, Proceedings in Informatics, pages 51–64. Carleton Scientific, 2002.
12. M. R. Garey and D. S. Johnson. *Computers and Intractability. A Guide to the Theory of NP-Completeness*. W. H. Freeman and Company, New York-San Francisco, 1979.
13. R. A. Guérin and A. Orda. Networks with advance reservations: The routing perspective. In *Proceedings of the 19th Annual Joint Conference of the IEEE Computer and Communications Societies (INFOCOM)*, pages 118–127, Los Alamitos, CA, USA, 2000. IEEE Computer Society Press.
14. V. Guruswami, S. Khanna, R. Rajaraman, B. Shepherd, and M. Yannakakis. Near-optimal hardness results and approximation algorithms for edge-disjoint paths and related problems. In *Proceedings of the 31st Annual ACM Symposium on Theory of Computing (STOC)*, pages 19–28, 1999.
15. V. Kann. Maximum bounded 3-dimensional matching is MAX SNP-complete. *Information Processing Letters*, 37:27–35, 1991.
16. J. Kleinberg. *Approximation algorithms for disjoint paths problems*. PhD thesis, Department of Electrical Engineering and Computer Science, Massachusetts Institute of Technology, 1996.
17. S. Leonardi. On-line network routing. In A. Fiat and G. J. Woeginger, editors, *Online Algorithms: The State of the Art*, LNCS 1442. Springer-Verlag, Berlin, 1998.
18. L. Lewin-Eytan, J. S. Naor, and A. Orda. Routing and admission control in networks with advance reservations. In *Proceedings of the 5th International Workshop on Approximation Algorithms for Combinatorial Optimization (APPROX)*, LNCS 2462, pages 215–228. Springer-Verlag, 2002.
19. S. Micali and V. V. Vazirani. An $O(\sqrt{|V|} \cdot |E|)$ algorithm for finding maximum matchings in general graphs. In *Proceedings of the 21st Annual IEEE Symposium on Foundations of Computer Science (FOCS)*, pages 17–27, 1980.

Joint Base Station Scheduling*

Thomas Erlebach[1], Riko Jacob[2], Matúš Mihaľák[3], Marc Nunkesser[2],
Gábor Szabó[2], and Peter Widmayer[2]

[1] Department of Computer Science, University of Leicester
t.erlebach@mcs.le.ac.uk
[2] Department of Computer Science, ETH Zürich
{jacob, nunkesser, szabog, widmayer}@inf.ethz.ch
[3] Department of Information Technology and Electrical Engineering, ETH Zürich
mihalak@tik.ee.ethz.ch

Abstract. Consider a scenario where base stations need to send data to users with wireless devices. Time is discrete and slotted into synchronous rounds. Transmitting a data item from a base station to a user takes one round. A user can receive the data item from any of the base stations. The positions of the base stations and users are modeled as points in Euclidean space. If base station b transmits to user u in a certain round, no other user within distance at most $\|b - u\|_2$ from b can receive data in the same round due to interference phenomena. The goal is to minimize, given the positions of the base stations and users, the number of rounds until all users have their data.

We call this problem the Joint Base Station Scheduling Problem (JBS) and consider it on the line (1D-JBS) and in the plane (2D-JBS). For 1D-JBS, we give a 2-approximation algorithm and polynomial optimal algorithms for special cases. We model transmissions from base stations to users as arrows (intervals with a distinguished endpoint) and show that their conflict graphs, which we call arrow graphs, are a subclass of the class of perfect graphs. For 2D-JBS, we prove *NP*-hardness and discuss an approximation algorithm.

1 Introduction

We consider different combinatorial aspects of problems that arise in the context of load balancing in time division networks. These problems turn out to be related to interval scheduling problems and interval graphs.

The general setting is that users with mobile devices are served by a set of base stations. In each time slot (round) of the time division multiplexing each base station serves at most one user. Traditionally, each user is assigned to a single base station that serves him until he leaves its cell or his demand is satisfied. The amount of data that a user receives depends on the strength of the signal

---

* Research partially supported by TH-Project TH-46/02-1 (Mobile phone antenna optimization: theory, algorithms, engineering, and experiments).

G. Persiano and R. Solis-Oba (Eds.): WAOA 2004, LNCS 3351, pp. 225–238, 2005.

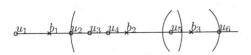

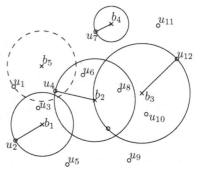

(a) This figure describes a possible situation in some time slot (round). Base station b_2 serves user u_2, b_3 serves user u_6. Users u_3, u_4 and u_5 are blocked and cannot be served. Base station b_1 cannot serve u_1 because this would create interference at u_2

(c) A possible situation in some time slot in the 2D case. Users u_2, u_4, u_7 and u_{12} are served. Base station b_5 cannot serve user u_1 here, because this would create interference at u_4 as indicated by the dashed circle

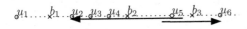

(b) Arrow representation of 1(a)

Fig. 1.1. The JBS-problem in one and two dimensions

that he receives from his assigned base station and on the interference, i.e. all signal power that he receives from other base stations. In [4], Das et al. propose a novel approach: Clusters of base stations jointly decide which users they serve in which round in order to increase network performance. Intuitively, this approach increases throughput, when in each round neighboring base stations try to serve pairs of users such that the mutual interference is low. We turn this approach into a discrete scheduling problem in one and two dimensions (see Figure 1.1), the Joint Base Station Scheduling problem (JBS).

In one dimension (see Figure 1.11(a)) we are given a set of n users as points $\{u_1, \ldots, u_n\}$ on a line and we are given positions $\{b_1, \ldots, b_m\}$ of m base stations. Note that such a setting could correspond to a realistic scenario where the base stations and users are located along a straight road. In our model, when a base station b_j serves a user u_i this creates interference in an interval of length $2|b_j - u_i|$ around the midpoint b_j. In each round each base station can serve at most one user such that at the position of this user there is no interference from any other base station. The goal is to serve all users in as few rounds as possible. In two dimensions (see Figure 1.11(c)), when base station b_j serves user u_i this creates interference in a disk with radius $\|b_j - u_i\|_2$ and center b_j.

The one-dimensional problem is closely related to interval scheduling problems, except that the particular way how interference operates leads to directed intervals (arrows). For these we allow that their tails can intersect (intersecting tails correspond to interference that does not affect the users at the heads of the arrows). We present results on this special interval scheduling problem. Similarly, the problem is related to interval graphs, except that we have conflict graphs of arrows together with the conflict rules defined by the interference (arrow graphs).

The problem of scheduling data transmissions in the smallest number of discrete rounds can be expressed as the problem of coloring the corresponding arrow graph with the smallest number of colors, where the colors represent rounds. In this paper, we prove that arrow graphs are perfect and can be colored optimally in $\mathcal{O}(n \log n)$ time. For the one-dimensional JBS problem with evenly spaced base stations we give a polynomial-time dynamic programming algorithm. For the general one-dimensional JBS problem, we show that for any fixed k the question whether all users can be served in k rounds can be solved in polynomial time. From the perfectness of arrow graphs and the existence of a polynomial-time algorithm for computing maximum weighted cliques in these graphs we derive a 2-approximation algorithm for JBS based on an LP relaxation and rounding. For the two-dimensional JBS problem, we show that it is NP-complete, and we discuss an approximation algorithm for a constrained version of the problem.

1.1   Related Work

Das et al. [4] propose an involved model for load balancing that takes into account different fading effects and calculates the resulting signal to noise ratios at the users for different schedules. In each round only a subset of all base stations is used in order to keep the interference low. The decision which base stations to use is taken by a central authority. The search for this subset is formulated as a (nontrivial) optimization problem that is solved by complete enumeration and that assumes complete knowledge of the channel conditions. The authors perform simulations on a hexagonal grid, propose other algorithms, and reach the conclusion that the approach has the potential to increase throughput.

There is a rich literature on interval scheduling and selection problems (see [6, 12] and the references given there for an overview). Our problem is more similar to a setting with several machines where one wants to minimize the number of machines required to schedule all intervals. A version of this problem where intervals have to be scheduled within given time windows is studied in [3]. Inapproximability results for the variant with a discrete set of starting times for each interval are presented in [2].

1.2   Problem Definitions and Model

We fully define the problems of interest in this section. Throughout the paper we use standard graph-theoretic terminology, see e.g. [14]. In the one-dimensional case we are given positions of base stations $B = \{b_1, \ldots, b_m\}$ and users $U = \{u_1, \ldots, u_n\}$ on the line in left-to-right order. Conceptually, it is more convenient to think of the interference region that is caused by some base station b_j serving a user u_i as an *interference arrow* of length $2|b_j - u_i|$ with midpoint b_j pointing to the user as shown in Figure 1.11(b). The interference arrow for the pair (u_i, b_j) has its head at u_i and its midpoint at b_j. We denote the set of all arrows resulting from pairs $P \subseteq U \times B$ by $\mathcal{A}(P)$. If it is clear from the context, we call the interference arrows just *arrows*. If two users are to be scheduled in the same round, then each of them must not get any interference from any other base station. Thus, two arrows are *compatible* if no head is contained in the other arrow; otherwise, we say that they are in *conflict*. (Formally, the head u_i of the arrow for (u_i, b_j) is contained in the arrow for (u_j, b_k) if u_i is contained in the closed interval $[b_k - |u_j - b_k|, b_k + |u_j - b_k|]$.) If we want to emphasize which user is affected by the interference from another transmission, we use the term *blocking*, i.e. arrow a_i blocks arrow a_j if a_j's head is contained in a_i. For each user we have to decide from which base station she is served. This corresponds to a selection of an arrow for her. Furthermore, we have to decide in which round each selected arrow is scheduled under the side constraint that all arrows in one round must be compatible. For this purpose it is enough to label the arrows with colors that represent the rounds.

For the two-dimensional JBS problem we have positions in $\mathbb{R}^2$ and *interference disks* $d(b_i, u_j)$ with center b_i and radius $\|b_i - u_j\|_2$ instead of arrows. We denote the set of interference disks for the user base-station pairs from a set P by $\mathcal{D}(P)$. Two interference disks are in conflict if the user that is served by one of the disks is contained in the other disk; otherwise, they are compatible. The problems can now be stated as follows:

1D-JBS
Input: A set of user positions $U = \{u_1, \ldots, u_n\} \subset \mathbb{R}$ and base station positions $B = \{b_1, \ldots, b_m\} \subset \mathbb{R}$.
Output: A set P of n user base-station pairs such that each user is in exactly one pair, and a coloring $C : \mathcal{A}(P) \to \mathbb{N}$ of the set $\mathcal{A}(P)$ of corresponding arrows such that any two arrows $a_i, a_j \in \mathcal{A}(P)$, $a_i \neq a_j$, with $C(a_i) = C(a_j)$ are compatible.
Objective: Minimize the number of colors used.

2D-JBS
Input: A set of user positions $U = \{u_1, \ldots, u_n\} \subset \mathbb{R}^2$ and base station positions $B = \{b_1, \ldots, b_m\} \subset \mathbb{R}^2$.
Output: A set P of n user base-station pairs such that each user is in exactly one pair, and a coloring $C : \mathcal{D}(\mathcal{P}) \to \mathbb{N}$ of the set $\mathcal{D}(\mathcal{P})$ of corresponding disks such that any two disks $d_i, d_j \in \mathcal{D}(\mathcal{P})$, $d_i \neq d_j$, with $C(d_i) = C(d_j)$ are compatible.
Objective: Minimize the number of colors used.

From the problem definitions above it is clear that both the 1D- and the 2D-JBS problems consist of a *selection problem* and a *coloring problem*. In the selection problem we want to select one base station for each user in such a way that the arrows (disks) corresponding to the resulting set P of user base-station pairs can be colored with as few colors as possible. We call a selection P *feasible* if it contains exactly one user base-station pair for each user. Determining the cost of a selection is then the coloring problem. This can also be viewed as a problem in its own right, where we no longer make any assumption on how the set of arrows (for the 1D problem) is produced. The conflict graph $G(A)$ of a set A of arrows is the graph in which every vertex corresponds to an arrow and there is an edge between two vertices if the corresponding arrows are in conflict. We call such conflict graphs of arrows *arrow graphs*. The *arrow graph coloring problem* asks for a proper coloring of such a graph. It is similar in spirit to the coloring of interval graphs. As we will see in Section 2.1, the arrow graph coloring problem can be solved in time $\mathcal{O}(n \log n)$. We finish this section with a simple lemma that leads to a definition:

Lemma 1. *For each 1D-JBS instance there is an optimal solution in which each user is served either by the closest base station to his left or by the closest base station to his right.*

Proof. This follows by a simple exchange argument: Take any optimal solution that does not have this form. Then exchange the arrow where a user is not served by the closest base station in some round against the arrow from the closest base station on the same side (which must be idle in that round). Shortening an arrow without moving its head can only resolve conflicts. Thus, there is also an optimal solution with the claimed property.   □

The two possible arrows by which a user can be served according to this lemma are called *user arrows*. It follows that for a feasible selection one has to choose one user arrow from each pair of user arrows.

2   Case on the Line—1D-JBS

As mentioned above, solving the 1D-JBS problem requires selecting an arrow for each user and coloring the resulting arrow graph with as few colors as possible.

2.1   Relation of Arrow Graphs to Other Graph Classes

In order to gain a better understanding of arrow graphs, we first discuss their relationship to other known graph classes.[1] We refer to [1, 13] for definitions and further information about the graph classes mentioned in the following.

First, it is easy to see that arrow graphs are a superclass of interval graphs: Any interval graph can be represented as an arrow graph with all arrows pointing in the same direction.

An arrow graph can be represented as the intersection graph of triangles on two horizontal lines $y = 0$ and $y = 1$: Simply represent an arrow with left endpoint ℓ and right endpoint r that points to the right (left) as a triangle with

---

[1] The connections between arrow graphs and known graph classes such as PI* graphs, trapezoid graphs, co-comparability graphs, AT-free graphs, and weakly chordal graphs were observed by Ekki Köhler, Jeremy Spinrad, Ross McConnell, and R. Sritharan at the seminar "Robust and Approximative Algorithms on Particular Graph Classes", held in Dagstuhl Castle during May 24–28, 2004.

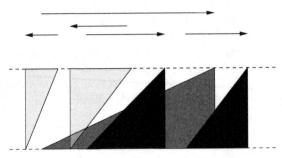

Fig. 2.1. An arrow graph (top) and its representation as a PI* graph (bottom)

corners $(\ell, 0)$, $(r, 0)$, and $(r, 1)$ (with corners $(r, 1)$, $(\ell, 1)$, and $(\ell, 0)$). It is easy to see that two triangles intersect if and only if the corresponding arrows are in conflict. See Figure 2.1 for an example.

Intersection graphs of triangles with endpoints on two parallel lines are called PI* graphs. They are a subclass of trapezoid graphs, which are the intersection graphs of trapezoids that have two sides on two fixed parallel lines. Trapezoid graphs are in turn a subclass of co-comparability graphs, a well-known class of perfect graphs. Therefore, the containment in these known classes of perfect graphs implies the perfectness of arrow graphs. Consequently, the size of a maximum clique in an arrow graph always equals its chromatic number.

As arrow graphs are a subclass of trapezoid graphs, we can apply known efficient algorithms for trapezoid graphs to arrow graphs. Felsner et al. [7] give algorithms with running-time $\mathcal{O}(n \log n)$ for chromatic number, weighted independent set, clique cover, and weighted clique in trapezoid graphs with n vertices, provided that the trapezoid representation is given. Their algorithm for chromatic number leads to a simple greedy coloring algorithm for arrow graphs (see [5]).

We sum up the discussed properties of arrow graphs in the following theorem.

Theorem 1. *Arrow graphs are perfect. In arrow graphs chromatic number, weighted independent set, clique cover, and weighted clique can be solved in time $\mathcal{O}(n \log n)$.*

One can also show that arrow graphs are AT-free (i.e., do not contain an asteroidal triple) and weakly chordal.

2.2   1D-JBS with Evenly Spaced Base Stations

Now we consider the 1D-JBS problem under the assumption that the base stations are evenly spaced. We are given m base stations $\{b_1, \ldots, b_m\}$ and n users $\{u_1, \ldots, u_n\}$ on a line, where the distance between any two neighboring base stations is the same. This assumption can be viewed as an abstraction of the fact that in practice, base stations are often placed in regular patterns and not in a completely arbitrary fashion.

Let d denote the distance between two neighboring base stations. The base stations partition the line into two *rays* and a set of *intervals* $\{v_1, \ldots, v_{m-1}\}$. In this section we additionally require that no user to the left of the leftmost base station be further away from it than distance d, and that the same holds for the right end. We define a solution to be *non-crossing* if there are no two users u and w in the same interval such that u is to the left of w, u is served from the right, and w from the left.

Lemma 2. *There is an optimal solution that is non-crossing.*

Proof. Take any optimal solution s that is not non-crossing. We show that such a solution can be transformed into another optimal solution s' that is non-crossing. Let u and w be two users such that u and w are in the same interval, u is to the left of w, and u is served by the right base station b_r in round t_1 by arrow a_r and w is served by the left base station b_l in round t_2 by arrow a_l; trivially, $t_1 \neq t_2$. Modify s in such a way that at t_1 base station b_r serves w and at t_2 base station b_l serves u. This new solution is still feasible because first of all both the left and the right involved arrows a_l and a_r have become shorter. This implies that both a_l and a_r can only block fewer users. On the other hand, the head of a_l has moved left and the head of a_r has moved right. It is impossible that they are blocked now because of this movement: In t_1 this could only happen if there were some other arrows containing w, the new head of a_r. This arrow cannot come from the left, because then it would have blocked also the old arrow. It cannot come from b_r because b_r is busy. It cannot come from a base station to the right of b_r, because such arrows do not reach any point to the left of b_r (here we use the assumption that the rightmost user is no farther to the right of the rightmost base station than d, and that the base stations are evenly spaced). For t_2 the reasoning is symmetric. $\square$

The selection of arrows in any non-crossing solution can be completely characterized by a sequence of $m - 1$ *division points*, such that the ith division point is the index of the last user that is served from the left in the ith interval. (The case where all users in the ith interval are served from the right is handled by choosing the ith division point as the index of the rightmost user to the left of the interval, or as 0 if no such user exists.) A brute-force approach could now enumerate over all possible $\mathcal{O}(n^{m-1})$ division point sequences (*dps*) and color the selection of arrows corresponding to each dps with the greedy algorithm.

Dynamic Programming

We can solve the 1D-JBS problem with evenly spaced base stations more efficiently by a dynamic programming algorithm that runs in polynomial time. The idea of the algorithm is to consider the base stations and thus the intervals in left-to-right order. We consider the cost $\chi_i(d_{i-1}, d_i)$ of an optimal solution up to the ith base station conditioned on the position of the division points d_{i-1} and d_i in the intervals v_{i-1} and v_i, respectively, see Figure 2.2.

Definition 1. *We denote by $\chi_i(\alpha, \beta)$ the minimum number of colors needed to serve users u_1 to u_β using the base stations b_1 to b_i under the condition that base station b_i serves exactly users $u_{\alpha+1}$ to u_β and ignoring the users $u_{\beta+1}, \ldots, u_n$.*

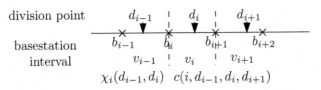

Fig. 2.2. Dynamic programming approach

Let $\Lambda(v_i)$ denote the set of potential division points for interval v_i, i.e., the set of the indices of users in v_i and of the rightmost user to the left of v_i (or 0 if no such user exists). The values $\chi_1(d_0, d_1)$ for $d_0 = 0$ (all users to the left of b_1 must be served by b_1 in any solution) and $d_1 \in \Lambda(v_1)$ can be computed directly by using the coloring algorithm from [7]. For $i \geq 1$, we compute the values $\chi_{i+1}(d_i, d_{i+1})$ for $d_i \in \Lambda(v_i)$, $d_{i+1} \in \Lambda(v_{i+1})$ from the table for $\chi_i(\cdot, \cdot)$. If we additionally fix a division point d_{i-1} for interval v_{i-1}, we know exactly which selected arrows intersect interval v_i regardless of the choice of other division points. Observe that this only holds for evenly spaced base stations and no "far out" users. For this selection, we can determine with the coloring algorithm from [7] how many colors are needed to color the arrows intersecting v_i. Let us call this number $c(i, d_{i-1}, d_i, d_{i+1})$ for interval v_i and division points d_{i-1}, d_i and d_{i+1}. We also know how many colors we need to color the arrows intersecting intervals v_0 to v_{i-1}. For a fixed choice of division points d_{i-1}, d_i and d_{i+1} we can combine the two colorings corresponding to $\chi_i(d_{i-1}, d_i)$ and $c(i, d_{i-1}, d_i, d_{i+1})$: Both of these colorings color all arrows of base station b_i, and these arrows must all have different colors in both colorings. No other arrows are colored by both colorings, so $\chi_i(d_{i-1}, d_i)$ and $c(i, d_{i-1}, d_i, d_{i+1})$ agree up to redefinition of colors. We can choose the best division point d_{i-1} and get

$$\chi_{i+1}(d_i, d_{i+1}) = \min_{d_{i-1} \in \Lambda(v_{i-1})} \max\{\chi_i(d_{i-1}, d_i), c(i, d_{i-1}, d_i, d_{i+1})\}$$

The running time is dominated by the calculation of the $c(\cdot)$ values. There are $\mathcal{O}(m \cdot n^3)$ such values, and each of them can be computed in time $\mathcal{O}(n \log n)$ using the coloring algorithm from [7]. The optimal solution can be found in the usual way by tracing back where the minimum was achieved from $\chi_m(x, n)$. Here the x is chosen among the users of the interval before the last base station such that $\chi_m(x, n)$ is minimum. For the traceback it is necessary to store in the computation of the χ values where the minimum was achieved. The traceback yields a sequence of division points that defines the selection of arrows that gives the optimal schedule. Altogether, we have shown the following theorem:

Theorem 2. *The base station scheduling problem for evenly spaced base stations can be solved in time* $\mathcal{O}(m \cdot n^4 \log n)$ *by dynamic programming.*

Note that the running time can also be bounded by $\mathcal{O}(m \cdot u_{\max}^4 \log u_{\max})$, where $u_{\max}$ is the maximum number of users in one interval.

2.3   Exact Algorithm for the k-Decision Problem

In this section we present an exact algorithm for the decision variant k-1D-JBS of the 1D-JBS problem: For given k and an instance of 1D-JBS, decide whether all users can be served in at most k rounds. We present an algorithm for this problem that runs in $\mathcal{O}(m \cdot n^{2k+1} \log n)$ time.

We use the result from Section 2.1 that arrow graphs are perfect. Thus the size of the maximum clique of an arrow graph equals its chromatic number.

The idea of the algorithm, which we call $A_{k-\text{JBS}}$, is to divide the problem into subproblems, one for each base station, and then combine the partial solutions to a global one.

For base station b_i, the corresponding subproblem S_i considers only arrows that intersect b_i and arrows for which the alternative user arrow[2] intersects b_i. Call this set of arrows A_i. We call S_{i-1} and S_{i+1} *neighbors* of S_i. A solution to S_i consists of a feasible selection of arrows from A_i of cost no more than k, i.e. the selection can be colored with at most k colors. To find all such solutions we enumerate all possible selections that can lead to a solution in k rounds. For S_i we store all such solutions $\{s_i^1, \ldots, s_i^l\}$ in a table T_i. We only need to consider selections in which at most $2k$ arrows intersect the base station b_i. All other selections need more than k rounds, because they must contain more than k arrows pointing in the same direction at b_i. Therefore, the number of entries of T_i is bounded by $\sum_{j=0}^{2k} \binom{n}{j} = \mathcal{O}(n^{2k})$. We need $\mathcal{O}(n \log n)$ time to evaluate a single selection with the coloring algorithm from [7]. Selections that cannot be colored with at most k colors are marked as irrelevant and ignored in the rest of the algorithm. We build up the global solution by choosing a set of feasible selections $s_1, \ldots, s_m$ in which all neighbors are compatible, i.e. they agree on the selection of common arrows. It is easy to see that in such a global solution all subsolutions are pairwise compatible.

We can find such a set of compatible neighbors by going through the tables in left-to-right order and marking every solution in each table as *valid* if there is a compatible, valid solution in the table of its left neighbor, or as *invalid* otherwise. A solution s_i marked as valid in table T_i thus indicates that there are solutions $s_1, \ldots, s_{i-1}$ in $T_1, \ldots, T_{i-1}$ that are compatible with it and pairwise compatible. In the leftmost table T_1, every feasible solution is marked as valid. When the marking has been done for the tables of base stations $b_1, \ldots, b_{i-1}$, we can perform the marking in the table T_i for b_i in time $\mathcal{O}(n^{2k+1})$ as follows. First, we go through all entries of the table T_{i-1} and, for each such entry, in time $\mathcal{O}(n)$ discard the part of the selection affecting pairs of user arrows that intersect only b_{i-1} but not b_i, and enter the remaining selection into an intermediate table $T_{i-1,i}$. The table $T_{i-1,i}$ stores entries for all selections of arrows from pairs of user arrows intersecting both b_{i-1} and b_i. An entry in $T_{i-1,i}$ is marked as valid if at least one valid entry from T_{i-1} has given rise to the entry. Then, the entries of T_i are considered one by one, and for each such entry s_i the algorithm looks

---

[2] For every user there are only two user arrows that we need to consider (Lemma 1). If we consider one of them, the other one is the *alternative user arrow*.

up in time $\mathcal{O}(n)$ the unique entry in $T_{i-1,i}$ that is compatible with s_i to see whether it is marked as valid or not, and marks the entry in T_i accordingly. If in the end the table T_m contains a solution marked as valid, a set of pairwise compatible solutions from all tables exists and can be retraced easily.

The overall running time of the algorithm is $\mathcal{O}(m \cdot n^{2k+1} \cdot \log n)$. There is a solution to k-1D-JBS if and only if the algorithm finds such a set of compatible neighbors. In the technical report [5] we give a formal proof of this statement.

Theorem 3. *Problem k-1D-JBS can be solved in time $\mathcal{O}(m \cdot n^{2k+1} \cdot \log n)$.*

2.4   Approximation Algorithm

In this section we present an approximation algorithm for 1D-JBS that relies on the properties of arrow graphs from Theorem 1. Let A denote the set of all user arrows of the given instance of 1D-JBS. From the perfectness of arrow graphs it follows that it is equivalent to ask for a feasible selection $A_{\mathrm{sel}} \subseteq A$ minimizing the chromatic number of its arrow graph $G(A_{\mathrm{sel}})$ (among all feasible selections) and to ask for a feasible selection A_{sel} minimizing the maximum clique size of $G(A_{\mathrm{sel}})$ (among all feasible selections). Exploiting this equivalence, we can express the 1D-JBS problem as an integer linear program as follows. We introduce two indicator variables l_i and r_i for every user i that indicate whether she is served by the left or by the right base station, i.e. if the user's left or right user arrow is selected. Moreover, we ensure by the constraints that no cliques in $G(A_{\mathrm{sel}})$ are large and that each user is served. The ILP formulation is as follows:

$$\min\ k \tag{2.1}$$

$$\text{s.t.}\ \sum_{l_i \in C} l_i + \sum_{r_i \in C} r_i \le k \qquad \forall\ \text{cliques } C \text{ in } G(A) \tag{2.2}$$

$$l_i + r_i = 1 \qquad \forall i \in \{1, \ldots, |U|\} \tag{2.3}$$

$$l_i, r_i \in \{0, 1\} \qquad \forall i \in \{1, \ldots, |U|\} \tag{2.4}$$

$$k \in \mathbb{N} \tag{2.5}$$

The natural LP relaxation is obtained by allowing $l_i, r_i \in [0, 1]$ and $k \ge 0$. Given a solution to this relaxation, we can use a rounding technique to get an assignment of users to base stations that has cost at most twice the optimum, i.e., we obtain a 2-approximation algorithm. Let us denote by *opt* the optimum number of colors needed to serve all users. Then $opt \ge k$, because the optimum integer solution is a feasible fractional solution. Construct now a feasible solution from a solution to the relaxed problem by rounding $l_i := \lfloor l_i + 0.5 \rfloor$, $r_i := 1 - l_i$. Before the rounding the size of every (fractional) clique is at most k; afterwards the size can double in the worst case. Therefore, the cost of the rounded solution is at most $2k \le 2opt$. We remark that there are examples where the cost of an optimal solution to the relaxed program is indeed smaller than the cost of an optimal integral solution by a factor of 2.

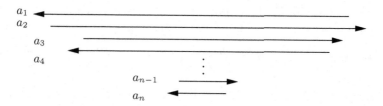

Fig. 2.3. Example of an arrow graph with an exponential number of maximum cliques. For every choice of arrows from a compatible pair (a_{2i-1}, a_{2i}) we get a clique of size $n/2$, which is maximum. The arrow graph can arise from a 1D-JBS instance with two base stations in the middle and $n/2$ users on either side

One issue that needs to be discussed is how the relaxation can be solved in time polynomial in n and m, as there can be an exponential number of constraints (2.2). (Figure 2.3 shows that this can really happen. The potentially exponential number of maximal cliques in arrow graphs distinguishes them from interval graphs, which have only a linear number of maximal cliques.) Fortunately, we can still solve such an LP in polynomial time with the ellipsoid method of Khachiyan [11] applied in a setting similar to [10]. This method only requires a separation oracle that provides us for any values of l_i, r_i with a violated constraint, if one exists. It is easy to check for a violation of constraints (2.3) and (2.4). For constraints (2.2), we need to check if for given values of l_i, r_i the maximum weighted clique in $G(A)$ is smaller than k. By Theorem 1 this can be done in time $\mathcal{O}(n \log n)$. Summarizing, we get the following theorem:

Theorem 4. *There is a polynomial-time 2-approximation algorithm for the 1D-JBS problem.*

3   General Case in the Plane—2D-JBS

We analyze the two-dimensional version (2D-JBS) of the base station scheduling problem. We show that the decision variant k-2D-JBS of the 2D-JBS problem is *NP*-complete and we present a constant factor approximation algorithm for a constrained version of it. The k-2D-JBS problem asks for a given k and an instance of 2D-JBS whether the users can be served in at most k rounds.

3.1   *NP*-Completeness of the 2D-JBS Problem

Here we briefly sketch our reduction from the general graph k-colorability problem [8] to 2D-JBS; the complete proof can be found in the technical report [5]. Our reduction follows the methodology presented in [9] for unit disk k-colorability.

Given any graph G, it is possible to construct in polynomial time a corresponding 2D-JBS instance that can be scheduled in k rounds if and only if G is k-colorable. We use an embedding of G into the plane which allows us to replace the edges of G with suitable base station chains with several users in a systematic way such that k-colorability is preserved. Our main result is the following:

Theorem 5. *The k-2D-JBS problem in the plane is NP-complete for any fixed $k \geq 3$.*

In the k-2D-JBS instances used in our reduction, the selection of the base station serving each user is uniquely defined by the construction. Hence, our reduction proves that already the coloring step of the 2D-JBS problem is *NP-complete*.

Corollary 1. *The coloring step of the k-2D-JBS problem is NP-complete for any fixed $k \geq 3$.*

3.2   Approximation Algorithms

Bounded Geometric Constraints. We consider instances where the base stations are at least a distance Δ from each other and have limited power to serve a user, i.e., every base station can serve only users that are at most $R_{\max}$ away from it. We also assume that for every user there is at least one base station that can reach the user. We present a simple algorithm achieving an approximation ratio depending only on the parameters Δ and $R_{\max}$.

Tiling the plane into a grid of squares of size $2R_{\max} \times 2R_{\max}$ and labelling the grid as in Figure 3.1 we get sets of squares S_a, S_b, S_c and S_d, where S_x is the set of squares with label x. We can place the grid in such a way that no base station lies on the boundary of a square. Note that if two base stations b_i and b_j are in different squares of the same label, their distance is greater than $2R_{\max}$ and, therefore, their transmissions cannot interfere. Now the algorithm proceeds as follows. While not all users are served, it goes in four steps through labels a, b, c and d. For each square of the current label, it repeatedly chooses an arbitrary base station from that square that can serve some user (i.e., the user

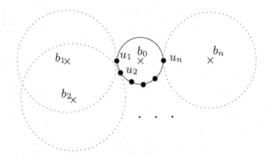

d	c	d	c	d
b	a	b	a	b
d	c	d	c	d
b	a	b	a	b
d	c	d	c	d

Fig. 3.1. Tiling of the plane into a grid of squares of size $2R_{\max} \times 2R_{\max}$ and labelling of the squares

Fig. 3.2. A greedy approach serves n users placed on a common interference disk in n time steps. An optimum algorithm can serve the users in one time step by assigning u_i to base station b_i, which lies on a halfline determined by b_0 and u_i

is at distance at most $R_{\max}$ from the base station and there is no interference at the user in the current round) and schedules the transmission from the chosen base station to that user in the current round. It keeps choosing base stations in the current square in this way as long as possible. If, after executing this step for all squares of the current label, there are still some unserved users, the algorithm proceeds to the next label and starts a new round.

We analyze the algorithm as follows. For every grid square s, let k_s denote the number of rounds in which a base station in s serves a user (in the solution computed by the algorithm). Let u_s be the number of users served by base stations in s. Note that $u_s \geq k_s$. Choose a square s for which k_s is maximum. It is clear that the solution of the algorithm uses at most $4k_s$ rounds, since the squares with the label of s are considered at least once every four rounds. Now we derive a lower bound on the number of rounds in the optimum solution. The u_s users served by the algorithm from base stations in s are contained in a square with side length $4R_{\max}$, as the maximum transmission radius is $R_{\max}$. The base stations that the optimum solution uses to serve these users must then be contained in a square of side length $6R_{\max}$ for the same reason. As disks of radius $\Delta/2$ centered at different base stations are interior-disjoint by assumption, an easy area argument shows that there can be at most $\rho = (6R_{\max} + \Delta)^2/\pi(\Delta/2)^2$ base stations in such a square. Therefore, even the optimal algorithm cannot serve more than ρ of the u_s users in one round. Hence, the optimum solution needs at least k_s/ρ rounds. This establishes the following theorem.

Theorem 6. *There exists an approximation algorithm with approximation ratio $\frac{16}{\pi}(\frac{6R_{\max}+\Delta}{\Delta})^2$ for 2D-JBS in the setting where any two base stations are at least Δ away from each other and every base station can serve only users within distance at most $R_{\max}$ from it.*

General 2D-JBS. In the technical report [5] we also discuss lower bounds on three natural greedy approaches for the general 2D-JBS problem: serve a maximum number of users in each round (*max-independent-set*), or repeatedly choose an interference disk of an unserved user with minimum radius (*smallest-disk-first*), or repeatedly choose an interference disk containing the fewest other unserved users (*fewest-users-in-disk*). In [5] we prove the following theorem.

Theorem 7. *There are instances (U, B) of 2D-JBS in general position (i.e., with no two users located on the same circle centered at a base station) for which the maximum-independent-set greedy algorithm, the smallest-disk-first greedy algorithm, and the fewest-users-in-disk greedy algorithm have approximation ratio $\Omega(\log n)$, where $n = |U|$.*

For instances of 2D-JBS that are not in general position, the smallest-disk-first greedy algorithm can have approximation ratio n, as shown in Figure 3.2.

4   Conclusion and Open Problems

In this paper we study the 1D- and 2D-JBS problems that arise in the context of coordinated scheduling in packet data systems. These problems can be split into a selection and a coloring problem. In the one-dimensional case, we have shown that the coloring problem leads to the class of arrow graphs, for which we have discussed its relation to other graph classes and algorithms. For the selection problem we propose an approach based on LP relaxation and rounding. For the 2D-problem, we have shown its NP-completeness. Several problems remain unsolved. In particular, it is open whether the 1D-JBS problem is *NP*-complete. For 2D-JBS it would be interesting to design approximation algorithms whose approximation ratio does not depend on the ratio $\frac{R_{\max}}{\Delta}$. Moreover, algorithmic results for more refined models would be interesting.

References

1. A. Brandstädt, V. B. Le, and J. P. Spinrad. *Graph classes: A survey.* SIAM Monographs on Discrete Mathematics and Applications. Society for Industrial and Applied Mathematics, Philadelphia, PA, 1999.
2. J. Chuzhoy and S. Naor. New hardness results for congestion minimization and machine scheduling. In *Proceedings of the 36th Annual ACM Symposium on the Theory of Computing (STOC'04)*, pages 28–34, 2004.
3. M. Cielibak, T. Erlebach, F. Hennecke, B. Weber, and P. Widmayer. Scheduling jobs on a minimum number of machines. In *Proceedings of the 3rd IFIP International Conference on Theoretical Computer Science*, pages 217–230. Kluwer, 2004.
4. S. Das, H. Viswanathan, and G. Rittenhouse. Dynamic load balancing through coordinated scheduling in packet data systems. In *Proceedings of Infocom'03*, 2003.
5. T. Erlebach, R. Jacob, M. Mihal̆ák, M. Nunkesser, G. Szabó, and P. Widmayer. Joint base station scheduling. Technical Report 461, ETH Zürich, Institute of Theoretical Computer Science, 2004.
6. T. Erlebach and F. C. R. Spieksma. Interval selection: Applications, algorithms, and lower bounds. *Algorithmica*, 46:27–53, 2001.
7. S. Felsner, R. Müller, and L. Wernisch. Trapezoid graphs and generalizations, geometry and algorithms. *Discrete Applied Mathematics*, 74:13–32, 1997.
8. M. R. Garey and D. S. Johnson. *Computers and Intractability.* Freeman, 1979.
9. A. Gräf, M. Stumpf, and G. Weißenfels. On coloring unit disk graphs. *Algorithmica*, 20(3):277–293, March 1998.
10. M. Grötschel, L. Lovász, and A. Schrijver. The ellipsoid method and its consequences in combinatorial optimization. *Combinatorica*, 1:169–197, 1981.
11. L. Khachiyan. A polynomial algorithm in linear programming. *Doklady Akademii Nauk SSSR*, 244:1093–1096, 1979.
12. F. C. R. Spieksma. On the approximability of an interval scheduling problem. *Journal of Scheduling*, 2:215–227, 1999.
13. J. P. Spinrad. *Efficient Graph Representations*, volume 19 of *Field Institute Monographs*. AMS, 2003.
14. D. West. *Introduction to Graph Theory.* Prentice Hall, 2nd edition, 2001.

Universal Bufferless Routing

Costas Busch[1], Malik Magdon-Ismail[1], and Marios Mavronicolas[2]

[1] Department of Computer Science,
Rensselaer Polytechnic Institute,
Troy, NY 12180, USA
[2] Department of Computer Science,
University of Cyprus,
Nicosia CY-1678, Cyprus

Abstract. Given an arbitrary network, and a routing problem with congestion C and dilation D, a long standing open problem is to show the existence of *bufferless* routing algorithms with optimal performance guarantees (routing time close to the lower bound $\Omega(C+D)$). Our main result is a new *deterministic* technique that constructs a universal bufferless algorithm by emulating a universal buffered algorithm. The heart of the emulation is to replace packet buffering with packet circulation on regions of the network. The cost of the emulation on the routing time is proportional to the square of the node buffer size used by the buffered algorithm. We apply this emulation to a simple randomized buffered algorithm to obtain a *distributed*, universal bufferless algorithm with routing time $O((C + D) \cdot \log^3(n + N))$, which is within poly-logarithmic factors from the optimal, where n is the size of the network and N is the number of packets. The *bufferless competitive ratio* is the ratio of the best achievable bufferless routing time, to the best achievable buffered routing time. We give the first non-trivial bound of $O(\log^3(n + N))$ for the bufferless competitive ratio for arbitrary routing problems.

1 Introduction

Packet Routing. has received a large amount of attention over the past decade on account of its importance to applications ranging from parallel and distributed algorithms to communication networks. The task is to deliver packets from their sources to their destinations along specified paths in a given network. A packet routing algorithm is *universal* if it can be applied to any routing problem on any network topology. For a given set of paths, the *routing time* (denoted rt) is the time at which the last packet reaches its destination. Universal algorithms with optimal or near-optimal routing time are known if packets may be buffered along their paths, [20, 23, 24].

A long standing and important open problem is to give universal *bufferless* routing algorithms with near optimal performance guarantees. In this paper, we will present a distributed bufferless routing algorithm that is optimal up to

G. Persiano and R. Solis-Oba (Eds.): WAOA 2004, LNCS 3351, pp. 239–252, 2005.

poly-logarithmic factors. We introduce a new technique for developing bufferless algorithms based upon emulating buffered algorithms. Applying this technique to a simple randomized buffered protocol gives the advertised result.

Preliminaries. A routing problem $Q = (G, \Pi, P)$ on the graph G with n nodes consists of a set of N packets $\Pi = \{\pi_1, \pi_2, \ldots, \pi_N\}$ that are to be routed on their respective paths $P = \{p_1, p_2, \ldots, p_N\}$, where p_i is a path in G. We will represent paths either as a sequence of edges, or as a sequence of nodes, and the length of a path $|p|$ is the number of edges in the path. The *edge-congestion* C is the maximum number of packets that use an edge in G, the *node-congestion* $\overline{C}$ is the maximum number of packets that use a node in G, and the *dilation* D is the maximum path length in P.

We assume a synchronous routing model, in which time is divided into a sequence of discrete time steps. An edge may be traveresed by at most one packet in either direction during a time step. A well known lower bound on the routing time in this model is given by $\Omega(C + D)$, and so the optimal routing time $rt^* = \Omega(C + D)$. In a buffered algorithm, packets may either traverse edges or be buffered at a node. In a bufferless algorithm, a packet must traverse an available edge at every time step.

An Impossibility Result. If all packets must follow the paths specified in P, without collisions or buffering, then the only degree of freedom for a bufferless routing algorithm is the injection times of the packets. Such a routing paradigm is known as *direct routing*, [3, 13]. In this case, it is shown in [13] that there exist routing problems for which bufferless routing times better than a $\sqrt{N}$ factor from optimal are not possible. Thus, if the paths remain unchanged, then near-optimal universal bufferless algorithms do not exist (where near optimal means within poly-logarithmic factors from the lower bound $C + D$). Thus, to obtain near-optimal bufferless schedules, we must allow packets to deviate from their paths. However, we still measure performance with respect to C and D of the original paths. The justification of this is that if the paths P themselves are optimal, i.e., they minimize $C + D$, then we obtain bufferless routing times that are near-optimal for the given sources and destinations. We do not discuss how to obtain the optimal paths, but rather how to send the packets to their destinations given the paths.

Contributions. Our main result is a deterministic technique for bufferless emulation of buffered algorithms. Given a near-optimal universal buffered algorithm that routes problems with simple paths, and uses buffers of size γ, we give a universal bufferless algorithm, which emulates the buffered algorithm. The cost of the emulation on the routing time is $O(\gamma^2 \cdot \log n)$.

We apply this emulation result to a simple randomized buffered algorithm that uses $O(\log(n + N))$ buffers to obtain a bufferless routing algorithm with routing time $O((C+D) \cdot \log^3(n+N))$ with high probability, which approximates within poly-logarithmic factors the optimal routing time for the given paths. If all the nodes know the network topology, and the values of C and N, then the bufferless algorithm is distributed, i,e., routing decisions are made locally at each node.

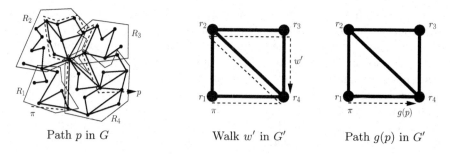

Path p in G — Walk w' in G' — Path $g(p)$ in G'

Fig. 1. An example of a region graph

Overview of the Approach. The main idea behind the bufferless emulation of a buffered algorithm is to use regions in the network in order to emulate buffer space. We decompose the graph into connected regions each containing approximately γ edges. The regions form a region graph, on which the nodes are regions. Now, a buffered algorithm executes as if the regions were the nodes. In the buffered algorithm, a packet is either buffered in a node (region) or "hops" from node to node. The path of each packet in the original graph is translated to a path on the region graph. The buffer needed at each node is at most γ. Figure 1 illustrates the general idea of decomposing the graph into regions and then mapping a packet's path to the graph in which every node corresponds to a region.

The buffered algorithm on the region graph is emulated by a bufferless algorithm on the original graph. If in the buffered algorithm a packet needs to be buffered in a node (region), then, in the emulation the packet "circulates" in the respective region by moving from one edge of the region to the next. A packet circulates until the buffered algorithm prescribes that the packet makes its next hop, in which case the packet moves to the respective adjacent region. Since the buffered algorithm requires γ buffer space per node (region), there is enough room to circulate all the packets in the γ edges of the region in a buffereless fashion.

Related Work. There are no previously known results for universal bufferless routing with near-optimal routing time guarantees. However, near-optimal bufferless routing has been obtained for specific bufferless routing models and architectures, which we summarize. In *hot-potato routing*, packets are deflected along available links in a collision [5]. Our model of bufferless routing is essentially the hot-potato routing model, with packets being deflected along particular available edges specified by the emulation (i.e., not on an arbitrary available edge as is typically done in hot-potato algorithms). Hot-potato routing algorithms have been extensively studied for a variety of architectures such as the mesh and torus [4, 6, 10, 12, 16, 19], hypercubes [9, 16, 18], trees [14, 25], vertex-symmetric networks [21], and leveled networks [8, 11]. Typically, by allowing packets to deviate from their paths slightly, one obtains routing times that are within polylogarithmic factors of optimal. In *direct routing*, packets follow their paths without buffering and without any collisions, [3, 13]. Wormhole routing is similar to direct routing, but here, packets occupy more than one edge [15, 17]. A dual to direct routing is *time constrained routing*, where the task is to schedule as many packets as possible within a given time frame [1]. In *matching routing*, packets are swapped at adjacent nodes, and permutation problems on trees have been studied in [2, 26].

There are two variants of buffered algorithms. Those that use buffers on every edge (*edge-buffers*) and those that use buffers in every node (*node-buffers*). For non-bounded degree networks, these variants are distinct. The existence of optimal, universal *buffered* routing algorithms using constant size edge-buffers was first established by Leighton, Maggs and Rao [20]. Thereafter, the main focus has been on constructive algorithms with optimal, $O(C + D)$, routing time, [7, 22, 23, 24]. These algorithms use large (proportional to the congestion C) buffers. Leighton *et al.* [20] improve this result, requiring only edge-buffers of size $O(\log ND)$ to obtain routing time $O(C + D \log ND)$. Cypher *at al.* [15] give an algorithm with edge-buffers of size $O(\log CD)$ and slightly better routing time. Our bufferless algorithm is based on emulating a universal buffered algorithm. However, the existing results, though powerful, do no suit our purpose because we need algorithms where the node-buffers are small (logarithmic), and so we offer a simple randomized algorithm that satisfies the conditions for bufferless emulation.

Paper Outline. We first discuss how to decompose a graph into connected regions of approximately a given size (Section 2). We then show how these regions are used for bufferless emulation of a buffered algorithm (Section 3). Finally we apply the emulation to a randomized buffered algorithm (Section 4) to obtain near-optimal universal bufferless routing (Section 5). We conclude with a discussion (Section 6). Due to lack of space, several proofs have moved to the full version of the paper.

2   Regions

We first discuss how to decompose a connected graph G into connected components of approximately a specified size. Such a decomposition will be required by the bufferless emulation algorithm. Specifically, let $G = (V, E)$ be an undirected connected graph. Let F be a subset of the edges in E. The subgraph induced by F is the graph $H = (U, F)$, where U is the union of all vertices in V that are incident with edges in F. We say that the edge set F is *connected* if the induced subgraph H is connected. A *connected decomposition* of G is a partition of the edges in E into disjoint sets $E_1, E_2, \ldots, E_k$ such that $\cup_{i=1}^{k} E_i = E$ and every E_i is connected. We refer to the E_i's as the *connected edge sets* or *regions* in the decomposition, and denote the number of edges in E_i as the size of E_i, $|E_i|$. Notice that the subgraphs, $H_1 = (V_1, E_1), \ldots, H_k = (V_k, E_K)$ induced by the edge sets may have overlapping vertex sets. We say that E_i is *connected to* E_j if and only if $V_i \cap V_j \neq \emptyset$. Notice that if E_i is connected to E_j, then $E_i \cup E_j$ is a connected edge set.

An $[\alpha, \beta]$-*partition* of G (if it exists) is a connected decomposition of G, $\{E_i, \ldots, E_k\}$, such that $\alpha \leq |E_i| \leq \beta$ for $i = 1, \ldots, k$. Notice that if $\alpha \approx \beta$, then an $[\alpha, \beta]$-partition decomposes G into connected edge sets of size approximately equal to α. We now show that such approximate decompositions are possible for any connected graph.

Theorem 1 (Existence of a $[k, 3k - 3]$-partition). *Let $G = (V, E)$ be a connected graph. For any k, where $1 < k \leq |E|$, there exists a $[k, 3k - 3]$-partition of G.*

The following example proves that the result of Theorem 1 is tight. For a given k, let G be any connected graph with $k - 2$ edges, and connect 3 such graphs in a wheel configuration as shown on the right. It is easy to see that the only decomposition in which every edge set has $\geq k$ edges is the entire graph itself, which has $3k - 3$ edges.

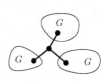

The proof in Theorem 1 is constructive, hence it can be directly converted to an algorithm. One can show that the time complexity of the algorithm to compute a decomposition is in $O(|E|^2)$.

Region Graph. Consider a connected graph $G = (V, E)$, with n nodes. Take an $[\alpha, \beta]$-partition of G, which gives regions (connected edge sets) $R_1, R_2, \ldots, R_k$. Let the subgraphs induced by these regions have vertex sets $U_1, U_2 \ldots, U_k$. The *region graph* $G' = (V', E')$, has a vertex set $V' = \{r_1, r_2, \ldots, r_k\}$ where each vertex r_i corresponds to the region R_i of G. Two vertices r_i, r_j are adjacent in G, i.e., $(r_i, r_j) \in E'$ if and only if $U_i \cap U_j \neq \emptyset$, i.e., the corresponding regions have intersecting vertex sets. An example of a region graph is given in Figure 1.

Routing Problems on Region Graph. Let $Q = (G, \Pi, P)$ denote a routing problem with edge-congestion C, node-congestion $\overline{C}$ and dilation D. Let $\{R_1, \ldots, R_k\}$ be an $[\alpha, \beta]$-partition of G. Every edge in G belongs to exactly one region. Let $G' = (V', E')$ be the corresponding region graph. The mapping $f : E \to V'$ is defined for every $e \in E$ by $f(e) = r_i$ if and only if $e \in R_i$. Consider a path $p \in P$, with $p = (e_1, e_2, \ldots, e_l)$. We define a function g which maps a path in G to a path in G' as follows. For any path $p = (e_1, e_2, \ldots, e_l)$ in G, consider the walk in G' given by $w' = (f(e_1), f(e_2), \ldots, f(e_l))$. $g(p)$ is the path obtained after removing all the cycles in w', $g(p) = (f(e_{i_1}), f(e_{i_2}), \ldots, f(e_{i_l}))$.

We now transform the routing problem Q on the original graph into a routing problem $Q' = (G', \Pi, P')$ on the region graph, in which the paths in G' are given by the transformed paths, $P' = \{p'_1, p'_2, \ldots, p'_N\}$ where $p'_i = g(p_i)$, $\forall p_i \in P$. Let $C', \overline{C'}$ and D' denote the edge-congestion, the node-congestion and the dilation of the paths in P'. For any routing problem, the edge-congestion is bounded by the node-congestion. A path uses node r_i only if it contains edges in R_i. By construction, $|R_i| \leq \beta$, so the number of edges in P that use R_i is at most βC, thus $\overline{C'} \leq \beta C$. Since $|g(p)| \leq |p|$ for any path p in G, we have the following lemma.

Lemma 1. $C' \leq \overline{C'} \leq \beta C$; $D' \leq D$.

Euler Tours on Regions. We define Euler tours with respect to the directed representation $G^D = (V, E^D)$ of the undirected graph G: each (undirected) edge $(u, v) \in E$ is replaced by two directed edges $(u, v), (v, u) \in E^D$. Let R_i^D denote the region of G^D that corresponds to the region R_i in G. Since the in-degree equals the out-degree of every node in R_i^D, R_i^D has an Euler tour. Let $\psi_i = (v_1, v_2, \ldots, v_1)$ denote an Euler tour in R_i^D. Note that ψ_i is walk in R_i. We will refer to ψ_i as the "Euler tour" of R_i (an abuse of notation, since ψ_i is not an Euler tour of R_i). Note that for an $[\alpha, \beta]$-partition of G, every Euler tour ψ_i satisfies $2\alpha \leq |\psi_i| \leq 2\beta$.

3   Emulation

Let $G = (V, E)$ be a connected graph with n nodes and let $\{R_1, \ldots, R_k\}$ be an $[\alpha, \beta]$-partition of G with corresponding region graph $G' = (V', E')$. For routing problem $Q = (G, \Pi, P)$ in G, we obtain the corresponding routing problem $Q' = (G', \Pi, P')$ in G'. Let (s_i, d_i) denote the source and destination of each packet $\pi_i \in \Pi$, and let $S = \{(s_1, d_1), (s_2, d_2), \ldots, (s_N, d_N)\}$. Let $Q_s = (G, \Pi, S)$ denote the routing problem in G in which the packets need to be delivered from their sources to their destination, without necessarily following the paths in P.

The general idea behind our approach is to design a bufferless routing Algorithm B to solve the routing problem Q_s. The bufferless algorithm will depend on a buffered Algorithm A to solve the routing problem Q' in G'. The bufferless algorithm will then *emulate* the running of Algorithm A in G' to solve Q_s in G.

3.1   Buffered Routing in G' – Algorithm A

Our bufferless algorithm in G will emulate a buffered algorithm A in G'. Algorithm A solves routing problem Q' in G' and uses node-buffers of size at most γ to do so. We require algorithm A to receive at most γ packets at every time step. It is then possible to divide the execution of Algorithm A, into a sequence of *phases*, in which each phase has the following two properties:

(i) Each phase is a fixed time period consisting of at least one time step;
(ii) During each phase, each packet traverses at most one edge in G', and each node receives at most γ packets from adjacent nodes or through injection.

A trivial division of the execution of Algorithm A into phases that satisfies these two properties is to take each phase to be a single time step. In Section 4, we give a specific buffered Algorithm A_1 in which each phase contains $O(\log(n + N))$ time steps. During a single phase of Algorithm A, a packet π may perform one of four actions (in G'):

(i) Remain in the buffer of its current node.   **[Buffering]**
(ii) Move from its current node to a neighboring node. **[Packet Transfer]**
(iii) Be injected into the network at its source node.   **[Injection]**
(iv) Move to and be absorbed in its destination node.   **[Absorbtion]**

3.2   Bufferless Routing in G – Algorithm B

Algorithm B emulates the phases of Algorithm A (which is faster than emulating the individual time steps of Algorithm A). Algorithm B emulates the buffering of packets and their transfer from node to node using an $[\alpha, \beta]$-partition of G, where $\alpha = 2\gamma$. (We assume that $2\gamma \leq |E|$ and by Theorem 1, we can set $\beta = 6\gamma - 3$.) In Algorithm A, when a packet is buffered in a node r_i of G', then Algorithm B emulates this by letting the packet circulate in the edges of region R_i in G. When in Algorithm A a packet is transferred from node r_i to node r_j of G', in Algorithm B the packet is transferred from region R_i to region R_j in G. Similarly, algorithm B handles the packet injection and absorbtion. Next we describe the emulation in more detail.

Phases and Rounds. Let Φ denote the number of phases in Algorithm A. In Algorithm B, time is divided into Φ phases. Each phase of B emulates a phase of A. In order to perform the emulation of a phase, Algorithm B further divides each phase into Σ *rounds*, where Σ is defined below. The duration of each round is $T_r = 4\beta^2 + 4\beta$ time steps. Thus the bufferless algorithm runs for $\Phi \cdot \Sigma \cdot T_r$ time steps in total.

For the duration of a round, a region is either in the *sending* or the *receiving* state – we say that the region is sending, or receiving. In the emulation, when a packet has to be transferred from one region to the next, the first region should be sending while the other receiving. We guarantee that for any pair of adjacent nodes there is a round in each phase in which one region is sending and the other is receiving (and vice-versa), as follows.

In order to determine if a region is sending or receiving, we first obtain a vertex coloring of G'. Let δ_i denote the color (non-negative integer in binary representation) assigned to node r_i in G' (which will also be the color of region R_i), and let δ denote the maximum color we obtain from the vertex coloring. Note that $\delta \leq n'$, where $n' = |V'| \leq |E|/\alpha$. Let σ denote the number of bits in δ, $\sigma = \lceil \log \delta \rceil \leq \lceil \log n' \rceil$. By pre-padding with zeros, we assume that every δ_i has σ bits. We define the *state parameter* $\mathbf{x}_i$ for region R_i to be the 2σ-bit integer $\bar{\delta}_i \delta_i$, where $\bar{\delta}_i$ is the binary complement of δ_i. We use the notation $\mathbf{x}_i(k)$ to denote the k-th bit of $\mathbf{x}_i$. We set $\Sigma = 2\sigma \leq 2\lceil \log n' \rceil$, i.e., each phase in Algorithm B, consists of 2σ rounds, $\omega_1, \omega_2, \ldots, \omega_{2\sigma}$. During round ω_k, if $\mathbf{x}_i(k) = 0$ then region R_i is sending, otherwise, if $\mathbf{x}_i(k) = 1$, then region R_i is receiving. Our assignment of colors ensures that during every phase, a region can send or receive from each of its neighbors.

Lemma 2. *If R_i and R_j are adjacent, then during every phase ϕ, there is at least one round ω_s (ω_r) in which R_i is sending (receiving) and R_j is receiving (sending).*

Proof. Since R_i and R_j are adjacent, δ_i and δ_j must differ at some bit s, $0 \leq s \leq \sigma - 1$. Thus, rounds s and $s + \sigma$ satisfy the requirements, since $\overline{\mathbf{x}_i(s + \sigma)} = \mathbf{x}_i(s) = \overline{\mathbf{x}_j(s)} = \mathbf{x}_j(s + \sigma)$.

Packet Circulation. Packet circulation is a basic function for the emulation. During packet circulation, a packet π repeatedly follows the Euler tour of the region R_i that it is in: at each time step, packet π follows the next edge in the Euler tour; when π reaches the end of the Euler tour it continues from the beginning of the tour, and so on. At the time step in which packet π traverses an edge $e \in \psi_i$, we say that e is the *current* edge of π.

At each round of a phase, a region is either sending or receiving. The speed at which a packet circulates in its region depends on whether the region is sending or receiving. If the region is receiving, then the packet follows the Euler tour in the normal fashion.

If the region is sending, then the packet moves at an effectively slower speed as follows. At time step 0 (the beginning of the round), suppose that π is at node u with current edge $e = (u, v) \in \psi_i$. At time step 0, packet π follows its current edge (u, v) and at time step 1, π appears in node v. At time step 1, suppose that its new current edge in ψ_i is (v, w); the packet *does not* follow its new current edge in ψ_i, but instead it follows edge (v, u) from v back to u, and thus at time step 2, it appears back in node u. Thus after two time steps, the packet has effectively not moved. We call such an operation an *oscillation*, and we say that packet π oscillates on its current edge in the Euler path. The time period of the oscillation is 2 time steps, The packet continues in this fashion for subsequent time steps, so at even time steps $t = 2i$, it appears in node u, and at odd time steps $t = 2i + 1$ it appears in node v, for $i \geq 0$. The packet performs β such oscillations on its current edge e, and so after 2β time steps, the packet appears at u and follows edge e for the last time. At time step $T_s = 2\beta + 1$, the packet is now at v and at this point it stops oscillating on edge e and begins oscillating on its new current edge $(v, w) \in \psi_i$. Thus, after T_s time steps, the packet advances by one edge in the Euler path of ψ_i. Consequently, since $|\psi_i| \leq 2\beta$, after $2\beta T_s = 4\beta^2 + 2\beta$ time steps, a packet circulating in region R_i has oscillated at least once on every edge of ψ_i.

Lemma 3. *After $4\beta^2 + 2\beta < T_r$ time steps, a packet circulating in a sending region R_i has oscillated at least once on every edge in ψ_i.*

Suppose that the directed edge $e = (u, v) \in \psi_i$, is an edge in the Euler path of a receiving region R_i. If at time step t, no packet has edge e as its current edge, then we say that e is *empty*. At each time step, we say that an empty edge is associated with an *empty slot*. Empty slots are similar to packets in that they too circulate – as the packets in a receiving region circulate (forward) in ψ_i, the empty slots circulate (backward) in ψ_i at the same rate. They continue to circulate until some packet occupies the empty edge.

Emulation of Buffering. Suppose that packet π is buffered at node r_i of G' during the execution of phase ϕ of Algorithm A. Assume that in Algorithm B, packet π is in region R_i of G. Packet π will circulate in R_i through the entire phase ϕ.

Lemma 4. *If packet π is in R_i at the end of phase $\phi - 1$ of bufferless Algorithm B, and in phase ϕ of buffered Algorithm A it is buffered in node r_i, then in phase ϕ of bufferless Algorithm B, it can be buffered in region R_i using circulation.*

Emulation of Packet Transfer. Suppose that in phase ϕ of Algorithm A, packet π moves from node r_i to node r_j. Assume that at the beginning of phase ϕ in Algorithm B, packet π is in region R_i. During phase ϕ in Algorithm B, π will move from R_i to R_j as follows. Packet π will circulate in R_i until a round ω of ϕ in which R_i is sending and R_j is receiving (the existence of such a round is guaranteed by Lemma 2).

Since r_i and r_j are adjacent in G', there exists a node u which is common to R_i and R_j. Since node u is in R_i, there exists an edge $e_i = (u_i, u) \in \psi_i$ on the Euler path of R_i. Similarly, there exists an edge $e_j = (u, u_j) \in \psi_j$ on the Euler tour of R_j. During round ω, packet π circulates (in slow mode) in region R_i along the Euler tour ψ_i. At some particular slow time step τ of the round, the current edge of π will be e_i. During the course of its $T_s > \beta$ oscillations on edge e_i, the packet will appear at the common node u at the $\beta + 1$ times $\tau + 1, \tau + 3, \ldots, \tau + 2\beta + 1$. If at any of these times, the edge $e_j \in \psi_j$ is an empty slot, i.e., not the current edge of any packet circulating (in normal mode) in R_j, then π switches from oscillation on edge e_i, making e_j its new current edge. π now continues to circulate in R_j at normal speed. Note that π will have completed its circulation on edge e_i in at most $4\beta^2 + 2\beta$ time steps, thus π will enter R_j within the first $4\beta^2 + 2\beta$ time steps of round ω.

We now show that during round ω, for at least one of the time steps $\tau + 1, \tau + 3, \ldots, \tau + 2\beta + 1$, the edge $e_j \in \psi_j$ will be an empty slot. Remember that empty slots circulate in R_j at the rate of one edge per time-step. Thus, if an empty slot is not occupied by any packet during its circulation, then every edge in ψ_j will become an empty slot at least once during a consecutive 2β time steps. In particular, edge e_j will become an empty slot at least once in the time steps $\tau + 1, \tau + 2, \tau + 3, \ldots, \tau + 2\beta + 1$. A problem arises if e_j becomes empty at time $\tau + k$ where k is even, because then packet π will not be at node u, able to utilize this edge. This problem is solved if there is a second *consecutive* empty slot in R_j that will also not be occupied by any other packet during its circulation. This second empty slot must also appear at least once in the time steps $\tau + 1, \tau + 2, \tau + 3, \ldots, \tau + 2\beta + 1$, and since both these empty slots cannot appear at $\tau + k$ for k even, we are assured that π will be able to transfer into R_j.

From the previous phase, suppose that there are at most γ packets circulating in R_j. During the current phase, at most γ more packets will enter R_j, by definition of the buffered Algorithm A. In the worst case, all the $\gamma - 1$ packets other than π that will enter have already entered, and none of the packets that are to leave this region in this phase have left yet. In this case there are at most $2\gamma - 1$ packets that could be circulating in R_j during round ω. Since $\alpha = 2\gamma$ and there are at least in $2\alpha = 4\gamma$ edges ψ_j, we conclude that there are at least $2\gamma + 1$ empty slots during round ω. By the pigeonhole principle, at least two of these empty slots must be consecutive, and we have the following lemma.

Lemma 5. *Suppose that in phase $\phi - 1$ of bufferless Algorithm B, at most γ packets are circulating in region R_j, and that packet π is circulating in the adjacent region R_i. Suppose that in buffered Algorithm A, packet π moves from*

r_i to r_j in phase ϕ. Then during phase ϕ of bufferless Algorithm B, packet π can be transferred (using circulation) from region R_i to R_j.

Emulation of Injection. Suppose that π is a packet that is to be injected into the network in Algorithm A. Let p be the path of π in G, and let e be the first edge in this path, and u the injection node. Suppose that $e \in R_i$ – note also that $u \in R_i$. In this case, π is injected into node r_i in G'. Suppose that π is injected into r_i during phase ϕ of buffered Algorithm A. Then π will be injected into R_i in phase ϕ of bufferless Algorithm B during the last round in which R_i is receiving. After injection, it will circulate in R_i until the end of phase ϕ. Let $e = (u, v)$ be an edge on the Euler path ψ_i of R_i. We know that from the previous analysis of packet transfer that if R_i had at most γ packets circulating in phase $\phi - 1$, then e will be an empty slot at least $2\gamma + 1$ times during every receiving round. At the time that e becomes empty, π is injected into the network and e becomes its current edge. π then continues to circulate in R_i. Note that at least γ packets could be injected into R_i from the *same* injection node during a single receiving round.

Lemma 6. *Suppose that in phase $\phi - 1$ of bufferless Algorithm B, at most γ packets are circulating in region R_i. Suppose that packet π has first edge $e \in R_i$ and that during phase ϕ of buffered Algorithm A, packet π is injected into node r_i. Then during phase ϕ of bufferless Algorithm B, packet π can be injected into R_i. Further, at least γ packets can be injected into the same node during a single receiving round.*

Emulation of Absorbtion. Suppose that packet π moves from node r_i to its destination node r_j in phase ϕ in buffered Algorithm A. We use the packet transfer emulation to first move the packet from region R_i to R_j in phase ϕ. This takes at most $4\beta^2 + 2\beta$ time steps. Then the packet circulates in the receiving region at normal speed until it reaches its destination node, at which point it is absorbed. Since the packet completes the Euler tour for R_j in at most 2β time steps, the number of time steps to move and be absorbed is $4\beta^2 + 4\beta \leq T_r$, giving the following lemma.

Lemma 7. *Suppose that in phase $\phi - 1$ of bufferless Algorithm B, at most γ packets are circulating in region R_j, and that packet π is circulating in the adjacent region R_i. Suppose that in phase ϕ of buffered Algorithm A, packet π is absorbed in r_j. Then, during phase ϕ of bufferless Algorithm B, packet π can be absorbed at its destination node in region R_j.*

3.3   Analysis of Emulation by Bufferless Algorithm B

First, we prove that Algorithm B correctly emulates Algorithm A. We then analyse the routing time of Algorithm B in G in terms of the routing time of Algorithm A in G'.

Correctness. Assume that $\alpha = 2\gamma \leq |E|$ in order to guarantee the existence of the $[\alpha, \beta]$-partition. Algorithm B correctly emulates algorithm A if at the end of every phase ϕ:

i. In Algorithm A, packet π is in node r_i iff in Algorithm B it is circulating in region R_i
ii. In algorithm A packet π is injected (absorbed) at node r_i, if and only if in Algorithm B packet π is injected (absorbed) into region R_i.

We show by induction on ϕ that Algorithm B correctly emulates Algorithm A. Observe that when $\phi = 1$, Algorithm A can only inject packets into nodes. The conditions of Lemma 6 are satisfied, and since at most γ packets are injected into a node in G', Algorithm B can succesfully inject these packets into the corresponding regions. Suppose that Algorithm B correctly emulates Algorithm A up to phase $\phi_0 \geq 1$. At the end of phase ϕ_0, there are at most γ packets circulating in any region R_i since every packet π in node r_i in the execution of Algorithm A is in region R_i in the execution of Algorithm B. Thus, the conditions of Lemmas 4, 5, 6, and 7 are satisfied for every packet π. Every action that π could make in phase $\phi_0 + 1$ of Algorithm A can now be emulated in phase $\phi_0 + 1$ of Algorithm B. By induction, we have the following theorem.

Theorem 2 (Correctness of Emulation). *Algorithm B correctly emulates in G every phase in the execution of Algorithm A in G'. Each packet in Algorithm B follows a path from its source to destination, hence Algorithm B solves routing problem Q_s without buffers.*

Routing Time. Let $rt_B(Q_s)$ be the routing time for Algorithm B to solve routing problem Q_s. Let $\Phi_A(Q')$ be the number of phases used by Algorithm A to solve routing problem Q'. Since Algorithm B emulates Algorithm A phase for phase, the number of phases of algorithm B is also $\Phi_A(Q')$. The routing time is therefore given by $\Phi_A \cdot \Sigma \cdot T_r$. Since $T_r = 4\beta^2 + 4\beta$, $\beta = 6\gamma - 3$ and $\Sigma = 2\lceil \log \delta \rceil$, we obtain:

Theorem 3 (Routing Time of Emulation). $rt_B(Q_s) = \Theta(\Phi_A(Q') \cdot \gamma^2 \cdot \log \delta)$.

Since $\delta \leq |E|/\alpha = O(n^2)$, from Theorem 3 we obtain an alternative upper bound on the routing time: $rt_B(Q_s) = O(\Phi_A(Q') \cdot \gamma^2 \cdot \log n)$.

4   A Randomized Buffered Algorithm

We give a buffered algorithm that can be used to obtaining bufferless routing on arbitrary networks. Since the per-node buffer size enters into the routing time of the bufferless emulation, it is necessary to have buffered algorithms that limit the amount of per-node buffering. We refer to this algorithm as Algorithm A_1.

Algorithm A_1 is a randomized routing algorithm for routing porblems with simple paths, in arbitrary networks. Let $Q' = (G', \Pi, P')$ be a routing problem

---

Algorithm 1 Buffered Algorithm A_1

1: Divide time into phases of length γ time steps.
2: **for** Each packet π **do**
3:     π selects uniformly at random an injection phase ϕ_π between phases 1 and $12\overline{C'}/\gamma$;
4:     Packet π is injected at the first time step of phase ϕ_π;
5:     Packet π follows its path at the speed of one edge per phase;

---

with acyclic paths P' on an arbitrary graph $G' = (V', E')$. Let $\overline{C'}$ be the node-congestion and D' the dilation. Let N be the number of packets and n' the size of V'. Algorithm A_1 uses buffers of size $\gamma = 6\log(n' + 2N)$.

We show that with high probability, Algorithm A_1 succesfully routes the packets, and at the same time satisfies the requirements in Section 3.1.

Theorem 4 (Routing Time of Algorithm A_1). *With probability at least* $1 - O(1/(n' + 2N))$, *Algorithm* A_1 *solves routing problem* Q' *in at most* $12\overline{C'}/\gamma + D'$ *phases. The node-buffer size required is* $\gamma = 6\log(n' + 2N)$.

5   A Universal Bufferless Algorithm

We use buffered Algorithm A_1 to construct bufferless Algorithm B_1 for arbitrary networks. Algorithm B_1 emulates Algorithm A_1. The buffer size used by algorithm A_1 is $\gamma = 6\log(n' + 2N)$. Since $n' \le |E|/\alpha$, in order to guarantee the existence of an $[\alpha, \beta]$-partition, we assume that $\alpha \le |E|$. Since $\alpha = 2\gamma$, we assume that $12\log(|E|/\alpha + 2N) \le |E|$. It is sufficient that $2N \le 2^{|E|/12} - |E|$.

Suppose $2N \le 2^{|E|/12} - |E|$. Since $n' \le n^2/2$, $\gamma \le 6\log(n^2/2 + 2N)$, independent of G'. Combining Theorems 3 and 4, and the fact that in the emulation, $\Phi_{B_1}(Q_s) = \Phi_{A_1}(Q')$, we obtain that $rt_{B_1}(Q_s) = O((12\overline{C'}/\gamma + D') \cdot \log \delta \cdot \log^2(n + 2N))$. Using Lemma 1 and the facts that $\beta \le 6\gamma$ and $\delta \le n' = O(n^2)$, we obtain that $rt_{B_1}(Q_s) = O((C + D) \cdot \log n \cdot \log^2(n + N))$, with probability at least $1 - O(1/(n' + N))$.

Consider now the case when $2N \ge 2^{|E|/12} - |E|$. We can send the N packets of routing problem Q_s on G to their destinations one after the other. Each packet takes time $O(D)$ to be delivered to its destination, and thus the total routing time to send all the packets is $O(DN)$. Clearly, $C \ge N/|E|$, and thus $C \ge (2^{|E|/12} - |E|)/2|E|$. Since $|E| = O(\log(N))$ and $D \le |E|$, the routing time is $ND \le CD|E| = O(C\log^2(N))$. This simple algorithm can easily be converted to a distributed algorithm with the same routing time.

Combining the above results for both cases of the number of packets, we obtain the main result of this paper:

Theorem 5 (Routing Time of Buffereless Algorithm). $rt_{B_1}(Q_s) = O((C + D) \cdot \log n \cdot \log^2(n + N))$, *with probability at least* $1 - O(1/(n' + N))$.

6 Discussion

We have presented a distributed algorithm for routing packets in bufferless networks. Our algorithm is based on the emulation of algorithms with buffers. We partition the original graph into regions, and construct a respective region graph. Each region serves the purpose of a buffer. We then consider an algorithm with buffers on the region graph, and emulate this algorithm by circulating the packets in the regions, and thus avoiding the need of buffers. With this technique, the resulting routing time of our algorithm is $O\left((C + D) \cdot \log^3(n + N)\right)$, which is poly-logarithmic factors away from the optimal for the given paths.

For a particular (source-destination) routing problem, we can define the *bufferless competitive ratio* as the ratio between the best possible routing time of a bufferless algorithm and the best possible routing time of a buffered algorithm. Our result shows that the bufferless competitive ratio is at most $\log^3(n + N)$ for *any* routing problem. A interesting open problem is to improve this bound.

References

1. Micah Adler, Sanjeev Khanna, Rajmohan Rajaraman, and Adi Rosen. Time-constrained scheduling of weighted packets on trees and meshes. In *Proceedings of 11th ACM Symposium on Parallel Algorithms and Architectures (SPAA)*, pages 1–12, 1999.
2. N. Alon, F.R.K. Chung, and R.L.Graham. Routing permutations on graphs via matching. *SIAM Journal on Discrete Mathematics*, 7(3):513–530, 1994.
3. Stephen Alstrup, Jacob Holm, Kristian de Lichtenberg, and Mikkel Thorup. Direct routing on trees. In *Proceedings of the Ninth Annual ACM-SIAM Symposium on Discrete Algorithms (SODA 98)*, pages 342–349, 1998.
4. A. Bar-Noy, P. Raghavan, B. Schieber, and H. Tamaki. Fast deflection routing for packets and worms. In *Proceedings of the Twelth Annual ACM Symposium on Principles of Distributed Computing*, pages 75–86, Ithaca, New York, USA, August 1993.
5. P. Baran. On distributed communications networks. *IEEE Transactions on Communications*, pages 1–9, 1964.
6. A. Ben-Dor, S. Halevi, and A. Schuster. Potential function analysis of greedy hot-potato routing. *Theory of Computing Systems*, 31(1):41–61, January/February 1998.
7. Petra Berenbrink and Christian Scheideler. Locally efficient on-line strategies for routing packets along fixed paths. In *Proceedings of the Tenth Annual ACM-SIAM Symposium on Discrete Algorithms*, pages 112–121, N.Y., January 17–19 1999. ACM-SIAM.
8. Sandeep N. Bhatt, Gianfranco Bilardi, Geppino Pucci, Abhiram G. Ranade, Arnold L. Rosenberg, and Eric J. Schwabe. On bufferless routing of variable-length message in leveled networks. *IEEE Trans. Comput.*, 45:714–729, 1996.
9. J. T. Brassil and R. L. Cruz. Bounds on maximum delay in networks with deflection routing. *IEEE Transactions on Parallel and Distributed Systems*, 6(7):724–732, July 1995.
10. A. Broder and E. Upfal. Dynamic deflection routing on arrays. In *Proceedings of the Twenty-Eighth Annual ACM Symposium on the Theory of Computing*, pages 348–358, May 1996.

11. C. Busch. Õ(Congestion + Dilation) hot-potato routing on leveled networks. In *Proceedings of the Fourteenth ACM Symposium on Parallel Algorithms and Architectures*, pages 20–29, August 2002.
12. C. Busch, M. Herlihy, and R. Wattenhofer. Hard-potato routing. In *Proceedings of the 32nd Annual ACM Symposium on Theory of Computing*, pages 278–285, May 2000.
13. Costas Busch, Malik Magdon-Ismail, Marios Mavronicolas, and Paul Spirakis. Direct routing: algorithms and complexity. In *Proceedings of the 12th Annual European Symposium on Algorithms ESA 2004*, Bergen, Norway, September 2004.
14. Costas Busch, Malik Magdon-Ismail, Marios Mavronicolas, and Roger Wattenhofer. Near-optimal hot-potato routing on trees. In *Proceedings of the 10th International Conference on Parallel and Distributed Computing (Euro-par)*, pages 820–827, September 2004.
15. Robert Cypher, Friedhelm Meyer auf der Heide, Christian Scheideler, and Berthold Vöcking. Universal algorithms for store-and-forward and wormhole routing. In *Proceedings of the 28th ACM Symp. on Theory of Computing (STOC)*, pages 356–365, 1996.
16. U. Feige and P. Raghavan. Exact analysis of hot-potato routing. In IEEE, editor, *Proceedings of the 33rd Annual Symposium on Foundations of Computer Science*, pages 553–562, Pittsburgh, PN, October 1992.
17. Ronald I. Greenberg and Hyeong-Cheol Oh. Universal wormhole routing. *IEEE Transactions on Parallel and Distributed Systems*, 8(3):254–262, 1997.
18. B. Hajek. Bounds on evacuation time for deflection routing. *Distributed Computing*, 1:1–6, 1991.
19. Ch. Kaklamanis, D. Krizanc, and S. Rao. Hot-potato routing on processor arrays. In *Proceedings of the 5th Annual ACM Symposium on Parallel Algorithms and Architectures*, pages 273–282, Velen, Germany, June 30–July 2, 1993.
20. F. T. Leighton, B. M. Maggs, and S. B. Rao. Packet routing and job-scheduling in $O(congestion + dilation)$ steps. *Combinatorica*, 14:167–186, 1994. (preliminary version appears in FOCS 1988).
21. Friedhelm Meyer auf der Heide and Christian Scheideler. Routing with bounded buffers and hot-potato routing in vertex-symmetric networks. In *Proceedings of the Third Annual European Symposium on Algorithms*, volume 979 of *LNCS*, pages 341–354, Corfu, Greece, 25–27 September 1995.
22. Friedhelm Meyer auf der Heide and Berthold Vöcking. Shortest-path routing in arbitrary networks. *Journal of Algorithms*, 31(1):105–131, April 1999.
23. Rafail Ostrovsky and Yuval Rabani. Universal $O(congestion+dilation+\log^{1+\varepsilon} N)$ local control packet switching algorithms. In *Proceedings of the 29th Annual ACM Symposium on the Theory of Computing*, pages 644–653, New York, May 1997.
24. Yuval Rabani and Éva Tardos. Distributed packet switching in arbitrary networks. In *Proceedings of the Twenty-Eighth Annual ACM Symposium on the Theory of Computing*, pages 366–375, Philadelphia, Pennsylvania, 22–24 May 1996.
25. Alan Roberts, Antonios Symvonis, and David R. Wood. Lower bounds for hot-potato permutation routing on trees. In *Proceedings of the 7th Int. Coll. Structural Information and Communication Complexity, SIROCCO*, pages 281–295, 20–22 June 2000.
26. L. Zhang. Optimal bounds for matching routing on trees. In *Proceedings of the 8th Annual ACM-SIAM Symposium on Discrete Algorithms*, pages 445–453, 1997.

Strong Colorings of Hypergraphs

Geir Agnarsson[1] and Magnús M. Halldórsson[2]

[1] Department of Mathematical Sciences, George Mason University,
MS 3F2, 4400 University Drive, Fairfax, VA 22030
geir@math.gmu.edu
[2] Department of Computer Science, University of Iceland,
Dunhaga 3, IS-107 Rvk, Iceland
mmh@hi.is

Abstract. A *strong vertex coloring* of a hypergraph assigns distinct colors to vertices that are contained in a common hyperedge. This captures many previously studied graph coloring problems. We present nearly tight upper and lower bound on approximating general hypergraphs, both offline and online. We then consider various parameters that make coloring easier, and give a unified treatment. In particular, we give an algebraic scheme using integer programming to color graphs of bounded composition-width.

Keywords: hypergraph, strong coloring, approximation, composition width.

1   Introduction

The purpose of this article is to discuss the properties of a special kind of vertex coloring of hypergraphs, where we insist on vertices that are contained in a common hyperedge receiving distinct colors. Such *strong colorings* capture a number of graph coloring problems that have been treated separately before.

A *hypergraph* $H = (V, \mathcal{E})$ consists of a finite set $V = V(H)$ of vertices and a collection $\mathcal{E} = \mathcal{E}(H) \subseteq \mathbb{P}(V)$ of subsets of V. A *strong coloring* of H is a map $\Psi : V(H) \to \mathbb{N}$ such that whenever $u, v \in e$ for some $e \in \mathcal{E}(H)$, we have that $\Psi(u) \neq \Psi(v)$. The corresponding *strong chromatic number* $\chi_s(H)$ is the least number of colors for which H has a proper strong coloring.

Strong coloring can be viewed as a regular vertex coloring problem of the *clique graph* $G_c(H)$ (also known as *2-section graph* or *representing graph* [6]) of the hypergraph H, defined on the same set of vertices, with edge set $E(G_c(H)) = \{\{u, v\} : u, v \in e \text{ for some } e \in \mathcal{E}(H)\}$. In this way, $\chi_s(H) = \chi(G_c(H))$, the ordinary chromatic number of the clique graph.

We consider both online and offline coloring algorithms. We analyze them in terms of their *competitiveness* or *approximation factor*, respectively, which in both cases is the maximum ratio between the number of colors used by the algorithm on an instance to the chromatic number of the instance. In the standard online graph coloring problem, the graph is presented one vertex at a time

G. Persiano and R. Solis-Oba (Eds.): WAOA 2004, LNCS 3351, pp. 253–266, 2005.

along with edges only to the previous vertices. Each time the algorithm receives a vertex, it must make an irrevocable decision as to its color.

Since the hypergraph H is our original input, we would like to use all the associated parameters that come with it, and use those to either obtain an optimal strong coloring or a good approximation. This is relevant because it is often difficult to deduce the hypergraph from the graph representation, or to find the "best" such hypergraph. Finding the smallest clique hypergraph H that is equivalent to a graph G, i.e. such that $G_c(H) = G$, is equivalent to finding the smallest Clique Cover (GT 17: Covering by Cliques in [13]), which is hard to approximate within $n^{2-\epsilon}$ factor, for any $\epsilon > 0$ [12].

One of the main objectives of this research is in opening a new line of research by unifying several coloring problems as strong coloring appropriate types of hypergraphs. We have gathered a host of results on these problems by modifying and sometimes slightly extending previous results on graph coloring. Finally, we have made the first step into a systematic treatment of parameters of hypergraphs and their clique graphs that make solution or approximation easier.

1.1   Instances of Strong Colorings Problems

Down-Coloring DAGs. Our original motivation to study strong hypergraph colorings stems from a digraph coloring problem that occurs when bounding storage space in genetic databases. A *down-coloring* of a DAG (acyclic digraph) $\overline{G}$ is coloring of the vertices so that vertices that share a common ancestor receive different colors. One motivation for such a coloring (see [2]) is to provide an efficient structure for querying relational tables referencing the digraph, including the retrieval of rows in a given table that are conditioned based on sets of ancestors from $\overline{G}$.

For a DAG $\overline{G}$, the binary relation $\leq$ on $V(\overline{G})$ defined by $u \leq v \Leftrightarrow u = v$, or there is a directed path from v to u in $\overline{G}$, is reflexive, antisymmetric, and transitive, and is therefore a partial order on $V(\overline{G})$. We denote by $\max\{\overline{G}\}$ the set of source vertices of $\overline{G}$, i.e., the maximal vertices with respect to this partial order $\leq$. For vertices $u, v \in V(\overline{G})$ with $u \leq v$, we say that u is a *descendant* of v. The *down-set* $D[u]$ of a vertex $u \in V(\overline{G})$ is the set of all descendants of u in $\overline{G}$, that is, $D[u] = \{x \in V(\overline{G}) : x \leq u\}$. As defined in [3], the *down-hypergraph* $H_{\overline{G}}$ of a DAG $\overline{G}$ contains the same set of vertices with the down-sets $\mathcal{E}(H_{\overline{G}}) = \{D[u] : u \in \max\{\overline{G}\}\}$ of sources of $\overline{G}$ as hyperedges. A down-coloring of a digraph corresponds to a strong coloring of the corresponding down-hypergraph.

Note that not all hypergraphs are down-hypergraphs of a DAG, but they are easily recognized: We say that a hypergraph has the *unique element property* if each hyperedge contains a vertex not contained in any other hyperedge. We can observe that a hypergraph H is a down-hypergraph of some DAG iff H has the unique element property [3].

Further properties between graphs, hypergraphs and the posets yielding them can be found in [30], where the corresponding clique graph associated with the poset is called an *upper bound graph* of the poset.

Distance-2 Coloring Graphs. A *distance-2 coloring* of a graph is a vertex coloring where vertices at distance two or less must receive different colors. This problem has received attention for applications in frequency allocation [37, 26], where two stations must use a different frequency if they are both to be able to communicate with a common neighbor. Another application given in [29] relates to the partition of the columns of a matrix for parallel solution so that columns solved in the same iteration do not share a non-zero element in the same row.

The *neighborhood hypergraph* N_G of a graph G consists of the same vertex set with a hyperedge consisting of the closed neighborhood $N[v] = \{u : u = v$ or $\{u, v\} \in E(G)\}$ of each vertex $v \in V(G)$. A strong coloring of N_G is equivalent to a distance-2 coloring of G.

The distance-2 coloring problem is also equivalent to an ordinary coloring problem on the *square graph* G^2 of the graph G. The k-th power G^k of a graph G is a graph on the same vertex set, with an edge between any pair of vertices of distance at most k in G. The square graph is indeed the clique graph of the neighborhood hypergraph. While it is easy to compute the power graph G^k from G, Motwani and Sudan [34] showed that it is NP-hard to compute the k-th root G of a graph G^k, for any $k \geq 2$. On the other hand, it is not hard to deduce the original graph when given its neighborhood hypergraph.

McCormick [29] was the first to show that the problem of coloring the power of a graph is NP-complete, for any fixed power. He gave a greedy algorithm with a $O(\sqrt{n})$-approximation for squares of general graphs, which was matched by the NP-hardness of an $\Omega(n^{1/2-\epsilon})$-approximation, for any $\epsilon > 0$ [4]. Several recent papers have studied distance-2 coloring planar graphs [26, 5], for which the current best upper bound is $1.66\Delta(G) + O(1)$ colors due to Molloy and Salavatipour [33].

1.2   Related Coloring Results

The best current upper bound for approximating ordinary graph coloring is $O(n(\lg \lg n / \lg n)^3)$ [15], while it is hard to approximate within $n^{1-\epsilon}$ factor, for any $\epsilon > 0$ [7]. For a survey on graph coloring approximations, see [36].

The *weak* hypergraph coloring problem is an alternative generalization of the graph coloring problem, where the vertices are to be colored so that no hyperedge is monochromatic. Several results are known about such approximations, including a $\Omega(n^{1-\epsilon})$ hardness [25].

Each color class in a strong coloring is called a *strong independent set* (strong stable set). The k-set packing problem is equivalent to the strong independent set problem in degree-k hypergraphs, by looking at the dual graph. This is NP-hard to approximate within factor $O(k/\lg k)$ [20]. This suggests, but does not guarantee, that coloring degree-k hypergraphs is hard to do within an asymptotic factor much smaller than k.

Strong coloring a hypergraph H is also equivalent to *edge coloring* the dual hypergraph H^*. Kahn [23] showed that $\chi'(H) \leq \Delta + o(\Delta)$, if no two hyperedges share many vertices. Further improvements were obtained by Molloy and Reed [32].

1.3   Overview of Paper

In the following section, we introduce the general version of the problem, which involves multicolorings, where a set of colors is to be assigned to each vertex. This follows naturally from some preprocessing of the hypergraph instance. In Section 3, we consider several parameters of graphs and hypergraphs and analyze their effect on approximability. In Section 4, we give bounds for online and offline strong coloring algorithms on general graphs. Finally, in Section 5, we present the technically most involved part of the paper, with a polynomial time coloring algorithm for the class of k-composite graphs.

2   Hypergraph Contractions and Multicolorings

To describe the strong coloring problem in its full generality, we must introduce *multicolorings*.

Multicoloring. For a simple graph G let $\nu : V(G) \to \mathbb{N}$ be a natural weight. By a *multicoloring* of (G, ν), we mean an assignment $\tilde{c} : V(G) \to \mathbb{P}(\mathbb{N})$ to the power set of $\mathbb{N}$, such that (i) $|\tilde{c}(u)| = \nu(u)$ for each $u \in V(G)$, and (ii) $\{u, v\} \in E(G) \Rightarrow \tilde{c}(u) \cap \tilde{c}(v) = \emptyset$. The corresponding *multichromatic number* $\tilde{\chi}(G, \nu)$ is then the smallest k which allows a legitimate multicoloring $\tilde{c} : V(G) \to \mathbb{P}(\{1, \ldots, k\})$.

In general, we may thus be given a *weighted hypergraph*, for which we seek a *strong multicoloring*. This corresponds to a multicoloring of the clique graph, whose weight function is identical to its corresponding hypergraph.

Contractions. One reason why it may be natural to generalize the problem to multicolorings is to handle certain contractions, or operations that simplify the instance. We consider particularly contractions that involve vertices with identical neighborhoods.

A *hypermodule* is a set S of vertices that appear identical to vertices outside S, i.e., for $u \in V \setminus S$ and $v, w \in S$, then $\{u, v\} \in E(G_c(H))$ iff $\{u, w\} \in E(G_c(H))$. A *contraction* takes a weighted hypergraph (H, ν) and a hypermodule S and produces a smaller reduced hypergraph (H', ν') where S has been replaced by a single vertex of weight $\chi(G_c(H[S]), \nu)$. One now can show that $\chi_s(H, \nu) = \chi_s(H', \nu')$.

Furthermore, degrees in the reduced hypergraph are no greater than before: for any vertex v we have $d_{H'}(v) \leq d_H(v)$. Thus, any result for approximation or time complexity involving degrees, number of vertices, or number of edges, carries over for the reduced hypergraph.

We may want to limit the kind of contractions that we seek. In particular, within our context, it is natural to search for *clique contractions*, where the clique graph $G_c(H[S])$ induced by S is a clique. In this case, degrees remain unchanged.

Note that a set of vertices that are contained in exactly the same hyperedges of H is a hypermodule that induces a clique in $G_c(H)$. By viewing each such hypermodule S of H as a single vertex u_S, and connecting two such vertices if, and only if, they are both contained in a common hyperedge of H, we obtain the *reduced graph* $G_r(H)$. This reduced graph $G_r(H)$ further has a natural weight $\nu : V(G_r(H)) \to \mathbb{N}$ given by $\nu(u_S) = |S|$. Hence, H yields a corresponding weighted reduced graph $(G_r(H), |\cdot|)$, something we will use in Section 5.

Hypergraph contraction preserves both *chordality* and *perfectness*. A polynomial time algorithm for multicoloring perfect graphs was given by Grötschel, Lovász, and Schrijver [8], under the problem name of *weighted coloring*.

3   Parameters of Graphs and Hypergraphs

The largest cardinality of a hyperedge of H will be denoted by $\sigma(H)$ and the largest cardinality of a hypermodule corresponding to a vertex of $G_r(H)$ will be denoted by $\mu(H)$. Clearly we have $\mu(H) \leq \sigma(H)$.

Maximum Degree. For a vertex $u \in V(H)$ of a hypergraph H, its *degree* $d_H(u)$ is the number of hyperedges that contain the vertex u. Note that the degree of u is usually much smaller than the number of neighbors of u (that is, the number of vertices contained in a common hyperedge with u.) The minimum (maximum) degree of a vertex in H is denoted by $\delta(H)$ ($\Delta(H)$).

Hypergraphs of degree at most t have the property that their clique graphs are $(t+1)$-claw free, i.e. contain no induced star on $(t+2)$-vertices. This ensures that almost any coloring obtains a ratio of at most t, even online.

Call an online coloring algorithm *frugal* if it does not introduce a new color unless it is forced to do so, i.e. if the corresponding vertex is already adjacent to vertices of all other colors. The First-Fit algorithm is clearly frugal. An offline algorithm is frugal if each vertex assigned a color i is adjacent to vertices of each color $1, \ldots, i-1$.

Lemma 1. *Any frugal coloring algorithm is at most* $\Delta(H)$*-competitive for the clique graph of a hypergraph* H.

Clique graphs of degree-2 hypergraphs contain the class of line graphs, and thus strong coloring such hypergraphs subsumes the edge coloring problem of multigraphs. This is hard to approximate within an absolute ratio of less than $4/3$, but can be done using at most $1.1\chi'(G) + 0.7$ colors [35], where χ' is the edge chromatic number.

We can obtain an incomparable bound in terms of the maximum degree of the clique graph, by extending an approach of [14] to multicolorings.

Theorem 1. *Multicoloring can be approximated within* $\lceil(\Delta(G) + 1)/3\rceil$.

This uses the following specialization of a lemma of Lovász [27]. It can be implemented in linear time [14] by first assigning the vertices greedily in order to the color class to which they have the fewest neighbors, followed by local improvement steps that move a vertex to a class with fewer neighbors.

Lemma 2. *Let G be a graph and let $t = \lceil(\Delta + 2)/3\rceil$. There is a partition of $V(G)$ into sets $V_1, \ldots, V_t$ such that the graph $G[V_i]$ induced by each V_i is of maximum degree at most 2.*

Given the graphs of maximum degree 2 promised by the lemma, we can color each of them optimally in linear time. (Details omitted.) Thus, we obtain a t-approximation to the multicoloring problem.

For unweighted graphs, a better approximation bound of $\lceil (\Delta + 1)/4 \rceil$ can be obtained [14], by using that graphs of maximum degree 3 can be colored optimally by way of Brooks theorem. It, however, does not apply for multicolorings.

Inductiveness. By the *inductiveness* (or the *degeneracy*) of H, denoted by $\text{ind}(H)$, we mean the parameter defined by $\text{ind}(H) = \max_{S \subseteq V(H)} \{\delta(H[S])\}$. Here, $H[S]$, for a vertex subset S denotes the *subhypergraph of H induced by S*, or the hypergraph with edge set $\mathcal{E}(H[S]) = \{X \cap S : X \in \mathcal{E}(H) \text{ and } |X \cap S| \geq 2\}$. Recall that the inductiveness naturally relates to a greedy coloring of a graph G that uses at most $\text{ind}(G) + 1$ colors (see [5]). The degree of a vertex v in $G_c(H)$ is at most $(\sigma(H) - 1)$ times its degree in H. Thus, the observation of [3] that $\text{ind}(G_c(H)) \leq \text{ind}(H)(\sigma(H) - 1)$. Hence, we have an $\text{ind}(H)$-factor approximation by the greedy algorithm. For ordinary graphs, we can obtain a simple contraction in the approximation by a factor of nearly 2. We observe here that this holds also for multicolorings. The following observation is from [16].

Theorem 2. *Suppose we are given a graph G with vertex weights w and a C-coloring of G (i.e., a partition of the vertex set into independent sets). Then we can approximate the multichromatic number within a factor of $\lceil C/2 \rceil$.*

Corollary 1. *Multicoloring can be approximated within $\lceil (\text{ind}(G) + 1)/2 \rceil$.*

The inductiveness measure is useful for bounding the performance of online algorithms. Irani [21] showed that the First-Fit coloring algorithm uses at most $O(\text{ind}(G) \lg n)$ to color an n-vertex graph G.

Corollary 2. *The First-Fit coloring algorithm is $O(\text{ind}(H) \lg n)$-competitive for coloring the clique graph of a hypergraph H.*

Composition Width. First we define a modular decomposition [22], which is also called *substitution decomposition* as in [31], and has been studied widely, since it is by many considered one of three most important hierarchical graph decomposition, the others being tree decomposition [38] and the graph decomposition upon which clique-width is defined [11].

Definition 1. *Let G be a graph with $V(G) = \{u_1, \ldots, u_k\}$ If $G_1, \ldots, G_k$ are graphs, then let $G' = G\langle G_1, \ldots, G_k \rangle$ denote the graph obtained by replacing each vertex u_i in G by the graph G_i, and connect each vertex in G_i to each vertex in G_j if, and only if, u_i and u_j are connected in G'. In this case we say that G is a modular decomposition of $G_1, \ldots, G_k$. The induced subgraphs G_i of G' are called* modules *of G'.*

Definition 2. *We call a graph G' k-composite if it is a null graph, or recursively, if there is a graph G on $\ell \leq k$ vertices and k-composite graphs $G_1, \ldots, G_\ell$, such that $G' = G\langle G_1, \ldots, G_\ell \rangle$. The composition-width of G', denoted $\text{cow}(G')$, is the least k for which G' is k-composite.*

REMARKS: (i) Every null graph is 1-composite and every clique is 2-composite. (ii) If G is k-composite, then G is k'-composite for each $k' \geq k$. (iii) Every graph on n vertices is n-composite.

Clique-Width. By a *labeled* graph G we mean a graph G provided with a labeling function $\iota : V(G) \to \mathbb{N}$. Consider the following four graph operations, introduced in [11]: (i) Create a new vertex u with a label i. (ii) Form the disjoint union of labeled graphs G_1 and G_2, denoted by $G_1 \oplus G_2$. (iii) Connect all i-labeled vertices with all the j-labeled vertices, where $i \neq j$ (iv) Relabel all vertices labeled i with the label j, where $i \neq j$. The *clique-width* of a graph G, denoted by $\mathrm{cw}(G)$ is the least number of labels needed so that G can be constructed by the above four graph operations.

Observation 3. *For a graph G we have $\mathrm{cw}(G) \leq \mathrm{cow}(G)$. If further $\mathrm{cw}(G) \in \{1,2\}$ then we have $\mathrm{cw}(G) = \mathrm{cow}(G)$.*

Clique-width was first defined and studied in [11] and it generalizes the notion of treewidth, introduced in [38], that is, a graph of bounded treewidth is necessarily also of bounded clique-width [10], but not conversely since a clique of arbitrary size has clique-width of two, while its treewidth is the number of vertices in the clique.

Problems definable in a certain variations of Monadic Second Order Logic, including maximum independent set, are solvable in polynomial time for graphs of bounded clique-width [9]. However, graph coloring has been shown to be not one of those (see [9]). Still, it has been shown that for a fixed $k \in \mathbb{N}$, the chromatic number of a graph G of clique-width of at most k, can be determined in time $O(2^{3k+1}k^2n^{2^{2k+1}+1})$ [24]. This however depends on that the expression that forms the graph using the four operations above is given.

4   Approximations for Down-Colorings

In this section we give bounds on the approximability of strong coloring general hypergraphs. The $O(\sqrt{m})$-approximation of strong independent sets of [18] leads to an equivalent approximation of strong coloring. However, we here obtain a bound on the inductiveness in terms of an arbitrary number m of edges, and obtain an approximation in terms of $\sigma(H)$, the largest hyperedge size.

Theorem 4. *For a hypergraph H with m edges, $\mathrm{ind}(G_c(H)) \leq \sqrt{m}\sigma(H)$.*

Proof. Let $k = \sqrt{m}$. Let S be a vertex subset inducing a subgraph of $G_c(H)$ of minimum degree $\mathrm{ind}(G_c(H))$. Let $H[S]$ be the subgraph of H induced by S, and let m_S be the number of hyperedges in it. If there is a vertex of degree at most k in $H[S]$, then its degree in $G_c(H[S])$ (which is at least $\mathrm{ind}(G_c(H))$) is at most $k\sigma(H)$, in which case the theorem follows. Otherwise, each vertex is of degree at least k and the number of edge-vertex incidences is at least $k|S|$. It follows that the average edge size in $H[S]$ is at least $k|S|/m_S$, and thus $\sigma(H) \geq k|S|/m_S \geq k|S|/m = |S|/k$.

Now, since the inductiveness of $G_c(H)$ is equal to the minimum degree of $G_c(H[S])$ which has at most $|S|$ vertices, we have that $\mathrm{ind}(G_c(H[S])) \leq |S|-1 < \sigma(H) \cdot k$ and the theorem follows.

Corollary 3. *There is a greedy algorithm that approximates the strong coloring of hypergraphs within a factor of $\sqrt{m}$. This yields a $\sqrt{M}$ approximation for the down-coloring of DAGs, where $M = |\max\{\overline{G}\}|$ is the number of source vertices in $\overline{G}$.*

Observe that we bounded the number of colors used by the algorithm in terms of the maximum edge size $\sigma(H)$. Thus, we have shown that the strong chromatic number of a hypergraph H differs from $\sigma(H)$ by a factor of at most $\sqrt{|\mathcal{E}(H)|} = \sqrt{m}$. In terms of down-graphs and hypergraphs, we obtain the following bound.

Corollary 4. $\chi(G_c(H_{\overline{G}})) = \chi_s(H_{\overline{G}}) \leq \sqrt{|\max\{\overline{G}\}|} \cdot \sigma(H_{\overline{G}}) \leq \sqrt{n} \cdot \sigma(H_{\overline{G}})$.

Compare the above corollary with [3, Obs. 2, p. 306], which shows that the bounds given can be obtained.

4.1   Approximation Hardness

We now give a reduction from the ordinary coloring problem that shows that the approximation of the greedy algorithm is close to best possible. The $\Omega(n^{1/2-\epsilon})$-hardness result for distance-2 coloring of [4] already yields the same hardness for strong coloring hypergraphs with n vertices and n edges. We show here similar result for down-hypergraphs and restricted types of DAGs.

Given a graph G_0, we construct a DAG $\overline{G}$ of height two by letting $V(\overline{G}) = E(G_0) \cup V(G_0)$ and $E(\overline{G}) = \{(e,v) : v \in e, v \in V(G_0), e \in E(G_0)\}$. The digraph has a source vertex for each edge in G_0, a leaf vertex for each node in G_0, and an edge from a source to a leaf if the leaf corresponds to a vertex incident on the edge corresponding to the source vertex.

Let $H = H_{\overline{G}}$ be the corresponding down-hypergraph. Note that the subhypergraph $H[V(G_0)]$ induced by the leaves is a graph and is exactly the graph G_0. The source nodes of H induce an independent set. Thus, $\chi(G_0) \leq \chi_s(H) \leq \chi(G_0)+1$. In fact, $\chi_s(H) = \chi(G_0)$ if $\chi(G_0) \geq 3$. By the results of Feige and Kilian [7], the chromatic number problem cannot be approximated within a factor of $|V(G)|^{1-\epsilon}$, for any $\epsilon > 0$, unless NP $\subseteq$ ZPP, i.e. unless there exist polynomial-time randomized algorithms for NP-hard problems. Here, we have $|V(G_0)| = \Omega(\sqrt{|V(H)|})$ and hence the following.

Theorem 5. *It is hard to approximate the down-coloring of DAGs within a factor of $n^{1/2-\epsilon}$, for any $\epsilon > 0$. This holds even for digraphs of height two.*

We may now ask if it is possible to give a better approximation for important special cases of the down-coloring problem. In particular, digraphs arising from pedigrees (i.e. records of ancestry for people) have some special properties; in particular, each vertex has in-degree at most 2, and normally a fairly small out-degree. We can show that even in this case, we cannot do better. (Proof omitted.)

Theorem 6. *It is hard to approximate the down-coloring of DAGs within a factor of $n^{1/2-\epsilon}$, for any $\epsilon > 0$, even when restricted to DAGs of in-degree and out-degree two.*

4.2 Online Coloring

In the standard online graph coloring problem, the graph is presented one vertex at a time along with edges only to the previous vertices. Each time the algorithm receives a vertex, it must make an irrevocable decision as to its color [19, 17].

The hypergraph model might lead to a different model, e.g. where all the vertices contained in edges incident on previous vertices are given. It can be inferred from the arguments below that this does not produce a great advantage.

Applying the result of Irani cited earlier, we obtain the following upper bound for online coloring.

Corollary 5. *First-Fit is $O(\sqrt{m}\lg n)$-competitive for the clique graph of a hypergraph H with m edges and n vertices. More generally, it is $O(\sqrt{m}\lg n)$-competitive for graphs that can be covered with m cliques.*

In the case of distance-2 coloring, one can argue a better bound. Namely, let G be the underlying graph and H_G be its neighborhood hypergraph. Then, $\Delta(H) = \Delta(G) + 1$ and $\sigma(H) = \Delta(G)$. Thus, the competitive ratio of any any frugal online coloring algorithm is at most $\min(\Delta(H), n/\sigma(H)) \le \sqrt{n}$.

Proposition 1. *Any frugal online coloring algorithm is $\sqrt{n}$-competitive for the distance-2 coloring graphs.*

On the hardness side, lower bounds for online graph coloring carry over for clique graphs, simply by viewing the graphs as hypergraphs. Halldórsson and Szegedy [19] showed that for any online algorithm, there is a $\lg n$-colorable graph on n vertices for which the algorithm uses $\Omega(n/\lg n)$ colors. This holds also for randomized algorithms against an oblivious adversary. This was later extended to a *known graph* model, where a graph isomorphic to the (fixed) input graph is given in advance [17]. Thus, there is one particular graph that is hard for any online coloring algorithm. By padding this graphs with isolated vertices (or small cliques), we can have this hold for graphs of any density.

Lemma 3. *For any n and m, there is a particular graph with n vertices and at most m cliques such that any online coloring algorithm is at least $\Omega(\sqrt{m}/\lg^2 m)$ competitive.*

5   Multicoloring, an Algebraic Approach

In this section we will consider an algebraic approach to determine the strong chromatic number of a given hypergraph by using integer programming. We then show how the same method can recursively yield an improved poly-time algorithm to obtain an optimal coloring for k-composite graphs.

Observation 7. *For a hypergraph H and its weighted reduced graph $(G_r(H),$ $|\cdot|)$, we have $\chi_s(H) = \tilde\chi(G_r(H), |\cdot|)$.*

Here we view a given hypergraph H as weighted graph $(G_r(H), |\cdot|)$, since by Observation 7 we have $\chi_s(H) = \tilde\chi(G_r(H), |\cdot|)$. Hence, we will here consider multicolorings of a weighted graph (G, ν), where $\nu : V(G) \to \mathbb{N}$ is a natural weight.

Consider our weighted graph (G, ν) where $V(G) = \{u_1, \ldots, u_k\}$. Let $G' = G\langle Q_1, \ldots, Q_k\rangle$ be the modular decomposition of the cliques $Q_1, \ldots, Q_k$, where each Q_i has $\nu(u_i)$ vertices. Clearly we have that $\tilde\chi(G, \nu) = \chi(G\langle Q_1, \ldots, Q_k\rangle)$, and so the computation of $\tilde\chi(G, \nu)$ can be trivially reduced to the computation of the chromatic number of a graph G'. However, taking further into the account the structure of $G\langle Q_1, \ldots, Q_k\rangle$, we can shorten the computations considerably, especially when the $\nu(u_i)$'s are large compared to k. We proceed as follows:

For each proper (i.e. nonempty) independent set $I \subseteq V(G)$ we form a variable x_I. We denote by $\mathcal{I}(G)$ the set of all independent sets of G, and for each $u \in V(G)$ we denote by $\mathcal{I}(G; u) \subseteq \mathcal{I}(G)$ the set of all independent sets of G that contain the vertex u. For each $u \in V(G)$ we form the following constraint

$$\sum_{I \in \mathcal{I}(G;u)} x_I = \nu(u). \tag{1}$$

Let us fix a listing of the elements of $\mathcal{I}(G)$: For an ordering $V(G) = \{u_1, \ldots, u_k\}$, note that each $U = \{u_{i_1}, \ldots, u_{i_\ell}\}$ yields a word $U \mapsto \mathrm{word}(U) = u_{i_1} \cdots u_{i_\ell}$, and hence the sets of $V(G)$ can be ordered lexicographically, viewing $u_1, \ldots, u_k$ as an ordered alphabet. We now can list the elements of $\mathbb{P}(V(G))$ *degree lexicographically*, or by *deglex* in short, in the following way [1]:

$$U_1 < U_2 \Leftrightarrow \begin{cases} |U_l| < |U_2| & \text{or} \\ |U_l| = |U_2| & \text{and } \mathrm{word}(U_1) < \mathrm{word}(U_2) \text{ lexicographically.} \end{cases} \tag{2}$$

With the deglex ordering (2), we can form the $|\mathcal{I}(G)|$-tuple $\mathbf{x}$ of the variables x_I, and the constraints from (1), determined by the k vertices of G, can be written collectively as $\mathsf{A}(G) \cdot \mathbf{x} = \mathbf{n}$, where $\mathbf{n} = (\nu(u_1), \ldots, \nu(u_k))$ and $\mathsf{A}(G)$ is a uniquely determined $|\mathcal{I}(G)| \times k$ matrix with only 0 or 1 as entries. Note that the sum Σ of all the variables x_I can be given by the dot-product $\Sigma = \mathbf{1} \cdot \mathbf{x}$, where $\mathbf{1}$ is the $|\mathcal{I}(G)|$-tuple with 1 in each of its entry.

Theorem 8. *For an integer weighted graph (G, ν) the multichromatic number $\tilde\chi(G, \nu)$ is given by the integer program*

$$\tilde\chi(G, \nu) = \min\{\mathbf{1} \cdot \mathbf{x} : \mathsf{A}(G) \cdot \mathbf{x} = \mathbf{n}, \ \mathbf{x} \in (\mathbb{N} \cup \{0\})^{|\mathcal{I}(G)|}\}, \tag{3}$$

where $\mathsf{A}(G)$ is uniquely determined by (1) and (2), and $\mathbf{n} = (\nu(u_1), \ldots, \nu(u_k))$.

It is well-known that the problem of solving an integer programming problem as (3) is NP-complete. However, considering the complexity in terms of $n = \sum_{u \in V(G)} \nu(u)$ (which corresponds to the number of vertices in the original hypergraph H) and assuming that $k = |V(G)|$ is fixed and "small" compared to n, it is worthwhile to discuss complexity analysis.

Lemma 4. *A connected graph G on k vertices has at most 2^{k-1} proper independent sets.*

By Lemma 4 the number of variables x_I in $\mathbf{x}$ is at most 2^{k-1}. Note also that by our deglex ordering, $\mathsf{A}(G)$ is already in reduced row echelon form with its first $k \times k$ submatrix being the $k \times k$ identity matrix I_k. If $N = \max_{u \in V(G)} \nu(u)$, then clearly each optimal solution $\mathbf{x}$ must satisfy $0 \leq x_I \leq N$ for each $I \in \mathcal{I}(G)$. Hence, we have at most $2^{k-1} - k$ free variables x_I in $\mathbf{x}$, namely those x_I with $|I| \geq 2$, each taking value in $\{0, 1, \ldots, N\}$. We now make some rudimentary computational observations, which are asymptotically tight for general G:

To check whether a set of i vertices is independent or not, we need $\binom{i}{2}$ edge-comparisons. Hence, $\mathcal{I}(G)$ can be obtained by at most $\sum_{i=1}^{k} \binom{k}{i}\binom{i}{2} = k(k-1)2^{k-3}$ comparisons. Having $\mathcal{I}(G)$, to determine $\mathcal{I}(G; u)$ for each $u \in V(G)$ we need at most $k2^{k-1}$ operations. In all, determining the linear program (3) we need at most $k(k-1)2^{k-3} + k^2 2^{k-1} < k^2 2^k$ operations. Further, for each given value of $\mathbf{x}$, the expressions $\mathsf{A}(G) \cdot \mathbf{x}$ and $\mathbf{1} \cdot \mathbf{x}$ can be evaluated in at most $k2^{k-1} + 2^{k-1} < k2^k$ steps. Hence, we have the following.

Observation 9. *For a connected integer weighted graph (G, ν) with $|V(G)| = k$ and $N = \max_{u \in V(G)} \nu(u)$, an optimal $\tilde{\chi}(G, \nu)$-multicoloring can be obtained by $k2^k(k + (N+1)^{2^{k-1}-k})$ operations, or in $O(k2^k(N+1)^{2^{k-1}-k})$ time for k fixed.*

With the notation from previous Section 2 we therefore have the following for a hypergraph.

Corollary 6. *For a hypergraph H with its reduced graph $G_r(H)$ on k vertices, the complexity of obtaining an optimal strong $\chi_s(H)$-coloring is given by $O(k2^k(\mu(H) + 1)^{2^{k-1}-k})$.*

Note that our complexities are polynomial expressions, only because we assume k to be fixed here.

We conclude this section by considering the complexity of obtaining an optimal coloring of a k-composite graph by using the integer program (3) recursively. The following is clear.

Proposition 2. *Let $k \in \mathbb{N}$ be given. The chromatic number $\chi(G')$ of a k-composite graph $G' = G\langle G_1, \ldots, G_\ell \rangle$ where $\ell \leq k$, is given by $\chi(G') = \tilde{\chi}(G, \nu)$, where the integer weight $\nu : V(G) \to \mathbb{N}$ is given by $\nu(u_i) = \chi(G_i)$ for each $i \in \{1, \ldots, \ell\}$.*

By Observation 9 we can obtain a $\chi(G)$-coloring by at most $k2^k(k + (N+1)^{2^{k-1}-k})$ operations, or in $O(k2^k(N+1)^{2^{k-1}-k})$ time, where $N = \max_i \chi(G_i)$, provided that we have an optimal coloring for each of the G_i's. If that is not the case however, we proceed recursively, but with more care, since we need at this point to keep track of the actual upper bound of arithmetic operations when applying the recursion. For a rooted tree (T, r) let T_u be the subtree rooted at $u \in V(T)$.

Lemma 5. *Let (T, r) be a rooted tree with n leaves, where each internal node has at least two children. If $f : \mathbb{N} \to \mathbb{R}$ is positive and non-decreasing, then*

$$\sum_{u \in V(T)} f(|V(T_u)|) \leq S_f(n) := (n-1)f(1) + \sum_{i=1}^{n} f(i).$$

REMARK: The bound of $S_f(n)$ from Lemma 5 is tight: Consider the degenerate binary tree T on $2n - 1$ vertices and n leaves, where each internal vertex has a leaf as a left child. In this case we have for any f that $\sum_{u \in V(T)} f(|V(T_u)|) = (n-1)f(1) + \sum_{i=1}^{n} f(i) = S_f(n)$.

An upper bound of the number of arithmetic operations need to obtain an optimal $\chi(G')$-coloring of a k-composite graph G', can now be obtained from the weighted rooted module tree $(T_{G'}, |\cdot|)$ of G' by $\sum_{u \in V(T_{G'})} f(w(u_{G''}))$, where each vertex $u_{G''}$ corresponds to an induced subgraph (strong module) G'' of G', and $w(u_{G''}) = |V(G'')|$, and the function f from Observation 9 is $f(n) = k2^k(k + (n+1)^{2^{k-1}-k})$, since $N = \max_{1 \leq i \leq k} \chi(G_i) \leq \sum_{i=1}^{k} |V(G_i)| = n$.

Theorem 10. *Provided the modular decompositions defining the k-composite connected graph G on n vertices are given, the number of arithmetic operations needed to obtain an optimal $\chi(G)$-coloring is $k2^{2^{k-1}+1}n + 4k(n+2)^{2^{k-1}-k+1}$ and hence can be obtained in $O(kn^{2^{k-1}-k+1})$ time, for k fixed.*

As mention above, the chromatic number of a connected graph G on n vertices of clique-width at most k, can be computed in $O(2^{3k+1}n^{2^{2k+1}+1})$ time [24], provided that the corresponding k-expression for G is given. By Observation 3 that also holds for k-composite graphs as well. However, the bound given in Theorem 10 is considerably better than the mentioned bound from [24]. In addition, it is not known, for $k \geq 4$, whether there exists a polynomial time algorithm to obtain a k-expression for a graph G of clique-width at most k.

For a graph G' known to have a module decomposition, a strong modular decomposition (where the modules do not have a nonempty intersection) which is unique, can be computed in $O(n^2)$ time, since such a decomposition of a graph is a special case of a modular decomposition a 2-structure (a slightly more general concept than a graph) [22]. However, in such a strong module decomposition $G' = G\langle G_1, \ldots, G_\ell \rangle$ it could be that $\ell > k$. But, if it is known that G' is k-composite, then the union of some of the strong modules $G_1, \ldots, G_\ell$ will make a single module of G'. To check this is the same as to check for modules in G, which will take at most $O(n^2)$ time as well. Hence, the module decomposition of G' constituting a k-decomposition will take at most $O(n^2)$ time at each step, which is much less than $f(n)$ from above, right before Theorem 10. Therefore, unlike for graphs of clique-width of k or less, we have the following corollary of Theorem 10, where we do not assume the modular k-decomposition.

Corollary 7. *The complexity of obtaining an optimal coloring for a k-composite graph on n vertices is $O(kn^{2^{k-1}-k+1})$.*

NOTE: A few words of warning are in order. If we do not have the modular decompositions, then we do need to compute the k-decomposition at each step of the recursive definition of a k-composite graph. A priori it could look as if the constant hidden in the $O(n^2)$ term might affect the overall complexity. However, a rooted tree with n leaves and no degree-2 internal vertex has at most $2n - 1$ vertices. So, the overall computation is at most $(2n - 1)O(n^2) = O(n^3)$ which is dominated by the expression in Theorem 10.

Acknowledgments

The authors are grateful to Walter D. Morris Jr. for many helpful discussions regarding integer programming.

References

1. W. W. Adams, P. Loustaunau. An Introduction to Gröbner Bases. *American Mathematical Society, Graduate Studies in Mathematics*, Volume 3, (1994).
2. G. Agnarsson and Á. Egilsson. On vertex coloring simple genetic digraphs. *Congressus Numerantium*, **161**, 117 – 127, (2004).
3. G. Agnarsson, Á. Egilsson and M. M. Halldórsson. Proper down-coloring simple acyclic digraphs. AGTIVE-2003, LNCS 3062, 299 – 312, Springer Verlag, (2004).
4. G. Agnarsson, R. Greenlaw, M. M. Halldórsson. On Powers of Chordal Graphs and Their Colorings. *Congressus Numerantium*, **142–147**, 2000.
5. G. Agnarsson and M. M. Halldórsson. Coloring Powers of Planar Graphs. *SIAM Journal of Discrete Mathematics*, **16**, No. 4, 651 – 662, (2003).
6. A. Brandstädt, V. B. Le, and J. P. Spinrad. Graph Classes: A Survey. *SIAM Monographs on Discrete Mathematics and Applications*, (1999).
7. U. Feige and J. Kilian. Zero Knowledge and the Chromatic number. *Journal of Computer and System Sciences*, **57**, no 2, 187–199, (1998).
8. M. Grötschel, L. Lovász, A. Schrijver. Polynomial algorithms for perfect graphs. *North-Holland Math. Stud.*, **88**, North-Holland, Amsterdam, (1984).
9. B. Courcelle, J.A. Makowsky, and U. Rotics. Linear Time Solvable Optimization Problems on Graphs of Bounded Clique Width. In WG '99.
10. B. Courcelle, S. Olariu. Upper bounds to the clique width of graphs. *Discrete Applied Mathematics*, **101**, 77 – 114, (2000).
11. B. Courcelle, J. Engelfriet, G. Rozenberg. Handle-rewriting hypergraph grammars. *J. Comput. System Sci.*, **46**, 218 – 270 , (1993).
12. P. Creszenci and V. Kann. A compendium of NP-optimization problems.
13. M. R. Garey and D. S. Johnson. *Computers and Intractability—A Guide to the Theory of NP-Completeness.* Freeman, San Francisco, CA, 1979.
14. M. M. Halldórsson and H.-C. Lau. Low-degree Graph Partitioning via Local Search with Applications to Constraint Satisfaction, Max Cut, and 3-Coloring. *Journal of Graph Algorithms and Applications*, 1(3):1–13, Nov. 1997.
15. M. M. Halldórsson. A still better performance guarantee for approximate graph coloring. *Information Processing Letters*, 45:19–23, 25 January 1993.
16. M. M. Halldórsson. Approximation via Partitioning. JAIST Research Report IS-RR-95-0003F, March 1995. Available at www.hi.is/~mmh/Publications.html
17. M. M. Halldórsson. Online coloring known graphs. *Electronic Journal of Combinatorics*, Feb 2000. www.combinatorics.org

18. M. M. Halldórsson, J. Kratochvíl, and J. A. Telle. Independent sets with domination constraints. *Discrete Applied Mathematics* **99**(1-3), 39–54, 17 Dec 1999. www.elsevier.nl/locate/jnlnr/05267

19. M. M. Halldórsson and Mario Szegedy. Lower bounds for on-line graph coloring. *Theoretical Computer Science*, 130:163–174, August 1994.

20. E. Hazan, S. Safra, and O. Schwartz. "On the Hardness of Approximating k-Dimensional Matching". Electronic Colloquium on Computational Complexity, TR03-020, 2003.

21. S. Irani. On-Line Coloring Inductive Graphs. *Algorithmica*, **11**:53–72, 1994.

22. Ö. Johansson. Graph Decomposition Using Node Labels. *Doctoral Dissertation, Royal Institute of Technology, Stockholm,* (2001).

23. J. Kahn. Asymptotically good list-colorings. *J. Combinatorial Th. (A)*, **73**:1–59, 1996.

24. D. Kobler, U. Rotics. Edge dominating set and colorings on graphs with fixed clique-width. *Discrete Applied Mathematics*, **126**, 197 – 221, (2003).

25. M. Krivelevich and B. Sudakov. Approximate coloring of uniform hypergraphs. *ESA '98.*

26. S. O. Krumke, M. V. Marathe, and S. S. Ravi. Approximation algorithms for channel assignment in radio networks. *2nd Dial M for Mobility*, 1998.

27. L. Lovász. On Decomposition of Graphs. *Stud. Sci. Math. Hung.*, **1**:237–238, 1966.

28. M. V. Marathe, H. Breu, H. B. Hunt III, S. S. Ravi and D. J. Rosenkrantz. Simple Heuristics for Unit Disk Graphs. *Networks*, **25**:59–68, 1995.

29. S. T. McCormick. Optimal approximation of sparse Hessians and its equivalence to a graph coloring problem. *Math. Programming*, 26(2):153–171, 1983.

30. F. R. McMorris, T. Zaslavsky. Bound graphs of a partially ordered set. *J. Combin. Inform. System Sci.*, **7**, 134–138, (1982).

31. R. H. Möhring, F. J. Radermacher. Substitution decomposition for discrete structures and connections with combinatorial optimization. *North-Holland Math. Stud.*, **95**, North-Holland, Amsterdam, 257 – 355, (1984).

32. M. Molloy and B. Reed. Near-optimal list colourings. *Random Structures and Algorithms*, **17**:376–402, 2000.

33. M. Molloy and M.R. Salavatipour. Frequency Channel Assignment on Planar Networks. In *ESA 2002*, LNCS 2461, pp. 736-747.

34. R. Motwani and M. Sudan. Computing roots of graphs is hard. *Discrete Appl. Math.*, 54:81–88, 1994.

35. T. Nishizeki and K. Kashiwagi. On the 1.1 edge-coloring of multigraphs. *SIAM Journal on Discrete Mathematics*, 3(3):391–410, 1990.

36. V. Th. Paschos. Polynomial approximation and graph-coloring. *Computing*, **70**, no. 1, 41–86, (2003).

37. S. Ramanathan and E. L. Lloyd. Scheduling algorithms for multi-hop radio networks. *IEEE/ACM Trans. on Networking*, 1(2):166–172, April 1993.

38. N. Robertson, P. Seymour. Graph minors V, Excluding a planar graph. *Journal of Combinatorial Theory (B)*, **41**, 92 – 114, (1986).

Deterministic Monotone Algorithms for Scheduling on Related Machines[*]

Pasquale Ambrosio and Vincenzo Auletta

Dipartimento di Informatica ed Applicazioni "R.M. Capocelli"
Universitá di Salerno
84081, Baronissi (SA), Italy
{ambrosio, auletta}@dia.unisa.it

Abstract. We consider the problem of designing monotone deterministic algorithms for scheduling tasks on related machines in order to minimize the makespan. Several recent papers showed that monotonicity is a fundamental property to design truthful mechanisms for this scheduling problem.

We give both theoretical and experimental results. For the case of two machines, when speeds of the machines are restricted to be powers of a given constant $c > 0$, we prove that algorithm LARGEST PROCESSING TIME is monotone for any $c \geq 2$ while it is not monotone for $c \leq 1.78$; algorithm LIST SCHEDULING, instead, is monotone only for $c > 2$.

For the case of m machines we restrict our attention to the class of "greedy-like" monotone algorithms defined in [AP04]. We propose the greedy–like algorithm UNIFORM_RR and we prove that it is monotone when speeds are powers of a given integer constant $c > 0$ and it obtains an approximation ratio that is not worse than algorithm UNIFORM, proposed in [AP04]. We also experimentally compare performances of UNIFORM, UNIFORM_RR, LPT, and several other monotone and greedy–like heuristics.

1   Introduction

In this paper we consider the problem of designing deterministic monotone algorithms for scheduling tasks on related machines in order to minimize the makespan (i.e. the maximum completion time). A classical result of game theory, recently rediscovered by [AT01], states that monotonicity is a necessary condition to design truthful (dominant strategies) mechanisms for this scheduling problem. Mechanisms are a classical concept of the theory of non-cooperative games [OR94]. In these games there are several independent agents that have to work together in order to optimize a global objective function. However, each player has its own private valuation function, maybe different from the global objective function, and may lie if this can improve its valuation of the game

---

[*] Work supported by the European Project IST-2001-33135, Critical Resource Sharing for Cooperation in Complex Systems (CRESCCO).

G. Persiano and R. Solis-Oba (Eds.): WAOA 2004, LNCS 3351, pp. 267–280, 2005.

output, even though this can produce a suboptimal solution. Non-cooperative games can be used to model problems that have to be solved in market environments where heterogenous entities have to cooperate in computing some global function but they compete for "resources" (e.g. the autonomous systems that regulate the routing of traffic in Internet) [Pa01].

The main idea of the *Mechanism Design* theory is to pay the agents to convince them to perform strategies that help the system to optimize the global objective function. A *Mechanism M=(A,P)* is a combination of two elements: an algorithm A computing a solution and a payment function P specifying the amount of "money" the mechanism should pay to each agent. A mechanism is *Truthful with Dominant Strategies* (in the sequel simply truthful) if its payments guarantee that agents are not stimulated to lie, whatever strategies other agents perform.

Recently, mechanism design has been applied to several optimization problems arising in computer science and networking that have been (re-)considered in the context of non-cooperative games [NR99, Ro00, Pa01].

State of the Art. The celebrated *VCG mechanism* [Cl71, Gr73, V61] is the prominent technique to derive truthful mechanisms for optimization problems. However, this technique applies only to *utilitarian* problems, that are problems where the objective function is equal to the sum of the agents valuation functions (e.g., shortest path, minimum spanning tree, etc.). In the seminal papers by Nisan and Ronen [NR99, NR00] it is pointed out that VCG mechanisms do not completely fit in a context where computational issues play a crucial role since they assume that it is possible to compute an optimal solution of the corresponding optimization problem (maybe a NP-hard problem). Scheduling, is a classical optimization problem that is not utilitarian, since we aim at minimizing the *maximum* over all machines of their completion times and it is NP-Hard. Moreover, scheduling models important features of different allocation problems and routing problems in communication networks. Thus, it has been the first problem for which not VCG based techniques have been introduced.

Nisan and Ronen [NR99, Ro00] give an m-approximation truthful mechanism for the problem of scheduling tasks on m *unrelated* machines, when each machine is owned by a different agent that declares the processing times of the tasks assigned to his/her machine and the algorithm has to compute the scheduling based on the values declared by the agents. In [AT01] it is considered the simpler variant of the task scheduling on *related* machines (in short $Q||C_{max}$), where each machine i has a speed s_i and the processing time of a task is given by the ratio between the weight of the task and the speed of the machine. They show that a mechanism $M = (A, P)$ for the $Q||C_{max}$ problem is truthful if and only if algorithm A is monotone. Intuitively, monotonicity means that increasing the speed of exactly one machine does not make the algorithm decrease the work assigned to that machine (see Section 2 for a formal definition). The result of [AT01] reduces the problem of designing a truthful mechanism for $Q||C_{max}$ to the algorithmic problem of designing a good algorithm which also satisfies the additional *monotonicity* requirement.

Several algorithms are known in literature for $Q||C_{max}$, but most of them are not monotone. Greedy algorithms were proposed by Graham in the '60s for the case of identical machines. He proves that algorithm LPT, which considers the tasks in non-increasing order by weight, is $(4/3 - 1/(3m))$-approximated [Gr69], while algorithm LIST SCHEDULING, which considers tasks in the same order as given in input, is $(2 - 1/m)$-approximated [Gr66]. Moreover, a PTAS can be constructed using LPT as a subroutine [Gr69]. The same algorithms can be used for $Q||C_{max}$. In particular, LPT is ϕ-approximated for two machines, where $\phi = \frac{1+\sqrt{5}}{2}$ is the golden ratio, and $\frac{2m}{m+1}$-approximated for m machines [GI80]. LS, instead, is $O(\log m)$-approximated [AA+93, CS80]. However, both these algorithms are not monotone [AT01]. Not greedy techniques have been used to provide a PTAS for $Q||C_{max}$ [HS88] and constant approximations for the online version of the problem [AA+93, BC97]. However, all these algorithms are intrinsically not monotone.

The first non-trivial monotone algorithm for $Q||C_{max}$ is given in [AT01], where a randomized 3-approximated mechanism for $Q||C_{max}$ is presented that is truthful in expectation. In [AP04] a technique is provided to construct a family of $(2 + \varepsilon)$-approximated monotone algorithms starting from a monotone allocation algorithm that is "greedy–like" (i.e. its cost is within an additive factor of $O(t_{max}/s_1)$ from the cost of LPT, where t_{max} is the largest task weight and s_1 is the smallest machine speed). The basic idea, derived by the PTAS of Graham, is to combine the optimal scheduling of the largest tasks with the schedule computed by a monotone "greedy–like" algorithm; however, in order to guarantee monotonicity, the scheduling of the large tasks and of the small tasks are computed independently. In [AP04] it is proposed the algorithm UNIFORM, which is greedy–like and it is monotone in the particular case where machine speeds are divisible (see Section 2 for a formal definition). Thus, they obtain a family of deterministic truthful $(2+\varepsilon)$-approximated mechanisms for the case of divisible speeds. This result, combined with payment functions of [AT01], implies the existence, for any fixed number of machines and any $\varepsilon > 0$, of deterministic truthful $(4 + \varepsilon)$-approximated mechanisms for the case of *arbitrary* speeds.

Our Results. In [AP04] two questions are left open. The first one is whether LPT is monotone in the particular case of divisible speeds; the second one is whether the algorithm UNIFORM is monotone also in the case of arbitrary speeds.

In this paper we try to answer to both the questions. With respect to the first question we give answers only for the case of 2 machines (see Section 3). We say that speeds are c–divisible if and only if they are powers of a given positive constant c. We prove that LPT is monotone for c–divisible speeds when $c \geq 2$ while it is not monotone when $c \leq 1.78$. We also prove that LS is monotone if $c > 2$. With respect to the second question, we prove that any "UNIFORM–like" algorithm is not monotone when speeds are not-divisible. It is possible to modify the algorithm to obtain monotonicity but this implies a much weaker approximation factor. We also describe a new algorithm UNIFORM_RR, based on UNIFORM, and prove that it is more efficient than UNIFORM and it is monotone for divisible speeds (see Section 4).

We experimentally evaluate the performances of algorithms UNIFORM and UNIFORM_RR, comparing it to LPT and to several other "UNIFORM–like" heuristics. The experiments show that algorithm UNIFORM_RR outperforms the other considered heuristics both with respect to the worst case and the average case approximation factor and it is very close to LPT.

2   The Problem

In this section we formally define the $Q||C_{max}$ problem. Consider m machines having speeds $s = \langle s_1, s_2, \ldots, s_m \rangle$, with $s_1 \leq s_2 \leq \ldots \leq s_m$, and n tasks of weights $\sigma = (t_1, t_2, \ldots, t_n)$. In the sequel we simply denote the i-th task with its weight t_i. A schedule is a mapping that associates each task to a machine. The amount of time to complete task j on machine i is t_j/s_i. The *work* of machine i, denoted as w_i, is given by the sum of the weights of the tasks assigned to i. The *load* (or completion time) of machine i is given by w_i/s_i. The cost of a schedule is the maximum load over all machines, that is, its *makespan*.

Given an algorithm A for $Q||C_{max}$, $A(\sigma, s) = (A_1(\sigma, s), A_2(\sigma, s), \cdots, A_m(\sigma, s))$ denotes the solution computed by A on input the task sequence σ and the speed vector s, where $A_i(\sigma, s)$ is the load assigned to machine i. The cost of this solution is denoted by $\text{COST}(A, \sigma, s)$. In the sequel we omit σ and s every time it is clear from the context. Following the standard notation of game theory, we denote by $s_{-i} = (s_1, s_2, \cdots, s_{i-1}, s_{i+1}, \cdots, s_m)$ the vector of the speeds of all machines except machine i and we write $s = (s_{-i}, s_i)$.

Definition 1 (Monotone Scheduling Algorithms). *A scheduling algorithm A is monotone iff for any machine i, fixed the speeds of the other machines s_{-i}, the work assigned to machine i is not decreasing with respect to s_i, that is for any $s_i' > s_i''$ it holds that $w_i(s_{-i}, s_i') \geq w_i(s_{-i}, s_i'')$.*

An *optimal algorithm* computes a solution of minimal cost $\text{OPT}(\sigma, s)$. Throughout the paper we assume that the optimal algorithm always produces the *lexicographically minimal* optimal assignment. As shown in [AT01], this algorithm is monotone.

An algorithm A is a *c-approximation algorithm* if, for every instance (σ, s), $\text{COST}(A, \sigma, s) \leq c \cdot \text{OPT}(\sigma, s)$. A *polynomial-time approximation scheme* (PTAS) for a minimization problem is a family $\mathcal{A}$ of algorithms such that, for every $\varepsilon > 0$ there exists a $(1 + \varepsilon)$-approximation algorithm $A_\varepsilon \in \mathcal{A}$ whose running time is polynomial in the size of the input.

LARGEST PROCESSING TIME (LPT) and LIST SCHEDULING (LS) are two greedy algorithms widely used for $Q||C_{max}$. LPT first sorts the tasks in nonincreasing order by weight and then process them assigning task t_j to machine i that minimizes $(w_i + t_j)/s_i$, where w_i denotes the work of machine i before task t_j is assigned; if more than one machine minimizing the above ratio exists then the machine with small index is chosen. LS uses the same rule as LPT to assign tasks to machines, but it processes the tasks in the same order as they appear in σ. For any fixed number of machines, there exists a PTAS for $Q||C_{max}$ that assigns the h largest tasks optimally, for h large enough, and the remaining tasks with LPT [Gr69].

We say that a scheduling algorithm A is greedy-close if for any speed vector s and for any task sequence σ we have that $\text{COST}(A, \sigma, s) \leq \text{COST}(\text{LPT}, \sigma, s) + O\left(\frac{t_{max}}{s_1}\right)$, where t_{max} is the largest task of σ and s_1 is the smallest speed in s. In [AP04] the monotone scheduling algorithm PTAS-GC is provided that uses a greedy-close algorithm GC as a subroutine. This algorithm splits the task sequence in two parts: the h largest tasks are scheduled optimally; the remaining tasks are scheduled by GC, independently from the schedule obtained in the first two phases. They prove that for any sequence of tasks and for any $\varepsilon > 0$ there exists an integer $h > 0$ such that PTAS-GC gives a solution that is within a factor of $(2 + \varepsilon)$ from the optimum.

We say that speeds of the machines are c–divisible, for any constant $c > 0$, if and only if each speed is a power of c. We say that the $Q||C_{max}$ problem is restricted to c–divisible speeds if speeds are c–divisible and each agent can declare only values that are powers of c.

3   Scheduling on Two Machines with c–Divisible Speeds

In this section we consider the case of two machines with c–divisible speeds and give upper and lower bounds on the values of c that guarantee the monotonicity of algorithms LPT and LS.

We start by proving two interesting properties of the schedules computed by LPT. The first lemma proves that the scheduling computed by LPT is such that if a task assigned to a machine (say i) is moved to another machine then it has a completion time that is not smaller than its completion time on machine i. This property, known as a Nash Equilibrium, is very important in the context of dynamic systems since it implies that the system is in a stable state and no entity has an incentive to move from its state.

Lemma 1. *Let $w_1, w_2, \cdots, w_m$ be the works assigned to the machines by LPT. For each task t, let $i(t)$ be the machine which t is assigned to. Then, for each $1 \leq j \leq m$, it holds that $\frac{w_j + t}{s_j} \geq \frac{w_{i(t)}}{s_{i(t)}}$.*

Lemma 2. *For each speed vector s and for each sequence of tasks σ, the schedule computed by LPT on input s and σ is such that for any i, j, if $s_i \leq s_j/2$ then $w_i \leq w_j$, where w_i is the work assigned by the algorithm to machine i.*

Let $c(A) > 0$ be the smallest real number such that for each $c \geq c(A)$ the algorithm A is monotone when restricted to c–divisible speeds. We briefly describe now the argument that we use to lower bound $c(A)$. Consider two speed vectors $s = \langle s_1, s_2 \rangle$ and $s' = \langle s'_1, s'_2 \rangle$, where s' differs from s only on speed of machine i and $s_i \leq s'_i$. For each sequence of tasks $\sigma = \langle t_1, t_2, \cdots, t_n \rangle$, we divide the tasks in σ in the following four sets with respect to the allocations computed by A with respect to s and s': $T_i(\sigma)$, for $i = 1, 2$, is the set of tasks of σ that are assigned to machine i both with respect to s and s'; $L(\sigma)$ is the set of tasks of σ that are assigned to machine 1 with respect to s and to machine 2 with respect to s'; $R(\sigma)$ is the set of tasks of σ that are assigned to machine 2 with respect to s and to machine 1 with respect to s'. In the following we omit the argument σ whenever it is clear from the context. It is easy to see that

$$w_1(s) = \frac{T_1 + L}{s_1}, \quad w_2(s) = \frac{T_2 + R}{s_2}, \quad w_1(s') = \frac{T_1 + R}{s_1'}, \quad w_2(s') = \frac{T_2 + L}{s_2'}.$$

$$(1)$$

Theorem 3. *For any $c \geq 2$, the algorithm* LPT *is monotone when restricted to the case of two machines with c–divisible speeds.*

Proof. Suppose by contradiction that LPT is not monotone for c–divisible speeds. Then, there exist two speed vectors $s = \langle s_1 \leq s_2 \rangle$ and $s' = \langle s_1' \leq s_2' \rangle$, where s' has been obtained from s by increasing only one speed, and a sequence of tasks $\sigma = \langle \sigma', t \rangle$ such that the scheduling of the tasks in σ computed by LPT with respect to s and s' is not monotone. Without loss of generality assume that σ is the shortest sequence that LPT schedules in a not monotone way. This means that the schedule of σ' is monotone while the allocation of t destroys the monotonicity. We distinguish three cases.

First of all, consider the case $s_2' \geq c \cdot s_2$. Since, by hypothesis, the schedule of σ' is monotone while the schedule of σ is not monotone we have that $w_2(\sigma', s) \leq w_2(\sigma', s')$ and $w_2(\sigma, s) > w_2(\sigma, s')$. By Eq. 1 it follows that

$$R(\sigma') \leq L(\sigma') < R(\sigma') + t.$$

$$(2)$$

Observe now that if LPT on input s assigns task t to machine 2 then $\frac{T_1(\sigma') + L(\sigma') + t}{s_1}$ $> \frac{T_2(\sigma') + R(\sigma') + t}{s_2}$, from which we obtain

$$T_2(\sigma') < \frac{s_2}{s_1}(T_1(\sigma') + L(\sigma') + t) - R(\sigma') - t.$$

$$(3)$$

Similarly, if LPT on input s' assigns task t to machine 1 then $\frac{T_1(\sigma') + R(\sigma') + t}{s_1'} \leq$ $\frac{T_2(\sigma') + L(\sigma') + t}{s_2'}$, from which we obtain

$$T_2(\sigma') + L(\sigma') + t \geq \frac{s_2'}{s_1'}(T_1(\sigma') + R(\sigma') + t)$$

$$(4)$$

Substituting Eq. 3 in Eq. 4 and making some algebraic manipulations we obtain that

$$L(\sigma') \geq \frac{s_2'}{s_1'}(T_1(\sigma') + R(\sigma') + t) - T_2(\sigma') - t$$

$$\geq \frac{s_2}{s_1}T_1(\sigma') + \frac{s_2}{s_1}(2R(\sigma') - L(\sigma')) + t(\frac{s_2}{s_1} - 1) + (R(\sigma') + t)$$

$$\geq (R(\sigma') + t)$$

where the last inequality holds since by hypothesis $\frac{s_2}{s_1} \geq 1$ and, since t is the smallest task, $R(\sigma') \geq L(\sigma')/2$. However, this contradicts Eq. 2 and therefore there is no instance σ for which the schedule computed by LPT is not monotone. The other two cases can be solved by reduction to the previous case.

A similar argument can be used to prove the following theorem.

Theorem 4. *For any $c > 2$, the algorithm LS is monotone when restricted to the case of two machines with c–divisible speeds.*

Intuitively, Theorem 3 states that LPT is monotone if a machine that wants to reduce its speed has to do it in a significant way (at least half in this case). It is interesting to study which is the value of $c(\text{LPT})$. Next Lemma gives a constructive lower bound on this value.

Lemma 5. *For any $c \leq 1.78$, the restriction of LPT to two machines and c–divisible speeds is not monotone.*

Proof. Consider the sequence of tasks $\sigma = \langle y \geq x \geq x/2 + 2\varepsilon \geq x/2 - \varepsilon \rangle$ and the speed vectors $s = \langle 1, c \rangle$ and $s' = \langle 1, c^2 \rangle$. Assume that x, y and ε are chosen in such a way that LPT, on input s, assign all tasks except for x to machine 2, while, on input s' it assigns the first two tasks to machine 2 and the other tasks to machine 1. Clearly, this schedule is not monotone since machine 2 receives a total load of $y + x + \varepsilon$ with speed c and a total load of $x + y$ with speed c^2. We observe that LPT produces the previous schedules when

$$\frac{y + x}{c} > x, \qquad \frac{y + x + \varepsilon}{c} < \frac{3}{2}x - \varepsilon \qquad (5)$$

and

$$\frac{y + x}{c^2} < x, \qquad \frac{y + x + x/2 - \varepsilon}{c^2} > x + \varepsilon. \qquad (6)$$

By trivial computations it can be seen that for any $c \leq 1.78$ it is possible to choose y, x and ε so that previous inequalities hold. In particular, for $c = 1.78$ we can take $y = 113.5, x = 68, \varepsilon = 0.005$.

The argument of the proof of Lemma 5 cannot be extended since for any $c \geq \frac{3 + \sqrt{17}}{4}$ it is not possible to choose y, x and ε in order to satisfy Eq. 5–6.

4   Algorithms UNIFORM–Like

In this section we prove that algorithm UNIFORM, proposed in [AP04], is not monotone with respect to not divisible speeds. In the sequel we assume that machine speeds are positive integers.

Algorithm UNIFORM works in two phases: first it uses LPT as a subroutine to compute a schedule of the tasks to $S = \sum_{i=1}^{m} s_i$ identical "virtual machines"; then, it assigns to each real machine i the work assigned to s_i virtual machines in such a way that each virtual machine is assigned to only one real machine. To guarantee the monotonicity of UNIFORM the mapping of the virtual machines to the real machines is such that $w_1/s_1 \leq w_2/s_2 \leq \cdots \leq w_m/s_m$. In particular, UNIFORM partitions the virtual machines into g blocks of the same size, where g is equal to $GCD(s_1, s_2, \ldots, s_m)$, in such a way that each virtual machine of block i has a work greater than any other machine in a block $j < i$. Finally, for each block i, for $i = 1, 2, \cdots, m$, it assigns s_i/g consecutive virtual machines to the real machine i, starting from the virtual machine with less work.

In [AP04] it is proved that UNIFORM is greedy-close and it is monotone in the particular case of divisible speeds. It is left open the question whether UNIFORM is monotone also when speeds are not divisible. Next Theorem gives a negative answer to this question.

Theorem 6. *Algorithm* UNIFORM *is not monotone with respect to not divisible speeds.*

Proof. We prove the Theorem by constructing an example where the allocation computed by algorithm UNIFORM is not monotone. Consider the task sequence $\sigma = \langle 2, 2, 2, 1, 1, 1 \rangle$ and the speed vectors $s = \langle 3, 8 \rangle$ and $s' = \langle 2, 8 \rangle$. Observe that on input (σ, s) algorithm UNIFORM partitions the virtual machines in only one block and assigns all the load to machine 2 (see Fig. 1(a)): thus, we have a work equal to 0 for machine 1 and a work equal to 9 for machine 2. On input (σ, s'), instead, algorithm UNIFORM splits the virtual machines in 2 blocks producing the schedule given in Fig. 1(b), where machine 1 obtains a work of 2. Thus, the algorithm is not monotone because machine 1 increases its load while reducing its speed.

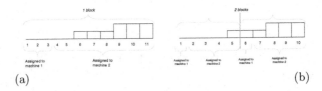

(a)   (b)

Fig. 1. An example of non monotone scheduling computed by UNIFORM. In (a) it is given the scheduling computed for $s = (3, 8)$; in (b) it is given the scheduling computed for $S' = (2, 8)$

The proof of Theorem 6 shows that any algorithm based on the partition of virtual machines in blocks will be not monotone if the number of blocks depends on the speeds of the machines. We can modify UNIFORM, so that it sets $g = 1$ and it considers all the virtual machines as in the same block. This new algorithm is monotone but it obtains a weak approximation since the assignment of the virtual machines to real machines is completely unbalanced (see Fig. 2(a)). We describe now a variation of UNIFORM that computes $g = GCD(s_1, s_2, \cdots, s_m)$ blocks but it makes a more clever assignment of the virtual machines of each block to the real machines.

Algorithm UNIFORM_RR, described in Alg. 1, uses a round-robin strategy to assign virtual machines of a block to real machines, starting from the virtual machine with lowest work that is assigned to the real machine with lowest speed. Fig. 2 shows the assignments computed by UNIFORM and UNIFORM_RR on an instance of 7 virtual machines and gives evidence of the more balanced scheduling computed by UNIFORM_RR. We state that algorithm UNIFORM_RR is monotone with respect to divisible speeds and it is $(2+\varepsilon)$-approximated. The proof of the monotonicity of UNIFORM_RR is a technical extension of the proof of monotonicity of UNIFORM given in [AP04] and we omit it from this extended abstract (a complete version of the paper can be found in [AA04]). In order to prove the bound on the approximation factor we show that for any speed vector s and any task sequence σ it holds that $\text{COST}(\text{UNIFORM\_RR}, \sigma, s) \leq \text{COST}(\text{UNIFORM}, \sigma, s)$.

Algorithm 1 UNIFORM_RR

Input: a task sequence σ, speed vector $s = \langle s_1, s_2, \ldots, s_m \rangle$, with $s_1 \leq s_2 \leq \ldots \leq s_m$
  1. Run algorithm LPT to allocate tasks of σ on $S = \sum_{i=1}^{m} s_i$ identical virtual machines.
  2. Order the virtual machines by nondecreasing load $l_1, \ldots, l_S$.
  3. Set $g := GCD(s_1, s_2, \ldots, s_m)$ and partition the virtual machines into g blocks $B_1, \ldots, B_g$, each consisting of S/g consecutive virtual machines. For $1 \leq i \leq g$ and $1 \leq k \leq S/g$, denote by B_{ik} the k-th virtual machine of the i-th block. Thus the virtual machine B_{ik} has load $l_{(i-1)\frac{S}{g}+k}$.
  4. For each block j

  (a) set $k_i = s_i/g$, for $1 \leq i \leq m$, and $x = 1$.
  (b) for $1 \leq k \leq S/g$

   – while $k_x = 0$ set $x = (x + 1) \bmod m$. Then, allocate the total load of the virtual machine B_{jk} to the real machine x and set $k_x = k_x - 1$.

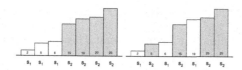

Fig. 2. Assignments computed by UNIFORM (left) and UNIFORM_RR (right) on an instance with two machines with speeds $s = \langle 3, 4 \rangle$. UNIFORM produces an assignment with makespan equal to 19.75; UNIFORM_RR produces an assignment with a makespan equal to 16.25

In [AP04] it is proved that the makespan of UNIFORM is obtained by machine m. We prove that a similar property holds also for UNIFORM_RR.

Lemma 7. *For any speed vector* $s = \langle s_1, s_2, \cdots, s_m \rangle$ *and any task sequence* σ *it holds that the makespan of the solution computed by* UNIFORM_RR *is equal to the completion time of the fastest machine.*

Proof. We prove the lemma by showing that for each block B the load assigned to machine m in block B is greater than or equal to the load assigned in the same block to any other machine.

Let $x_1 \leq x_2 \leq \cdots \leq x_{s_m/g}$ and $y_1 \leq y_2 \leq \cdots \leq y_{s_j/g}$ be the loads of the virtual machines assigned to machine m and j, for any $j < m$, respectively. We observe that $x_h \geq y_h$ for $1 \leq h \leq s_j/g$ and $x_h \geq y_{s_j/g}$ for $s_j/g < h \leq s_m/g$. Then,

$$\frac{1}{s_m} \sum_{h=1}^{s_m/g} x_h \geq \frac{1}{s_j} \sum_{h=1}^{s_j/g} y_h.$$

Lemma 8. *For any speed vector* $s = \langle s_1, s_2, \cdots, s_m \rangle$ *and any task sequence* σ *it holds that the cost of the solution computed by* UNIFORM_RR *is not greater than the cost of the solution computed by* UNIFORM.

Proof. By Lemma 7 it is sufficient to prove that UNIFORM_RR assigns to machine m a total load not greater than the load assigned by UNIFORM to the same machine.

Observe that the two algorithms compute the same assignment of tasks to the virtual machines and the same partition of virtual machines in blocks. Thus, it is sufficient to prove that the load assigned by algorithm UNIFORM_RR to machine m for each block B is not greater than the load assigned by UNIFORM to the same machine. Let $x_1 \leq x_2 \leq \cdots \leq x_{s_m/g}$ and $y_1 \leq y_2 \leq \cdots \leq y_{s_m/g}$ be the works of the virtual machines of block B assigned to machine m algorithms by UNIFORM and UNIFORM_RR, respectively. It can be easily seen that $x_h \geq y_h$ for $1 \leq h \leq s_m/g$ and the lemma follows.

Theorem 9. *For any speed vector* $s = \langle s_1, s_2, \cdots, s_m \rangle$ *and any task sequence* σ *it holds that*
$$\text{COST}(\text{UNIFORM\_RR}, \sigma, s) \leq (2 + \varepsilon)\text{OPT}(\sigma, s).$$

Proof. The theorem follows by Lemma 8 and Theorem 16 of [AP04].

5   Experimental Results

In this section we describe the results of an experimental analysis on the performances of several monotone scheduling algorithms. We have performed two different experiments: in the first experiment we have measured the approximation factors of several monotone heuristics, comparing them to the approximation of LPT; in the second experiment, instead, we have measured the approximation factors of the algorithms obtained by plugging different monotone greedy–like

Table 1. Algorithms considered in our testing and their theoretical approximation factors

Upper bounds to approximation factors of monotone scheduling algorithms	
LPT (1)	$\left(\frac{2n}{n+1}\right)$
LPT RESTRICTED(2)	$\left(\frac{2n}{n+1}\right)$
UNIFORM G=1 (3)	$(4 + \varepsilon)$
UNIFORM G=1 RESTRICTED (4)	$(4 + \varepsilon)$
UNIFORM RESTRICTED (5)	$(4 + \varepsilon)$
UNIFORM_RR G=1 (6)	$(4 - \varepsilon)$
UNIFORM_RR G=1 RESTRICTED (7)	$(4 - \varepsilon)$
UNIFORM_RR RESTRICTED (8)	$(4 - \varepsilon)$

Table 2. Summary of the number of instances performed in each run

Jobs	Machines	Instances					
		$\beta = 1$	$\beta = 2$	$\beta = 3$	$\beta = 4$	$\beta = 5$	$\beta = 6$
10	4	3690	7380	11070	14760	22140	29520
25	5	3690	11070	14760	19680	29520	39360
100	10	5538	17694	33232	54310	66466	88620

algorithms in the scheme described in [AP04]; in the third experiment we have measured the total quantity of money paid by the mechanisms induced from the algorithms UNIFORM and UNIFORM_RR. In our testing we have considered three basic algorithms: LPT, UNIFORM and UNIFORM_RR. We have also considered several variations of these three algorithms, obtained by changing the number of blocks, if used, or rounding the speeds of the machines. In particular, the restricted versions of the two algorithms take in input the machine speeds, round up the speeds to a power of 2 and then compute the scheduling. Table 1 summarizes the algorithms we have considered in our testing. We have performed experiments with respect to arbitrary speeds. We executed our measures on three different runs: in each run we fix the number of machines and the number of tasks and select speeds uniformly in a range $[1, 2^\beta]$ with $1 \leq \beta \leq 6$ and task weights uniformly in a range $[1, 2^\alpha]$ with $0 \leq \alpha \leq 8$. Table 2 gives a summary of the instances performed in each run. For each instance we have measured the makespan and the approximation factor. Then we have computed the average makespan, the average approximation factor and the worst case approximation factor in each run. We have also performed similar experiments for speeds and weights selected according to a normal distribution and for 2–divisible speeds. The results obtained are substantially equivalent and we omit them. Figure 3 shows the worst case approximation factors obtained in the three runs. Table 3, instead, shows the average approximation factors, where the average is computed on the set of all the instances. Experiments give evidence that UNIFORM_RR obtains the best results among the monotone algorithms considered in our testing.

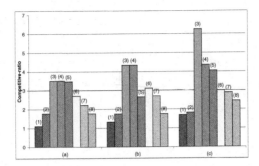

Fig. 3. Worst case approximation factors obtained in our testing. Algorithms are labeled as shown in Table 5. (a) 10 tasks and 4 machines (b) 15 tasks and 5 machines (c) 100 tasks and 15 machines

Table 3. Average Competitive ratio computed on all the instances

Average approximation factors of monotone scheduling algorithms	
LARGEST PROCESSING TIME	1.377031
LARGEST PROCESSING TIME RESTRICTED	1.777902
UNIFORM MCD=1	4.692374
UNIFORM MCD=1 RESTRICTED	4.062987
UNIFORM RESTRICTED	3.387385
UNIFORM ROUND-ROBIN MCD=1	2.935213
UNIFORM ROUND-ROBIN MCD=1 RESTRICTED	2.600026
UNIFORM ROUND-ROBIN RESTRICTED	1.988051

Its approximation factor is very close to LPT, both in the worst case and in the average case. UNIFORM, instead, obtains an approximation factor that is very close to the theoretical bound. An unexpected result is that the restricted version of UNIFORM_RR obtains better results than the unrestricted version of the same algorithm and its performances improve when the number of tasks increase. Our interpretation is that the restriction version of the problem uses more blocks and thus obtains a more balanced assignment of virtual machines to real machines, counterbalancing the approximation induced by the rounding of the machine speeds.

We have also experimentally measured the impact of the proposed greedy-close monotone algorithms on the performances of the PTAS-Gc algorithm defined in [AP04]. Notice that PTAS-Gc takes three inputs: the task sequence, the speed vector and a parameter h, that is the number of tasks that are allocated optimally in the first phase of the algorithm. Our testing is organized in two runs: the first run is performed on instances with 15 tasks and 4 machines; the second run is performed on instances with 25 tasks and 5 machines. For each instance of σ and s we run the algorithm with $h \in \{0, 3, 5, 8\}$ Our experiments point out two interesting aspects: the first one is that, since the largest tasks are

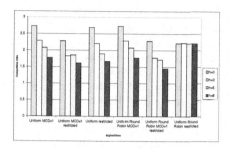

Fig. 4. Average approximation factor of PTAS-Gc, computed on all the instances of our testing, for different selections of Gc and h

assigned optimally, the difference in the performances between UNIFORM_RR and the other heuristics is significantly less; the second one is that all the considered heuristics, except for UNIFORM_RR, improve their performances when h increases. In particular, for h sufficiently large, algorithm UNIFORM outperforms UNIFORM_RR. However, this relatively small improvement in the approximatio factor is counterbalanced by a dramatic growing in the computatio time.

6   Conclusion

The contribution of this paper is twofold. From a theoretical point of view, we have proved that greedy algorithms like LPT and LS are monotone if we restrict to the case of 2 machines with c–divisible speeds, for c large enough. We think that this technique can be generalized to prove that greedy algorithms can be made monotone with a loss in the approximation factor even for the case of $m > 2$ machines. From an experimental point of view we have analyzed several heuristics, based on the algorithm UNIFORM, and proved that making a more clever assignments of virtual machines to real machines can significantly improve the performances of the algorithm. In particular, we have shown that in several cases rounding machine speeds can yield better results than solving the problem with respect to the original speeds. However, if we could prove that LPT is monotone for c-divisible speeds, for a small c, we could obtain even better approximations.

References

[AA04]   P. Ambrosio and V. Auletta. Deterministic monotone algorithms for scheduling on related machines. Technical Report Università di Salerno, 2004.

[AT01]   A. Archer and E. Tardos. Truthful mechanisms for one-parameter agents. In *Proc. of FOCS*, pages 482 - 491, 2001.

[AA+93] J. Aspnes, Y. Azar, A. Fiat, S. Plotkin and O. Warts. On-Line Routing of Virtual Circuits with Applications to Load Balancing and Machine Scheduling. In *Journal of ACM*, 44: 486 - 504.

[AD+04] V. Auletta, R. De Prisco, P. Penna and G. Persiano. The Power of Verification for One-Parameter Agents. In *Proc. of ICALP '04*, 2004.

[AP04] V. Auletta, R. De Prisco, P. Penna and G. Persiano. Deterministic Truthful Mechanisms for Scheduling on Selfish Machines. In *Proc. of STACS '04*, LNCS 2996, pages 608 - 619, 2004.

[BC97] P. Berman, M. Charikar and M. Karpinski. On-Line load Balancing for Related Machines. In *Proc. of WADS '97*, 1997.

[CS80] Y. Cho and S. Sahni. Bounds for list scheduling on uniform processors. *SIAM Journal of Computing*, vol.9 n.1, 1980.

[Cl71] E. H. Clarke. Multipart Pricing of Public Goods. *Public Choice*, pages 17 - 33, 1971.

[GI80] T. Gonzalez, O. Ibarra and S. Sahni. Bounds for LPT schedules on uniform processors. *SIAM Journal of Computing*, vol.6, 1977.

[Gr73] T. Groves. Incentive in Teams. *Econometrica*, 41 : 617 - 631, 1973.

[Gr66] R. L. Graham. Bounds for certain multiprocessing anomalies. *Bell System Tech. Journal*, 45 : 1563 - 1581, 1966.

[Gr69] R. L. Graham. Bound on multiprocessing timing anomalies. *SIAM Journal on Applied Mathematics*, 17:416-429, 1969.

[HS88] D. S. Hochbaum and D. B. Shmoys. A Polynomial Approximation Scheme for Scheduling on uniform processors: Using the Dual Approximation Approach. In *SIAM Journal of Computing*, vol. 17(3) pages 539 - 551, 1988.

[NR00] N. Nisan and A. Ronen. Computationally Feasible VCG Mechanisms. In *Proceedings of the 2nd ACM Conference on Electronic Commerce (EC)*, pages 242 - 252, 2000.

[NR99] N. Nisan and A. Ronen. Algorithmic Mechanism Design. In *Proc. of the 31st STOC*, pages 129 - 140, 1999.

[OR94] M. J. Osborne and A. Rubinstein. *A course in Game Theory*. MIT Press, 1994.

[Pa01] C. H. Papadimitriou. Algorithms, Games and the Internet. In *Proc. of the 33rd STOC*, 2001.

[Ro00] A. Ronen. *Solving Optimization Problems Among Selfish Agents*. PhD thesis, Hebrew University in Jerusalem, 2000.

[V61] W. Vickrey. Counterspeculation, Auctions and Competitive Sealed Tenders. *Journal of Finance*, 8 - 37, 1961.

Better Bounds for Minimizing SONET ADMs

Leah Epstein[1],[*] and Asaf Levin[2]

[1] School of Computer Science, The Interdisciplinary Center,
Herzliya, Israel
lea@idc.ac.il.
[2] Department of Statistics, The Hebrew University,
Jerusalem, Israel
levinas@mscc.huji.ac.il

Abstract. SONET add-drop multiplexers (ADMs) are the dominant cost factor in SONET /WDM rings. The number of SONET ADMs required by a set of traffic streams is determined by the routing and wavelength assignment of the traffic streams. Following previous work, we consider the problem where the route of each traffic stream is given as input, and we need to assign wavelengths so as to minimize the total number of used SONET ADMs. This problem is known to be NP-hard, and the best known approximation algorithm for this problem has a performance guarantee of $\frac{3}{2}$. We improve this result, and present a $\frac{10}{7}$-approximation algorithm. We also study some of the previously proposed algorithms for this problem, and give either tight or tighter analysis of their approximation ratio.

1 Introduction

WDM (Wavelength Division Multiplexing)/SONET (Synchronous Optical NETworks) rings form a very attractive network architecture that is being deployed by a growing number of telecom carriers. In this architecture each wavelength channel carries a high-speed SONET ring. The key terminating equipments are optical add-drop multiplexers (OADM) and SONET add-drop multiplexers (ADM). Each vertex is equipped with exactly one OADM. The OADM can selectively drop wavelengths at a vertex. Thus, if a wavelength does not carry any traffic from or to a vertex, its OADM allows that wavelength to optically bypass the vertex. Therefore, in each SONET ring a SONET ADM is required at a vertex if and only if it carries some traffic terminating at this vertex. In this paper we study the problem of minimizing the total cost incurred by the SONET ADMs.

Formally, we are given a set E of circular-arcs over the vertices $0, 1, \ldots, n-1$, where the vertices are ordered clockwise. A pair of arcs $(i, j), (k, l)$ is *non-intersecting* if the clockwise paths along the cycle $0, 1, \ldots, n-1, 0$ that connects i to j and the clockwise path that connects k to l do not share any arc of the

---

[*] Research supported by Israel Science Foundation (grant no. 250/01).

G. Persiano and R. Solis-Oba (Eds.): WAOA 2004, LNCS 3351, pp. 281–294, 2005.

cycle. A set of arcs is non-intersecting if each pair of arcs from this set is non-intersecting. A feasible solution is a partition of E into non-intersecting subsets of arcs $E_1, E_2, \ldots, E_p$. The cost of E_i is the number of different vertices of the ring that are end-points of the arcs of E_i. The cost of the solution is the sum of costs of E_i for all i. The goal is to find a minimum cost feasible solution.

For an arc (i, j), we define its *length* as $\ell(i, j) = j - i \mod n$. For a subset of arcs, the length of the subset is the total length of its arcs. Throughout the paper we often use vertex numbers x where $x \geq n$ to denote the vertex $x \mod n$. We omit the mod operation to simplify notations.

A *chain* is an open directed path of length at most $n-1$, and a *cycle* is a closed directed path of length exactly n. W.l.o.g. we can assume that the arcs in each E_i form a connected component (either a chain or a cycle). This is so because if the arcs in E_i are disconnected, then we can partition E_i to its connected components without increasing its total cost. Therefore, we ask for a partition of E into cycles and (open-)chains. The cost of a feasible solution equals the sum of $|E|$ and the number of chains in the solution.

Liu, Li, Wan and Frieder [3] proved that this problem is NP-hard. They also considered a set of heuristics, and tested them empirically. Gerstel, Lin and Sasaki [2] also designed some heuristics for this problem. Wan, Calinsecu, Liu and Frieder [4] proved that any nontrivial heuristic is a 7/4-approximation algorithm. I.e., any algorithm such that none of its chains can be united to form a larger chain (a local optimum) is a 7/4-approximation algorithm. Calinsecu and Wan [1] provided a 3/2-approximation algorithm, and analyzed the worst-case performance of the previously studied heuristics. Below, we further describe the results of [1].

Let OPT be a given optimal solution to our problem with cost opt. Assume that for $i = 2, 3, \ldots$, OPT has CY_i cycles with i arcs, and for $i = 1, 2, \ldots$, OPT has CH_i chains with i arcs. We further assume that CY_2 is maximized among all optimal solutions, and as noted in [1] that no feasible solution can have a higher value of CY_2. For an algorithm A, we also use A to denote the cost of its returned solution. We sometimes use APX to denote the cost of a solution returned by an approximation algorithm.

A feasible solution SOL induces a partition of the arcs into an Eulerian subgraph and a set of *mega-chains* as follows: We consider the set of cycles and chains used by SOL as a set of arcs in directed auxiliary graph over $\{0, 1, \ldots, n-1\}$ where cycles are loops and a chain is a directed arc from its starting vertex to its end vertex. In this directed graph we find a maximal subgraph in which the in-degree of each vertex equals its out-degree. The remaining arcs define a minimal set of chains such that each such chain is directed from a vertex whose out-degree is greater than its in-degree, towards a vertex whose in-degree is greater than its out-degree. Each such chain in the auxiliary graph corresponds to a *mega-chain* in the original graph (by replacing each arc in the auxiliary graph by its corresponding chain). Therefore, each mega-chain is composed of chains. The remaining arcs in the original graph are the Eulerian subgraph. Note that the Eulerian subgraph need not be connected. Note that the number of mega-chains in SOL is independent of SOL, and is common to all feasible solutions.

We now formalize Algorithm Iterative Matching (IM) (see [1]). The algorithm maintains a set of valid chains of arcs $\mathcal{P}$ that covers E throughout its execution. Initially, $\mathcal{P}$ consists of chains each of which is an arc in E. The fit graph $\mathcal{F}(\mathcal{P})$ is defined as follows: its vertex set is $\mathcal{P}$, and two of its vertices are connected by an edge if the two corresponding chains have a common end-point, and they can be concatenated to form a valid chain. The algorithm constructs $\mathcal{F}(\mathcal{P})$, and if its edge set is not empty, then it finds a maximum matching $\mathcal{M}$ in $\mathcal{F}(\mathcal{P})$. Then, it merges each matched pair of chains of arcs in $\mathcal{M}$ into a longer chain. When the edge set of $\mathcal{F}(\mathcal{P})$ is empty, $\mathcal{P}$ is the valid chain generation that is given as output. Calinescu and Wan [1] showed that the approximation ratio of Algorithm IM is at most $\frac{5}{3}$, and provided a negative example for the algorithm that shows that its approximation ratio is at least $\frac{3}{2}$. We improve the negative examples of the algorithm by presenting an example where the approximation ratio of the algorithm is at least 1.6. For a variant PPIM of algorithm IM with a preprocessing step that removes all cycles with two arcs each, we present a negative example that shows that the approximation ratio of PPIM is at least $\frac{14}{9}$. We further show that the approximation ratio of IM is strictly less than $5/3$.

Calinescu and Wan considered a variant of Algorithm IM: Algorithm Preprocessed Iterative Matching (PIM) defined as follows:

1. Preprocessing phase: repeatedly remove cycles consisting of remaining arcs until no more cycle can be obtained.
2. Matching phase: apply Algorithm IM to the arcs remaining after the first phase.

They showed that Algorithm PIM has an approximation ratio of at most $\frac{3}{2}$, and gave a negative example for PIM that shows that its approximation ratio is at least $\frac{4}{3}$. We show that the $\frac{3}{2}$ bound is tight. We also provide a better analysis of the approximation ratio of the algorithm. This improved analysis in Section 3 does not improve the worst case performance of the algorithm, but together with our Algorithm GPTS defined in Section 4 it provides Algorithm COMB, and the main result of this paper (shown in Section 5) is that Algorithm COMB is a 10/7-approximation algorithm. We show that the approximation ratio of algorithm COMB is at least $\frac{4}{3}$.

Algorithm Preprocessed Cut and Merge (PCM) is defined as follows:

1. Remove all cycles of two arcs each.
2. Choose a cycle's arc $(i, i+1)$ with minimum load, and let B_i denote the subset of E that pass through $(i, i+1)$. Partition $E \setminus B_i$ into an optimal set of chains using a greedy procedure. Let $\mathcal{P}$ be the obtained chains.
3. Construct a weighted bipartite graph with sides B_i and $\mathcal{P}$ as follows: if $a \in B_i$ can be merged with $P \in \mathcal{P}$, add an edge between a and P with weight equal to the number of their common end-vertices. Find a maximum-weight matching in the resulting graph. Merge each pair of arc and a chain into a larger chain. This step is repeated until no further merging can be obtained.

Calinescu and Wan [1] proved that the approximation ratio of PCM is between $3/2$ and $5/3$. In this paper we close this gap, and improve the lower bound on the approximation ratio of PCM by showing that the approximation ratio of PCM is exactly $5/3$.

Algorithm Preprocessed Eulerian Tour-Trail Splitting (PET-TS) is defined as follows:

1. Remove all cycles of two arcs each.
2. Eulerian tour phase: add a minimum size set of fake arcs E' to make the directed graph with arc set $E \cup E'$ Eulerian (and if it is disconnected, each connected component is Eulerian). Find an Eulerian tour in this graph, and remove all the fake arcs from this tour to obtain a set of trails.
3. Trail decomposing phase: Decompose each trail into simple paths and circuits.
4. Chain split phase: split each (open) path into valid chains by walking along the path from its first arc, and generating a valid chain whenever overlap occurs; split each invalid circuit into valid chains by walking along the circuit from each arc, generating a valid chain whenever overlap occurs, and then choose the one with the smallest number of open chains.
5. Chain merging phase: Repeatedly merge any pair of open chains into a larger valid chain until no more merging can occur.

Calinescu and Wan [1] proved that the approximation ratio of PET-TS is between $3/2$ and $7/4$. We narrow this gap by showing that the approximation ratio of PET-TS is at least $5/3$.

The paper [1] also considered MCC-TS that is a variation of PET-TS in which the Eulerian tour phase is replaced by the following: define a weighted directed graph $H(E)$ with vertex set E as follows. For any pair of non-intersecting arcs $e_1, e_2 \in E$, such that $e_1 = (i, j)$ and $e_2 = (k, l)$, add an arc from e_1 to e_2 and an arc from e_2 to e_1. If e_1, e_2 do not share any end-vertices, then the weights of both arcs are set to two. If $j = k$, then the weight of the arc from e_1 to e_2 is set to one, and the weight of the arc from e_2 to e_1 is set to two. Otherwise, the weight of the arc from e_1 to e_2 is set to two, and the weight of the arc from e_2 to e_1 is set to one. Now, find a minimum weight circuit cover of $H(E)$. Remove from it all the arcs of weight two to obtain a set of paths and circuits. Calinescu and Wan [1] proved that the approximation ratio of MCC-TS is between $3/2$ and $8/5$. We close this gap by showing that the approximation ratio of MCC-TS is exactly $14/9$. If the pre-processing step of two arcs cycles removal is not performed, we show that the bound $8/5$ is tight (we call this algorithm NMCC-TS).

Note that although we consider the absolute approximation ratio in this paper, all results are valid for the asymptotic approximation ratio as well. All negative examples can be easily magnified by taking multiple copies of each input arc, to form arbitrary large negative examples.

2   Negative Examples

In this section we give negative examples where the approximation ratio is at least 3/2. This will show that the upper bound of 3/2 on the performance of PIM given in [1] is tight i.e. that the following theorem holds.

Theorem 1. *The approximation ratio of PIM is exactly $\frac{3}{2}$. The approximation ratio of any algorithm that removes cycles in an arbitrary order (even if it removes the two arc cycles first) and then solves the remaining instance, is at least $\frac{3}{2}$.*

If we are interested in the design of a better approximation algorithm, the negative examples in this section exclude the option that a better analysis of PIM or a design of a similar algorithm that replaces the matching phase may be the answer.

Proof. We start with a very simple example showing that an algorithm which removes cycles in an arbitrary way cannot perform better than 3/2. Let $n = 3$ and the input arcs be $(0, 1), (0, 2), (1, 2), (1, 0), (2, 0), (2, 1)$. Clearly, OPT consists of three two arc cycles which are $(i, i + 1), (i + 1, i)$ for $i = 0, 1, 2$, and therefore $opt = 6$. However, if the algorithm removes the cycle $(0, 1), (1, 2), (2, 0)$, then it is left with three arcs of length $2 > n/2$ that cannot be combined. Therefore, we have $APX = 9$. This gives approximation ratio of at least 3/2.

The above input consists of two arcs cycles only. As it was already noticed in [1], it is easy to remove such cycles before processing any algorithm, and prevent the situation above. In the next example we show that even if there are no two arc cycles in the input, still an arbitrary removal procedure cannot reach smaller performance ratios. Moreover, we consider the following **exponential-time** algorithm: first, remove all cycles of two arcs, next, remove cycles one after the other until the remaining arcs do not contain a cycle, and finally solve optimally (in exponential time) the remaining arcs. We show that this algorithm has an approximation ratio of at least 3/2. Since this algorithm outperforms PIM, we conclude that it is a 3/2-approximation algorithm (however, not a polynomial-time).

For a given integer parameter $\alpha \geq 2$, consider $n = 2\alpha^2 - 4\alpha + 4$, and the arc set (with the optimal solution) is given by: for every $0 \leq i \leq \alpha - 3$ and every $0 \leq j \leq \alpha - 3$, we have the arcs $(\alpha j + i, \alpha j + i + 1)$, $(\alpha j + i + 1, n - \alpha i - j - 5)$, $(n - \alpha i - j - 5, n - \alpha i - j - 4)$ and $(n - \alpha i - j - 4, \alpha j + i)$. For every $0 \leq i \leq \alpha - 4$, we have the arcs $(n - \alpha(i + 1) - 4, n - \alpha(i + 1) - 3)$, $(n - \alpha(i + 1) - 3, n - \alpha(i + 1) - 2)$ and $(n - \alpha(i + 1) - 2, n - \alpha(i + 1) - 4)$. For every $1 \leq j \leq \alpha - 3$, we have the arcs $(\alpha j - 2, \alpha j - 1)$, $(\alpha j - 1, \alpha j)$ and $(\alpha j, \alpha j - 2)$. Finally, we have the twelve arcs of the following four triangles: $(\alpha^2 - 2\alpha - 2, \alpha^2 - 2\alpha - 1, \alpha^2 - 2\alpha)$, $(\alpha^2 - 2\alpha, \alpha^2 - 2\alpha + 1, \alpha^2 - 2\alpha + 2)$, $(n - 4, n - 3, n - 2)$ and $(n - 2, n - 1, 0)$. Then, OPT has $(\alpha - 2)^2$ cycles of four arcs and $2(\alpha - 2) + 4$ cycles of three arcs, and its total cost is exactly $4(\alpha - 2)^2 + 3[2(\alpha - 2) + 4] = 4\alpha^2 - 10\alpha + 16$.

We now argue that the instance contains the arc $(t, t + 1)$ for every t. The arcs $(0, 1), \ldots, (\alpha^2 - 2\alpha - 3, \alpha^2 - 2\alpha - 2)$ are given by the arcs $(\alpha j + i, \alpha j + i + 1)$

for $0 \le i \le \alpha - 3, 0 \le j \le \alpha - 3$. In this set there is a gap of two arcs every $\alpha - 2$ arcs which is filled by the arcs $(\alpha j - 2, \alpha j - 1), (\alpha j - 1, \alpha j)$ for $1 \le j \le \alpha - 3$. Similarly the arcs $(\alpha^2 - 2\alpha + 2, \alpha^2 - 2\alpha + 3), \ldots, (n - 5, n - 4)$ are given by the arcs $(n - \alpha i - j - 5, n - \alpha i - j - 4)$ for $0 \le i \le \alpha - 3, 0 \le j \le \alpha - 3$. In this set there is again a gap of two arcs every $\alpha - 2$ arcs which is filled by the arcs $(n - \alpha(i+1) - 4, n - \alpha(i+1) - 3), (n - \alpha(i+1) - 3, n - \alpha(i+1) - 2)$ for $0 \le i \le \alpha - 4$. The remaining eight arcs $(\alpha^2 - 2\alpha - 2, \alpha^2 - 2\alpha - 1), \ldots, (\alpha^2 - 2\alpha + 1, \alpha^2 - 2\alpha + 2)$ and $(n - 4, n - 3), \ldots, (n - 1, 0)$ are given by the arcs of the last four triangles (except for the last arc of each triangle). Assume that the cycle which consists of n arcs is exactly the cycle that our algorithm removes.

Note that in the remaining arc set S each arc has length at least 4. We next show that in the optimal solution for the remaining arcs each arc consists of its own chain. To see this it is enough to show that if there is a pair of arcs in S with a common end-vertex v, then their total length is at least $n + 1$ (this claim also shows that the original instance does not contain two arcs cycles). First, note that if one of the arcs incident at v occurs in one of the triangles of OPT, then its length is exactly $n - 2$, the other arc has length at least 4, and therefore their total length is greater than $n + 1$. Therefore, we can assume that the pair of arcs incident at v are from the four arcs cycles of OPT. Let $0 \le i, j \le \alpha - 3$:

- Assume that $v = \alpha j + i$. Then, the arcs incident at v are $(n - \alpha i - j - 4, v)$ and $(v, n - \alpha(i - 1) - j - 5)$, and their total length is $n + \alpha - 1 > n$ for all values of $\alpha \ge 2$.
- Assume that $v = n - \alpha i - j - 5$. Then, the arcs incident at v are $(\alpha j + i + 1, v)$ and $(v, \alpha(j + 1) + i)$, and their total length is $n + \alpha - 1 > n$ for all values of $\alpha \ge 2$.

Therefore, our optimal solution for S is a chain for each arc. Since $|S| = 2(\alpha - 2)^2 + 1[2(\alpha - 2) + 4]$ (S contains two arcs from each cycle of four arcs in OPT, and one arc from each triangle of OPT), we conclude that the cost of the approximation algorithm is $|E| + |S| = 6(\alpha - 2)^2 + 4[2(\alpha - 2) + 4] = 6\alpha^2 - 16\alpha + 24$. Therefore, the approximation ratio of the algorithm approaches $3/2$ as α goes to infinity (also n grows to infinity).

3   A Better Analysis of the Algorithm PIM

In this section we assume that the Preprocessing phase of Algorithm PIM first removes cycles with two arcs, and only if such cycles do not exist, other cycles are removed.

The proof of the next theorem is similar to the proof of Lemma 19 in [1].

Theorem 2. *Algorithm PIM returns a solution whose cost is at most*
$$1 \cdot 2CY_2 + \frac{4}{3} \cdot 3CY_3 + \frac{7}{5}5CY_5 + 1 \cdot (2CH_1 + 3CH_2) + \frac{5}{4}(4CH_3 + 5CH_4) + \frac{3}{2}4CY_4$$
$$+ \frac{3}{2}\left(\sum_{i=6}^{n} iCY_i + \sum_{i=5}^{n-1}(i+1)CH_i\right).$$

Proof. To prove the claim we assign the cost of the solution obtained by Algorithm PIM to the arcs, such that the following properties hold:

1. The total cost assigned to the arcs that belong to a cycle in OPT of two arcs is exactly the cost paid by OPT to this cycle, i.e. 2.

2. The total cost assigned to the arcs that belong to a cycle in OPT of three (resp. five) arcs is at most $\frac{4}{3}$ (resp. $\frac{7}{5}$) times the cost paid by OPT to this cycle, i.e. 4 (resp. 7). The total cost assigned to the arcs that belong to a cycle in OPT of four arcs or at least six arcs is at most $\frac{3}{2}$ times the cost paid by OPT to this cycle.

3. The total cost assigned to the arcs that belong to a chain in OPT of at most two arcs is exactly the cost paid by OPT to this chain. The total cost assigned to the arcs that belong to a chain in OPT of three or four arcs is at most $\frac{5}{4}$ times the cost paid by OPT to this chain. The total cost assigned to the arcs that belong to a chain in OPT of at least five arcs is at most $\frac{3}{2}$ times the cost paid by OPT to this chain.

To prove property 1, note that the preprocessing phase take out exactly all cycles of OPT of exactly two arcs, and therefore their cost in the solution obtained by PIM is exactly their cost in OPT.

We prove the other properties by considering not the solution obtained by PIM, but an alternative solution that is no better than PIM in terms of cost. We replace the solution of PIM with the solution obtained by PIM after the first iteration of the matching phase. This is clearly an upper bound on the cost of PIM. We further replace the solution by a solution that does not create an optimal matching, but some feasible matching that we construct. The matching is constructed by uniting matchings that may be created from the remaining arcs (after cycle removal by PIM) from each component of OPT (cycle or chain) separately. Note that a remaining path of s arcs contributes a matching of size $\lfloor s/2 \rfloor$.

We now prove property 2. Each cycle of OPT loses at least one arc in the preprocessing phase. Consider a cycle of OPT which contains k arcs ($k \geq 3$). Let $\ell \geq 1$ be the number of arcs that the cycle loses in the preprocessing phase. There is a matching of the arcs from this cycle of size $\lfloor \frac{k-1}{2} \rfloor - (l-1)$ (after the first arc from this cycle is removed, we have a matching of size $\lfloor \frac{k-1}{2} \rfloor$. Every other arc that is removed destroys at most one arc in the matching). The cost of the cycle depends on the number of chains created from it. This number is at most $\lceil \frac{k-1}{2} \rceil \leq \frac{k}{2}$. I.e., for three arcs cycles we have at most one chain, and five arcs cycles we have at most two chains. This gives the costs 4 and 7 for cycles of three and five arcs (respectively). For a cycle of k arcs we get at most $k/2$ chains, which proves the case of four arc cycles, and cycles of at least six arcs.

Next, we prove property 3. For a chain in OPT of k arcs such that l arcs are taken out in the preprocessing phase, there is a matching of the arcs from this chain of size $\lfloor \frac{k}{2} \rfloor - l$ (before the preprocessing phase we have a matching of size $\lfloor \frac{k}{2} \rfloor$, and every arc that is removed, destroys at most one arc in the matching). Therefore, the number of chains created after the first matching step equals the number of chains OPT has if the number of arcs in the chain is one or two. If the number of arcs in the chain is three or four, then the number of chains after the first matching phase is at most two (whereas OPT has one chain). Therefore, in this case we have a ratio of at most $\frac{5}{4}$ between the cost OPT pays for this chain and the cost PIM pays for this chain. For longer chains, we get at most $\lceil \frac{k}{2} \rceil - 1$ new chains (compared to OPT), and so PIM pays at most $\frac{3}{2}$ times the cost OPT pays.

Taking the optimal matching instead of the matching we describe above, only decreases the cost of the solution, and therefore the claim holds also for the solution obtained by PIM.

4   Algorithm GPTS

In this section we study a different approximation algorithm for the problem. Given a set of input arcs we apply a certain greedy clean-up preprocessing phase that is composed of five steps. First, we remove all cycles of two arcs (the number of the cycles that we remove in this step is exactly CY_2). Then, we remove greedily certain subgraphs (cycles or chains with a certain number of arcs and certain length), by removing each time a single such subgraph as long as the remaining graph contains a subgraph with the desired property. E.g., in Step 2 we remove cycles of three arcs each one at a time until the remaining arc-set does not contain a cycle with exactly three arcs.

Algorithm Greedy-Preprocessing Trail-Split (GPTS):
1. Remove all cycles of two arcs.
2. Remove greedily cycles of three arcs until there are no such cycles.
3. Remove greedily mega-chains of at most three arcs with length in the interval $[\frac{3n}{4}, n - 1]$ until there are no such mega-chains.
4. Remove greedily cycles of four arcs until there are no such cycles.
5. Remove greedily cycles of five arcs until there are no such cycles.
6. Cover the rest of the arcs with chains in the following way:
   (a) Find a set of mega-chains (with arbitrary lengths) that connect the odd-vertices and remove them. For each such mega-chain of length greater than n, decompose it into chains of length at most n.
   (b) Partition the rest of the arcs (these are from the Eulerian subgraph) into chains (or cycles) of length at most n.

Observation 3. *Consider a mega-chain with length in the interval $[kn, (k + 1)n - 1]$ that is created in step 6a. Then, the number of chains that result from it is at most $2k + 1$.*

Proof. We perform a greedy partition that chooses a maximum length chain at each step. Therefore, the total length of each pair of consecutive chains in a mega-chain is at least n (otherwise, they can be united).

Observation 4. *The chains obtained in step 6b have an average length of at least $\frac{n}{2}$.*

Proof. The claim follows because the total length of each pair of consecutive chains resulting from the Eulerian subgraph is at least n (otherwise, they can be united).

Notations: consider OPT. Partition it into mega-chains and an Eulerian subgraph. There may be several options to do that, therefore we fix an arbitrary partition. Denote by:

- CY - the number of cycles in OPT that contain at least six arcs.
- CH_E - the number of chains in OPT that are part of the Eulerian subgraph defined with respect to OPT.
- MC - the total number of mega-chains.
- CH_M - the number of chains in OPT that are part of mega-chains defined with respect to OPT.
- MC_3 - the number of mega-chains of at most three arcs with total length in the interval $[\frac{3n}{4}, n-1]$. Note that such a mega-chain is a chain of OPT.
- MC_s - the number of mega-chains with total length less than $\frac{3n}{4}$.
- MC_4 - the number of mega-chains with exactly four arcs and total length in the interval $[\frac{3n}{4}, n-1]$.
- MC_5 - the number of mega-chains with at least five arcs and total length in the interval $[\frac{3n}{4}, n-1]$.
- MC^i - the number of mega-chains with total length in the interval $[in, (i+1)n-1]$. MC^i is defined for all $i \geq 1$.
- CH_E^1 and CH_E^2 - the number of chains in OPT that are part of the Eulerian subgraph, of exactly one arc, and exactly two arcs, respectively.
- CH_M^1 and CH_M^2 - the number of chains in OPT that are part of mega-chains, of exactly one arc, and exactly two arcs, respectively.

Note that a mega-chain in OPT with total length in the interval $[in, (i+1)n-1]$ consists of at least $i+1$ chains (in OPT).

The total length of all arcs is at most $UB_L = (CY_2 + CY_3 + CY_4 + CY_5 + CY)n + CH_En + MC_3n + MC_s\frac{3n}{4} + MC_4n + MC_5n + \sum_{i=1}^{\infty}[MC^i(i+1)n]$.

Consider Algorithm GPTS, and denote by A the number of cycles removed in step 2, B- the number of chains removed in step 3, C- the number of cycles removed in step 4. D - the number of cycles removed in Step 5. Then, the following inequalities hold: $3A \geq CY_3$, $3A + 3B \geq CY_3 + MC_3$, $3A + 3B + 4C \geq CY_3 + MC_3 + CY_4$, and $3A + 3B + 4C + 5D \geq CY_3 + MC_3 + CY_4 + CY_5$. Subject to these constraints we want to minimize $nA + \frac{3nB}{4} + nC + nD$ (this is a lower bound on the total length that we gain by removing these subgraphs). An optimal solution of this (parametric) mathematical program is $A = \frac{CY_3}{3}$, $B = \frac{MC_3}{3}$, $C = \frac{CY_4}{4}$, and $D = \frac{CY_5}{5}$.

This proves that each feasible solution to the mathematical program has a cost of at least $\frac{n}{3}CY_3 + \frac{n}{4}MC_3 + \frac{n}{4}CY_4 + \frac{n}{5}CY_5$, and this is the cost of the (feasible) solution that we outlined above, and therefore it is optimal.

Therefore, the total length of the arcs that are left in the beginning of step 6 is at most $UB_L' = UB_L - \frac{CY_3}{3}n - \frac{MC_3}{3} \cdot \frac{3n}{4} - \frac{CY_4}{4}n - \frac{CY_5}{5}n$.

By Observations 3 and 4, the total number of chains obtained by our algorithm is at most $\frac{UB_L'}{n/2} + MC$ (note that this include also the chains from step 3).

In the remaining of this section we use the fact that Step 1 of the algorithm correctly identifies all the cycles of OPT that have two arcs. Therefore, we can assume that $CY_2 = 0$.

We use the fact that $MC = MC_3 + MC_s + MC_4 + MC_5 + \sum_{i=1}^{\infty} MC^i$ to obtain that the number of chains resulted by our algorithm is at most: $APX_{CH} = \frac{4}{3}CY_3 + \frac{3}{2}CY_4 + \frac{8}{5}CY_5 + 2CY + 2CH_E + \frac{5}{2}MC_3 + \frac{5}{2}MC_s + 3MC_4 + 3MC_5 + \sum_{i=1}^{\infty} MC^i[2(i+1)+1]$. We denote $OMC = MC_3 + MC_s + \sum_{i=1}^{\infty}[MC^i(i+1)]$. Since $2(i+1)+1 \le \frac{5}{2}(i+1)$ for all $i \ge 1$, we reduce the term $\sum_{i=1}^{\infty} MC^i[2(i+1)+1]$ into $\frac{5}{2}\sum_{i=1}^{\infty}[MC^i(i+1)]$.

Therefore, the total cost of the solution obtained by our algorithm is $apx = |E| + APX_{CH} \le |E| + \frac{4}{3}CY_3 + \frac{3}{2}CY_4 + \frac{8}{5}CY_5 + 2CY + 2CH_E + 3MC_4 + 3MC_5 + \frac{5}{2}OMC$, and the cost of OPT is $opt = |E| + CH_E + MC_4 + MC_5 + OMC$.

We consider the partition of E according to the roles of arcs in OPT to the following (disjoint) subsets: $E(CY_3)$: arcs that belong to cycles of three arcs in OPT. $E(CY_4)$: arcs that belong to cycles of four arcs in OPT. $E(CY_5)$: arcs that belong to cycles of five arcs in OPT. $E(CY)$: all the other arcs that belong to cycles in OPT. $E(CH_E)$: arcs that belong to the Eulerian subgraph but not to $E(CY_3) \cup E(CY_4) \cup E(CY_5) \cup E(CY)$. $E(MC_4)$: arcs that belong to mega-chains of exactly four arcs with total length in the interval $[\frac{3n}{4}, n-1]$. $E(MC_5)$: arcs that belong to mega-chains of at least five arcs with total length in the interval $[\frac{3n}{4}, n-1]$. $E(OMC)$: the rest of the arcs that belong to mega-chains.

The following equations and inequalities hold using the numbers of arcs in the subgraphs of OPT.

$$|E(CY_3)| + \frac{4}{3}CY_3 = \frac{13}{9}|E(CY_3)| \tag{1}$$

$$|E(CY_4)| + \frac{3}{2}CY_4 = \frac{11}{8}|E(CY_4)| \tag{2}$$

$$|E(CY_5)| + \frac{8}{5}CY_5 = \frac{33}{25}|E(CY_5)| \tag{3}$$

$$|E(CY)| + 2CY \le |E(CY)| + \frac{2}{6}|E(CY)| = \frac{4}{3}|E(CY)| \tag{4}$$

$$|E(CH_E)| \ge 3CH_E - 2CH_E^1 - CH_E^2 \tag{5}$$

$$|E(MC_4)| + 3MC_4 \le \frac{7}{5}(|E(MC_4)| + MC_4) \tag{6}$$

$$|E(MC_5)| + 3MC_5 \le \frac{4}{3}(|E(MC_5)| + MC_5) \tag{7}$$

$$|E(OMC)| \ge 3OMC - 2CH_M^1 - CH_M^2 \tag{8}$$

Using the inequalities (after re-considering the cycles of two arcs), it is possible to prove the following theorem. We omit the proof due to space constraints.

Theorem 5. *Algorithm GPTS returns a feasible solution whose cost is at most*

$$2CY_2 + \frac{13}{9}|E(CY_3)| + \frac{7}{5}\left(|E(CY_5)| + |E(MC_4)| + MC_4\right) + \frac{7}{4}\left(2CH_E^1 + 3CH_E^2\right.$$

$$\left. +2CH_M^1 + 3CH_M^2\right) + \frac{11}{8}\left(opt - 2CY_2 - |E(CY_3)| - |E(CY_5)| - |E(MC_4)|\right.$$

$$\left. -MC_4 - 2CH_E^1 - 3CH_E^2 - 2CH_M^1 - 3CH_M^2\right).$$

5   An $\frac{10}{7}$-Approximation Algorithm: Algorithm COMB

In this section we design a new approximation algorithm COMB. Algorithm COMB combines the two algorithms: PIM and GPTS. It simply applies both PIM and GPTS, and picks the better solution.

It is possible to adapt the bound on PIM as follows. $PIM \leq 2CY_2 + \frac{4}{3}|E(CY_3)| + \frac{7}{5}|E(CY_5)| + (2CH_1 + 3CH_2) + \frac{5}{4}(MC_4 + E(MC_4)) + \frac{3}{2}\Big(opt - 2CY - |E(CY_3)| - |E(CY_5)| - 2CH_1 - 3CH_2 - MC_4 - |E(MC_4)|\Big).$

We omit the proof of the following theorem.

Theorem 6. *The approximation ratio of Algorithm COMB is at most $\frac{10}{7}$, and at least $\frac{4}{3}$.*

The upper bound is proved by considering a convex combination of the cost of the two algorithms, and showing an upper bound of $10opt/7$ on it. The lower bound is shown by reconsidering example 15 from [1].

6   Other Algorithms

In this section we consider several previously known algorithms, and give tight or tighter bounds on their performance. Due to space restrictions, the analysis of algorithms PCM, PET-TS and IM is omitted.

6.1   MCC-TS

In [1] algorithm MCC-TS has a preprocessing step of two arcs cycles removal. However, the algorithm can be easily adapted to work without this step, and the analysis still works. While building the auxiliary graph the option of two arcs that form a cycle should be taken into account, and the arcs between those arcs both get weight one. It was shown [1] that the performance ratio for this algorithm is in the interval [1.5, 1.6]. We show that the upper bound is tight. To distinguish between the two versions we call them MCC-TS (the version with pre-processing) and NMCC-TS (without pre-processing).

The proof of the following theorem is omitted.

Theorem 7. *Algorithm NMCC-TS has approximation ratio of exactly 1.6.*

For algorithm MCC-TS (with two arc cycles removal), we can show a tight bound of 14/9. We prove it using the next two lemmas.

Lemma 1. *Algorithm MCC-TS has approximation ratio of at least 14/9.*

Proof. Let $n = 24m^4$ for an integer $m > 1$. The input arcs are described in Table 1. The input consists of five families of arcs. Each family has certain amount of parallel copies of arcs (this amount appears in the column Amount). The arc set of each family is parameterized by i or by i, s. For each value of the parameters in the Index range (that appears in the second column) we have the amount of parallel copies of the arcs that appear in the Arcs column.

Table 1. Input arcs

Amount	Index range	Arcs
$12m^3$	$0 \leq i < n$	$(i, i+n/2), (i+n/2, i-2m^2), (i-2m^2, i)$
$24m^2$	$0 \leq i < n, 1 \leq s \leq m$	$(i, i+n/3-sm^2)$
$12m^2$	$0 \leq i < n, 1 \leq s \leq m$	$(i, i+n/3+2(s+1)m^2)$
$12m^3$	$0 \leq i < n$	$(i, i+2), (i, i+3)$
$6m^3$	$0 \leq i < n$	$(i, i-4), (i, i-6)$

We give an upper bound on *opt* by the cost of the following solution. The solution has for every $0 \leq i < n$, $12m^3$ cycles which are $(i, i+n/2), (i+n/2, i-2m^2), (i-2m^2, i)$. For $0 \leq i < n$ we have $6m^3$ cycles of $(i, i+2), (i+2, i+4), (i+4, i)$ and $6m^3$ of $(i, i+3), (i+3, i+6), (i+6, i)$ for $0 \leq i < n$. Finally, for every $0 \leq i < n$ and for every $2 \leq s \leq m$ there are $12m^2$ identical cycles: $(i, i+n/3-sm^2), (i+n/3-sm^2, i+2n/3-2sm^2), (i+2n/3-2sm^2, i)$. The arcs $(i, i+n/3-m^2)$ and $(i, i+n/3+2(m+1)m^2)$, are not combined into cycles but into paths, $12m^2$ copies of $(i, i+n/3-m^2), (i+n/3-m^2, i+2n/3-2m^2)$ for every $0 \leq i < n$ and $6m^2$ of $(i, i+n/3+2(m+1)m^2), (i+n/3+2(m+1)m^2, i+2n/3+4(m+1)m^2)$ for every $0 \leq i < n$. Since $m > 1$, $4(m+1)m^2 < n/3$.

The MCC solution may consist of the following cycles (it manages to combine all arcs into long cycles). Note that each pair of consecutive arcs in each cycle is indeed valid for MCC as their combined length is less than n. This will hold due to the choice of $n = 24m^4$ which gives $(m+2)m^2 < n/3$ and $2(m+1)m^2 < n/6$. We have $12m^2$ copies of the following cycle $(i, i+n/2), (i+n/2, i+5n/6-sm^2), (i+5n/6-sm^2, i+n/3-(s+2)m^2), (i+n/3-(s+2)m^2, i+2n/3-(2s+2)m^2), (i+2n/3-(2s+2)m^2, i)$, for every $0 \leq i < n$ and for every $1 \leq s \leq m$. These cycles can be decomposed into three chains, no matter which arc is chosen to be first. We have the following cycle $6m^3$ times for every $0 \leq i \leq 4m^2 - 1$. The number of arcs in a cycle is $48m^2$, and no vertex is repeated until the cycle is closed. The cycle consists of $6m^2$ phases of eight arcs. For $0 \leq q \leq 6m^2-1$, we have the eight arcs $(i+4qm^2, i+2+4qm^2), (i+2+4qm^2, i+2+(4q+2)m^2), (i+2+(4q+2)m^2, i+4+(4q+2)m^2), (i+4+(4q+2)m^2, i+(4q+2)m^2), (i+(4q+2)m^2, i+3+(4q+2)m^2), (i+3+(4q+2)m^2, i+3+(4q+4)m^2), (i+3+(4q+4)m^2, i+6+(4q+4)m^2), (i+6+(4q+4)m^2, i+(4q+4)m^2)$. Since $m > 1$ is an integer, $2m^2 \geq 8$, and so vertices with different residues (indices mod $2m^2$) cannot coincide. Vertices with the same residue are distinct due to the different coefficients of m^2. The decomposition of each cycle creates $24m^2$ chains. We get that $opt \leq n(36m^3 + 18m^3 + 18m^3 + 36m^2(m-1) + 54m^2) = n(108m^3 + 18m^2)$. $APX = (12nm^3) \cdot 8 + 48m^2 \cdot 6m^3 \cdot 4m^2 \cdot 1.5 = nm^3(96 + 72) = 168nm^3$. This gives a ratio of $168m/(108m + 18)$ which tends to $14/9$ for large m.

Lemma 2. *Algorithm MCC-TS has approximation ratio of at most $14/9$.*

Proof. For every arc e, define a weight $w(e)$ in the following way. $w(e) = 1/3 + 2\ell(e)/(3n)$. We show the following properties.

1. The total sum of weights of arcs is at most $(5/9)$opt.
2. The number of new chains caused by decomposition is at most the total sum of weights.

The total cost for original chains and valid cycles constructed by MCC is bounded by opt, so the result of proving the properties would be $APX \leq 14opt/9$.

We start with proving property 1. Consider a cycle C in OPT which consists of k arcs. The total cost paid by OPT for C is k. The total weight of the arcs of C is exactly $k/3 + 2/3$, and $k/3 + 2/3 \leq 5k/9$ for $k \geq 3$. Consider a chain created by OPT which consists of k arcs. The total cost paid by OPT for this chain is $k + 1$. The total weight of the arcs of this chain is at most $k/3 + 2/3$, and $(k/3 + 2/3) \leq 5(k + 1)/9$ for $k \geq 1$. Next, we prove property 2. Consider a (not necessarily valid) cycle of $2k + 1$ arcs constructed by MCC which is of length sn for some integer s. Every such cycle can be split into at most $2s - 1$ chains in the following way. Let i be the starting vertex of an arc of the cycle, then i would be the first end-point of the first chain and the last end-point of the last chain (it can be the case where those two chains are combined into one). The distance to go from the first end-point to the last is sn. The length of two consecutive chains along the cycle is at least $n + 1$ (otherwise, they can be merged). If there are $2s$ chains, this means that the distance between the first and the last is more than sn, and therefore there are at most $2s - 1$ chains. On the other hand any pair of successive arcs can be combined in a chain due to the construction of the MCC graph, so $k + 1$ chains are always possible. We get that the number of new chains is at most $\min(k + 1, 2s - 1)$. The weight for these $2k + 1$ arcs is $(2k + 1)/3 + 2s/3 = (2k + 2)/3 + (2s - 1)/3 \geq (2/3 + 1/3) \min(k + 1, 2s - 1)$. Therefore, the weight of the cycle is at least the amount of additional cost caused by the decomposition.

Consider a cycle of $2k$ arcs which is of length sn for an integer s. We can get that the number of new chains is at most $\min(k, 2s - 1)$. The weight for these $2k$ arcs is $2k/3 + 2s/3 > 2k/3 + (2s - 1)/3 \geq (2/3 + 1/3) \min(k, 2s - 1)$.

Consider a chain of $2k$ or $2k + 1$ arcs with length in the interval $[sn, (s + 1)n)$. Note that the original connected component built by MCC is already a chain and not a cycle. There are at most $\min(k, 2s)$ new chains. The weight of the chain is at least $2k/3 + 2s/3 \geq \min(k, 2s)$.

Summarizing we proved the following theorem.

Theorem 8. *Algorithm MCC-TS has approximation ratio of exactly* $14/9$.

7   Conclusion

We introduced an approximation algorithm COMB for the problem of minimizing the number of SONET ADMs. COMB is a combination of two algorithms, one of them was introduced in this paper and the other was previously studied. Algorithm COMB is the current best approximation algorithm for this problem.

Table 2. Summary of results

Heuristic	Lower bound on the approximation ratio in [1]	Lower bound on the approximation ratio (this paper)	Upper bound on the approximation ratio (this paper)	Upper bound on the approximation ratio in [1]
COMB	–	4/3	**10/7**	–
PIM	4/3	3/2	–	3/2
PCM	3/2	5/3	–	5/3
MCC-TS	3/2	14/9	14/9	8/5
NMCC-TS	3/2	8/5	–	8/5
PET-TS	3/2	5/3	–	7/4
IM	3/2	8/5	< 5/3	5/3
PPIM	3/2	14/9	< 5/3	5/3

We showed that it is a 10/7 approximation algorithm, and we provided a lower bound on its worst-case performance of 4/3. Closing this gap, and finding a better approximation algorithm is left for future research. We also raise the following question: Is there a good approximation algorithm whose preprocessing step consists of cycle removal solely (without removal of chains)?

A summary of the results in the paper can be found in Table 2.

References

1. G. Calinescu and P.-J. Wan, "Traffic partition in WDM/SONET rings to minimize SONET ADMs," *Journal of Combinatorial Optimization*, **6**, 425-453, 2002.
2. O. Gerstel, P. Lin and G. Sasaki, "Wavelength assignment in a WDM ring to minimize cost of embedded SONET rings," *Proc. INFOCOM 1998*, **1**, 94-101, 1998.
3. L. Liu, X. Li, P.-J. Wan and O. Frieder, "Wavelength assignment in WDM rings to minimize SONET ADMs," *Proc. INFOCOM 2000*, **2**, 1020-1025, 2000.
4. P.-J. Wan, G. Calinescu, L. Liu and O. Frieder, "Grooming of arbitrary traffic in SONET/WDM BLSRs," *IEEE Journal on Selected Areas in Communications*, **18**, 1995-2003, 2000.

Author Index

Lecture Notes in Computer Science

For information about Vols. 1–3308

please contact your bookseller or Springer